AF389513

Advancements in Biomonitoring and Remediation Treatments of Pollutants in Aquatic Environments

Advancements in Biomonitoring and Remediation Treatments of Pollutants in Aquatic Environments

Editor

Elida Nora Ferri

MDPI • Basel • Beijing • Wuhan • Barcelona • Belgrade • Manchester • Tokyo • Cluj • Tianjin

Editor
Elida Nora Ferri
University of Bologna
Italy

Editorial Office
MDPI
St. Alban-Anlage 66
4052 Basel, Switzerland

This is a reprint of articles from the Special Issue published online in the open access journal *Applied Sciences* (ISSN 2076-3417) (available at: https://www.mdpi.com/journal/applsci/special_issues/Biomonitoring_Remediation_Pollutants).

For citation purposes, cite each article independently as indicated on the article page online and as indicated below:

LastName, A.A.; LastName, B.B.; LastName, C.C. Article Title. *Journal Name* **Year**, *Volume Number*, Page Range.

ISBN 978-3-0365-4505-9 (Hbk)
ISBN 978-3-0365-4506-6 (PDF)

Contents

About the Editor

Elida Nora Ferri

Elida Nora Ferri is a Senior Researcher currently working at the University of Bologna, Bologna, Italy. She holds a degree in biology from the same university. Her research interests include membrane biochemistry; analytical biochemistry; the development of bioluminescent, chemiluminescent, and flow-biosensor assays for metabolites and enzymes; luminescent studies on free radical production and antioxidant activities in biological specimens; and the biomonitoring of environmental pollution (e.g., honey bees and bioluminescent bacteria) and of remediation (photodegradation or AOPs) treatments' effectiveness by bioluminescent bacteria and microalgae toxicity tests. With approximately four decades of experience in academia, she has participated in several Italian, international, and applied research projects, publishing more than 130 papers in peer-reviewed international and national journals and books, 4 patents, and 2 books.

applied sciences

MDPI

Editorial

Special Issue: Advancement in Biomonitoring and Remediation Treatments of Pollutants in Aquatic Environments

Elida Nora Ferri

Department of Pharmacy and Biotechnology, University of Bologna, Via S. Donato 15, 40127 Bologna, Italy; elidanora.ferri@unibo.it

Citation: Ferri, E.N. Special Issue: Advancement in Biomonitoring and Remediation Treatments of Pollutants in Aquatic Environments. *Appl. Sci.* **2022**, *12*, 5091. https://doi.org/10.3390/app12105091

Received: 12 May 2022
Accepted: 18 May 2022
Published: 18 May 2022

Publisher's Note: MDPI stays neutral with regard to jurisdictional claims in published maps and institutional affiliations.

A series of negative, attendant circumstances threaten the ecological equilibrium, the quality, and even the existence of the different aquatic ecosystems nowadays more than ever. In both fresh and marine waters, it is possible to find traces or significant amounts of contaminants produced by and for human activities, unpredictably affecting the different trophic levels. Together with the excessive drawings on freshwater reserves, driven by the rise in population and the drought crisis linked to the climate changes, the pollution of the tiny fraction of freshwater available to us urgently requires effective, sustainable, and innovative monitoring and remediation tools. In light of the above, this special issue was proposed and so far, 16 papers have been published in this Special Issue, fairly equally distributed between monitoring activities and remediation proposals.

The first paper reported the application of a polar organic chemical integrative sampler (POCIS) to the monitoring, by passive sampling, of marine water employed in aquarium tanks [1]. The passive accumulation devices provide in situ integrative sampling, pre-concentration and purification steps, allowing the determination of time-weighted average (TWA) concentrations of trace levels in the supply water of hydrocarbons, DPS, 2,4-DTBP, BPA and the UV-filters OC and EHMC. The result highlighted the usefulness of pre-concentration in the passive sampling approach, which permitted enhanced sensitivity.

In the paper [2], the performance of conventional monitoring of PCBs by water sampling followed by GC-MS/MS determination was compared with biomonitoring using the Baikal omul, *Coregonus migratorius*. Comparative statistical analyses were carried out to study both the influence of the year, season, locations, layers of the water column on water samples values and that of the age and sex of omul on the ratio and concentrations of indicator congeners in fish tissues. The determination in water was simple, obtained by measuring only two indicator congeners. The PCB accumulation in the muscle tissue depended on their hydrophobicity and on the age of individuals. Hence, the ratio of congeners in fish tissues did not reflect their ratio in the water samples.

By using a manipulative feeding experiment, possible damages to the intestinal microbial community and morphology of *Micropterus salmoides* produced by virgin polypropylene microplastics (MPs) were investigated [3]. The tested MPs were fibres, fragments, and films added into the water or to the feed. After a four-week trial, no significant differences were detected in weight gain (WG%) and specific growth rate (SGR). MP exposure had no statistically significant effect on the composition or diversity of the intestinal microbial community, although it could partly influence the intestinal morphology and the recombination process of the intestinal microbial community.

In [4], the authors carried out a numerical study, reproducing the changes in 1,4-dioxane concentrations in groundwater observed in the monitoring wells of a waste dumping site. The quasi-3D flow and transport simulation model was useful in predicting the changes in pollutant amounts in the hydrologic environment, elucidating the transport under both natural and artificially induced water flows. The simulation revealed that the pumping actions could promote pollutant transport to areas where it would otherwise be difficult to reach. This phenomenon was unanticipated, and it would influence the groundwater treatment efficiency.

In light of the differences that exist among pollutant mechanisms and among organism sensitivity, a careful evaluation of the suitability of three freshwater and marine microalgae and of *Vibrio fisheri* bacteria in assessing the long-term toxicity of a set of benzo-fused nitrogen heterocycles was reported in [5]. No species-specific sensitivity for any whole family of pollutants was observed and the freshwater *R. subcapitata* was the only alga affected by most of the compounds at the lowest concentrations. The *Vibrio fisheri* assay showed low detrimental effects or even growth stimulation regarding the contaminant levels in surface and drinking waters.

Heavy metal (HMs) contamination in fish from aquaculture can represent a serious health risk. The patterns of HM accumulation (Cu, Zn, Cd, Hg, Pb, As) in the muscles and livers of *Cyprinus carpio* and *Ictalurus punctatus* grown in a net cage deployed in a lake, where the sediments were rich in heavy metals, were monitored over a 16 month period [6]. The water pH values (8.6–9.5) promoted the aggregation of metals, reducing their content in water. The fish raised in aquaculture that were provided only metal-free feed and with limited access to natural food sources do not represent a health risk for their HM content. Conversely, native fish in free contact with sediments and contaminated sources of food represented a moderate-to-high risk.

The groundwater quality and the origin of threats to it can be determined by analysing the physico-chemical parameters and the metal content, as it was assessed in [7] on samples collected from different countries. Various parameters were associated with anthropogenic activities. High EC conductivity can affect shallow depth groundwater. The concentration of nitrate can be connected to high agricultural waste and sewage contamination. A high concentration of chloride indicates a high degree of organic pollution, whereas ammonium is an indicator of recent organic pollution, both of natural and artificial origin. Turbidity in groundwater depends on the exploitation activities in the area. The statistical analysis of data can reliably indicate the pollution source.

The integrated biomarker response index (IBR), the bioeffects assessment index (BAI) and the principal components analysis (PCA) were evaluated for their sensitivity in revealing stress patterns in *Phaeodactylum tricornutum* under contaminant exposure [8]. The antioxidant enzymes CAT, SOD and APX, and the TBARS oxidative stress markers revealed an overall suitability, producing statistically significant dose–response relations. IBR appears as a useful tool, as it reflects the contaminant levels at the sites, despite the variability in biomarkers used for its calculation. The BAI index revealed low sensitivity against the oxidative stress data set, confirming that its applicability is more adequate for general toxicity biomarkers and non-specific stress parameters. PCA satisfactorily delivered a significant negative correlation for the overall experiment relative to compound concentration and index scores and it was positively correlated with the BAI index.

In [9], a model for the assessment of the health risk depending on the consumption of contaminated water is proposed, depending either on the hydraulic or on water quality models. The health impact without using water quality stations was compared to that obtained in the presence of ten water quality sensors, both deployed utilizing the volume of the contamination water (VCW) and the time detection (TD), respectively. The use of water sensors has shown a large difference in reducing the number of people exposed to the risk of using contaminated water; the result was quite optimal (the health impact was reduced by up to 98.37%) in the case of sensors deployed using TD.

A remedial investigation and feasibility study was carried out on groundwater contaminated by tetrachloroethene, trichloroethene, and vinyl chloride [10]. A double tracer approach was used to reassess the hydraulic conductivity and hydrodynamic dispersity. A soybean oil-based emulsion (EVO), with and without hydrogen gas infusion, was employed and the simple EVO treatment was the most suitable in supporting the native microbial consortia and in removing the substrate from the subsurface. Microbial consortia changes were monitored. Careful and frequent addition of substrate to the low permeability regions was key in successfully remediating them.

The work reported in [11] focused on the same topic. The chemical reduction by an organo-iron compound (liquid ferrous lactate), combined with emulsified vegetable oil, was employed to minimize the accumulation of the tetrachloroethene by-product in groundwater. Ferrous lactate (FL) forms aqueous solutions that are convenient for injection during in situ operations and that may have a longer distance of transmission than solid, zerovalent iron. The used mixture was effectively adsorbed into soil pores and continued to release electron donors; it had at least 90 days of slow-release effect and reached a minimum distance of 5 m. Moreover, the metabolites did not accumulate in a notable manner in areas where FL concentration was adequate.

The tolerance mechanisms developed by microorganisms that inhabit the impacted sites could be exploited in remediation processes, for example in [12], where it is reported that marine bacteria collected from seven Portuguese fishing ports were isolated and grown in the presence of Tributyltin (TBT). The bacteria that exhibited higher growth were evaluated for their capability to degrade this highly toxic, endocrine disrupting compound. The results were ascertained using ecotoxicological tests. Two strains were selected and identified as *Pseudomonas putida* and *Serratia marcescens*. *P. putida* showed the highest degradation potential. After 96 h of exposure to *S. marcescens*, the media was still toxic, suggesting the need for longer contact time or the production of toxic by-products.

An effective remediation of polluted water was obtained by removing the unwanted chemicals by physical absorption mechanisms on nanomaterials, thanks to the large specific surface area [13]. Multi-walled carbon nanotubes (MWCNTs) were synthesized, purified, functionalized, and used as an adsorbent for the removal of Ismate violet 2R dye from the contaminated water. The morphological structure, the physical, chemical, thermodynamic, and kinetics properties were evaluated throughout. The adsorption is a physisorption process and under optimal conditions, the percentage of removal of the MWCNTs from real wastewater reached 88%.

The in situ degradation of pesticide containing agro-wastewaters (remnant volumes from spraying activities or leftover pesticide waste) was realized by a portable photo-reactor prototype, employing UV lamps and H_2O_2 [14]. The degradation process was evaluated by the pesticides' absorbance decrease, the mineralization process by measuring the TOC values, and the toxic effects of irradiated and non-irradiated solutions by biological tests. The photo-reactor prototype was able to remove about 80% (mean value) of the active molecules in an acceptable time (8 h). The mean mineralization rate was about 44%. These results were extremely positive, since the photoproducts are more prone to inherent water remediation processes than the parent compounds.

The effectiveness of a TiO_2 photo-catalysis in removing glucocorticoids (GCs) under simulated and natural sunlight from solutions prepared in tap water or in river water at environmental concentrations has been evaluated [15]. The degradation profiles followed a pseudo-first order kinetics and the TiO_2-assisted photo-catalysis appeared to be independent of the chemical structure of the GCs. The photocatalytic process was from one to two orders of magnitude faster than the not-catalysed photolysis. The two main photo-oxidation paths, directly linked to OH radicals from the irradiated TiO_2 catalyst, were defined. The author concluded that the great stability of the titania crystals under irradiation, coupled with the good adsorption properties and catalytic activity, overcome any other depollution method.

The paper [16] reports the degradation of the highly toxic pesticide methiocarb (MC) by electrochemical oxidation in aqueous solutions, utilizing a boron-doped diamond anode and two different supporting electrolytes. The removal rate, total organic carbon (TOC), and total nitrogen (TN) were determined. The effects on the biotoxicity were assessed using the freshwater crustacean *Daphnia magna*. The hydroxyl radical-dependent degradation showed a first-order mechanism and the MC and TOC removal rates were higher for the solutions containing NaCl instead of Na_2SO_4. Moreover, in the presence of Na_2SO_4, no nitrogen removal occurred. The electrochemical oxidation effectively degraded methiocarb, reducing drastically the acute toxicity towards *D. magna*, especially in chloride-containing solutions. In those solutions, the electric charge increase must be avoided, since this increase in the ecotoxicity is probably due to the formation of perchlorate and organochlorinated compounds, which can exert a toxic effect on *Daphnia*.

Funding: This research received no external funding.

Conflicts of Interest: The author declares no conflict of interest.

References

1. Scapuzzi, C.; Benedetti, B.; Di Carro, M.; Chiesa, E.; Pussini, N.; Magi, E. Passive Sampling of Organic Contaminants as a Novel Approach to Monitor Seawater Quality in Aquarium Ocean Tanks. *Appl. Sci.* **2022**, *12*, 2951. [CrossRef]
2. Gorshkov, A.G.; Kustova, O.V.; Bukin, Y.S. Assessment of PCBs in Surface Waters at Ultratrace Levels: Traditional Approaches and Biomonitoring (Lake Baikal, Russia). *Appl. Sci.* **2022**, *12*, 2145. [CrossRef]
3. Zhang, C.; Wang, Q.; Wang, S.; Pan, Z.; Sun, D.; Cheng, Y.; Zou, J.; Xu, G. Effects of Virgin Microplastics on Growth, Intestinal Morphology and Microbiota on Largemouth Bass (Micropterus salmoides). *Appl. Sci.* **2022**, *11*, 11921. [CrossRef]
4. Pongritsakda, T.; Nakamura, K.; Wang, J.; Watanabe, N.; Komai, T. Prediction and Remediation of Groundwater Pollution in a Dynamic and Complex Hydrologic Environment of an Illegal Waste Dumping Site. *Appl. Sci.* **2022**, *11*, 9229. [CrossRef]
5. Canova, L.; Sturini, M.; Maraschi, F.; Sangiorgi, S.; Ferri, E.N. A Comparative Test on the Sensitivity of Freshwater and Marine Microalgae to Benzo-Sulfonamides,-Thiazoles and -Triazoles. *Appl. Sci.* **2021**, *11*, 7800. [CrossRef]
6. Alvarado, C.; Cortez-Valladolid, D.M.; Herrera-López, E.J.; Godínez, X.; Martín Ramírez, J. Metal Bioaccumulation by Carp and Catfish Cultured in Lake Chapala, and Weekly Intake Assessment. *Appl. Sci.* **2021**, *11*, 6087. [CrossRef]
7. Alina Soceanu, A.; Dobrinas, S.; Dumitrescu, C.I.; Manea, N.; Sirbu, A.; Popescu, V.; Vizitiu, G. Physico-Chemical Parameters and Health Risk Analysis of Groundwater Quality. *Appl. Sci.* **2021**, *11*, 4775. [CrossRef]
8. Leal Pires, V.; Novais, S.C.; Lemos, M.F.L.; Fonseca, V.; Duarte, B. Evaluation of Multivariate Biomarker Indexes Application in Ecotoxicity Tests with Marine Diatoms Exposed to Emerging Contaminants. *Appl. Sci.* **2021**, *11*, 3878. [CrossRef]
9. Essa, Q.; Shahra, E.Q.; Wu, W.; Gomez, R. Human Health Impact Analysis of Contaminant in IoT-Enabled Water Distributed Networks. *Appl. Sci.* **2021**, *11*, 3394. [CrossRef]
10. Guerra, P.; Bauer, A.; Reiss, R.A.; McCord, J. In Situ Bioremediation of a Chlorinated Hydrocarbon Plume: A Superfund Site Field Pilot Test. *Appl. Sci.* **2021**, *11*, 10005. [CrossRef]
11. Liu, M.-H.; Hsiao, C.-M.; Lin, C.-E.; Leu, J. Application of Combined In Situ Chemical Reduction and Enhanced Bioremediation to Accelerate TCE Treatment in Groundwater. *Appl. Sci.* **2021**, *11*, 8374. [CrossRef]
12. Hugo, R.; Monteiro, H.R.; Moutinho, A.B.; Campos, M.J.; Esteves, A.C.; Lemos, M.F.L. Insights into the Restoration of Tributyltin Contaminated Environments Using Marine Bacteria from Portuguese Fishing Ports. *Appl. Sci.* **2021**, *11*, 6411. [CrossRef]
13. Abualnaja, K.M.; Alprol, A.M.; Ashour, M.; Mansour, A.T. Influencing Multi-Walled Carbon Nanotubes for the Removal of Ismate Violet 2R Dye from Wastewater: Isotherm, Kinetics, and Thermodynamic Studies. *Appl. Sci.* **2021**, *11*, 4786. [CrossRef]
14. Esposito, B.; Riminucci, F.; Di Marco, S.; Metruccio, E.G.; Osti, F.; Sangiorgi, S.; Ferri, E.N. A Simple Device for the On-Site Photodegradation of Pesticide Mixes Remnants to Avoid Environmental Point Pollution. *Appl. Sci.* **2021**, *11*, 3593. [CrossRef]
15. Pretali, L.; Albini, A.; Cantalupi, A.; Maraschi, F.; Nicolis, S.; Sturini, M. TiO_2-Photocatalyzed Water Depollution, a Strong, yet Selective Depollution Method: New Evidence from the Solar Light Induced Degradation of Glucocorticoids in Freshwaters. *Appl. Sci.* **2021**, *11*, 2486. [CrossRef]
16. Fernandes, A.; Pereira, C.; Coelho, S.; Ferraz, C.; Sousa, A.C.; Pastorinho, M.R.; Pacheco, M.J.; Ciríaco, L.; Lopes, A. Ecotoxicological Evaluation of Methiocarb Electrochemical Oxidation. *Appl. Sci.* **2020**, *10*, 7435. [CrossRef]

applied sciences

MDPI

Article

Passive Sampling of Organic Contaminants as a Novel Approach to Monitor Seawater Quality in Aquarium Ocean Tanks

Chiara Scapuzzi [1], Barbara Benedetti [1], Marina Di Carro [1], Elvira Chiesa [2], Nicola Pussini [2] and Emanuele Magi [1,*]

[1] Department of Chemistry and Industrial Chemistry, University of Genoa, Via Dodecaneso 31, 16146 Genoa, Italy; chiara.scapuzzi@edu.unige.it (C.S.); barbara.benedetti@unige.it (B.B.); marina.dicarro@unige.it (M.D.C.)
[2] Acquario di Genova, Ponte Spinola, 16128 Genoa, Italy; echiesa@costaedutainment.it (E.C.); npussini@costaedutainment.it (N.P.)
* Correspondence: emanuele.magi@unige.it

Abstract: The determination of trace pollutants in seawater is challenging, and sampling is a crucial step in the entire analytical process. Passive samplers combine in situ sampling and preconcentration, thus limiting the tedious treatment steps of the conventional sampling methods. Their use to monitor water quality in confined marine environment could bring several advantages. In this work, the presence of organic contaminants at trace and ultra-trace levels was assessed in the Genoa Aquarium supply-and-treated water using Polar Organic Integrative Samplers (POCIS). Both untargeted gas chromatography-mass spectrometry and targeted liquid chromatography-tandem mass spectrometry were employed. The untargeted approach showed the presence of hydrocarbons, diphenyl sulfone and 2,4-di-tert-butyl-phenol. Only hydrocarbons were detected in all the samples. Nineteen emerging contaminants, belonging to different classes (pharmaceuticals, UV-filters, hormones and perfluorinated compounds), were selected for the target analysis. Thirteen analytes were detected, mainly in supply water, even though the majority of them were below the quantitation limit. It is worthy to note that two of the detected UV-filters had never been reported in seawater using the POCIS samplers. The comparison of the analytes detected in supply and treated water indicated a good performance of the Aquarium water treatment system in the abatement of seawater contaminants.

Keywords: passive sampling; trace contaminants; emerging contaminants; seawater; chromatography–mass spectrometry

Citation: Scapuzzi, C.; Benedetti, B.; Di Carro, M.; Chiesa, E.; Pussini, N.; Magi, E. Passive Sampling of Organic Contaminants as a Novel Approach to Monitor Seawater Quality in Aquarium Ocean Tanks. *Appl. Sci.* **2022**, *12*, 2951. https://doi.org/10.3390/app12062951

Academic Editors: Elida Nora Ferri and Simone Morais

Received: 18 January 2022
Accepted: 10 March 2022
Published: 14 March 2022

Publisher's Note: MDPI stays neutral with regard to jurisdictional claims in published maps and institutional affiliations.

1. Introduction

Coastal marine waters constitute the main receptor of human activities. They are subjected to different anthropogenic stressors, including wastewater treatment plant discharge, shipping traffic, runoff and atmospheric depositions [1–3]. The low concentration of seawater contaminants is due to dilution and to the complex hydrodynamics of the marine environment [3]. The detection and quantification of pollutants in marine environments represent an analytical challenge due to the matrix complexity and the trace levels of contaminants that require sensitive methodologies. Sampling is a crucial step in the analytical process. Conventional sampling methods usually require several liters of water, repetitive samplings and treatment steps. On the other hand, passive accumulation devices provide in situ integrative sampling, allowing the determination of time-weighted average (TWA) concentrations. TWA provides an estimation of the contamination level of the considered water environment that can be very advantageous to assess the exposure of living organisms and the associated toxicological risks. Passive samplers are useful for tracking spatial and temporal concentration profiles in waters [4] and to assess the occurrence of episodic events [5]. Moreover, these techniques combine sampling, preconcentration and purification steps, limiting sample manipulation. Among passive sampling devices, the Polar Organic Chemical Integrative Sampler (POCIS) was designed to sample polar compounds [5]. However, POCIS was applied to the sampling of chemicals characterized by a $logK_{ow}$ comprised between 0 and 5 [6]. It consists in a sorbent phase sandwiched between two microporous membranes, usually made of polyethersulfone (PES), which are kept together by two stainless rings.

POCIS devices generally show low affinity for highly hydrophobic and ionic compounds [7]. Indeed, low diffusion through the membranes was found for hydrophobic substances because the PES polymer can act as the primary receiving sorption phase for these compounds [8]. Nevertheless, the detection of some lipophilic substances using these samplers is reported [9,10]. The introduction and testing of new sorption materials and membranes in passive techniques are aimed at increasing the range of chemicals that can be sampled [6].

POCIS are mainly employed to monitor contaminants in freshwater and wastewater [10–14], while their use for seawater monitoring remains limited [9,15,16]. Furthermore, the use of this device in a confined marine environment was only reported for the detection of micro-pollutants in a Spanish fish farm located in a coastal area [17]. Nevertheless, their use to monitor water quality of circumscribed environments (such as swimming pools or aquarium tanks) could bring several advantages. Aquarium water is generally collected from the coastal zone and subjected to treatments. Then, renewal water systems allow the water in fish tanks to be replaced to some extent every day, thus making the passive sampling approach useful for verifying water quality over time.

The aim of this work was to identify the presence of polar organic contaminants in Aquarium water (Genoa, Liguria, Italy). POCIS samplers were employed to continuously monitor the quality of supply water (withdrawn from the sea), of treated water (filtered and disinfected) and of shark tank water, verifying the efficacy of the treatment system used before tank filling. Gas chromatography-mass spectrometry (GC-MS) and liquid chromatography-tandem mass spectrometry (LC-MS/MS) were employed to detect the presence of classical and emerging contaminants.

2. Materials and Methods

2.1. Standard and Reagents

Dichloromethane, 2-propanol, trifluoroacetic acid, acetonitrile (ACN) and methanol (CH_3OH), all of analytical or chromatographic grade, were obtained from Merck (Darmstadt, Germany). Water was purified by a Milli-Q system (Millipore, Watford, Hertfordshire, UK). A standard mixture containing even n-alkanes with chains from 10 to 40 carbon atoms (C10–C40) was purchased from Sigma-Aldrich (Milan, Italy). Analytical standards of diphenyl sulfone (DPS), 2,4-ditert butylphenol (2,4-DTBP), ibuprofen (IBU), perfluorooctanoic acid (PFOA), perfluorooctane sulfonate (PFOS), gemfibrozil (GBZ), estrone (E1), triclosan (TCS), carbamazepine (CBZ), bisphenol-A (BPA), β-estradiol (E2), 17-α-ethinylestradiol (EE2), benzophenone-3 (BP-3), ethylhexyl methoxycinnamate (EHMC), octocrylene (OC), octyl dimethyl p-aminobenzoic acid (OD-PABA), naproxen (NAP), diclofenac (DIC), mefenamic acid (MEF) and caffeine (CAF) were obtained from Sigma-Aldrich (Milan, Italy). All standards were of high purity grade (>97%).

2.2. Genoa Aquarium Water

Supply water, used in all Aquarium oceanic tanks, is directly taken from the sea through a submarine pipeline. This pipe draws marine water at about 10 m from the breakwater and 50 m depth. Before use, raw water is subjected to physico-chemical and microbiological analysis; furthermore, a toxicity test is performed using ephyrae. The ephyra is one of the jellyfish life cycle stages and ephyrae are used as bioindicators for toxic organic substances. If the mentioned tests ensure good water quality, a mechanical treatment is performed by two sand prefilters (each one of 250 m^3 per hour). Quarzifer sand granulometry is 0.4–0.8 mm. The filtration step is followed by UV disinfection, and the system (Panaque srl, Capranica, Italy) consists in eight 75-watt lamps of 156 cm height. After these treatments, water is stored in aerated accumulation tanks (treated water).

The multitaxa shark tank hosts various species of sharks from different environments and other fishes. Its size is 9.9×23.5 m and contains a water volume of approximately 1200 m^3. The tank presents an internal treatment system consisting of four sand filters (each one with a 300 m^3 per hour discharge) and O$_3$ disinfection. Every day, 150 m^3 of the tank water is replaced (approximately 10% of the total volume), and filters are backwashed.

2.3. Sampling Method

Two studies were carried out to investigate the presence of organic contaminants in raw and treated Aquarium waters using POCIS devices.

The initial monitoring was performed between March 2017 and January 2018. POCISs were used to monitor shark tank water, treated water and supply water.

In particular, the following samplings of two weeks were performed:

- Sampling I (spring): Two POCIS deployed in the shark tank (P1-P2) and two POCIS in a laboratory beaker filled with 4 L of Aquarium treated water (T1-T2);
- Sampling II (winter): Two POCIS deployed in the shark tank (P3-P4), two POCIS in a laboratory beaker filled with 4 L of Aquarium treated water (T3-T4) and two POCIS in a beaker filled with 4 L of supply water (S1-S2).

POCIS deployed in the shark tank were protected by stainless-steel grids, a Zn tablet was fixed on the protective system to avoid corrosion, and the devices were positioned at the bottom of the tank. Treated and supply waters, sampled in laboratory beakers, were subjected to magnetic agitation (Figure S1), and water was replaced as frequently as the shark tank water. This action allowed the minimization of differences between the two water samplings, although agitation conditions may still have an influence on the sampling rate of analytes [18].

The second study, carried out in January 2019, involved a sampling of two weeks:

- Sampling III (winter): Three POCIS deployed in a tank of 150 L (Figure S2) filled with supply water (S3, S4 and S5) or treated water (T5, T6 and T7).

POCISs were placed parallel to the water flow to allow the correct diffusion of analytes into the sampler [19]. Based on the results of the first study, no POCISs were deployed in fish tanks. Indeed, the focus of this last monitoring (sampling III) was to confirm the good performance of the filtration system and to evaluate the presence of emerging contaminants. To take into account the blank signals, for each study two POCISs were submerged in beakers filled with 1.5 L of milli-Q water for 7 days.

The average temperature and salinity of the sampled waters were 23 °C and 36 ng kg^{-1}, respectively.

2.4. POCIS Preparation and Extraction

Both commercial POCIS (E&H Services, Praha, Czech Republic) and devices assembled in our laboratories were used for passive sampling. The laboratory assembly was performed in accordance with the characteristics of commercial ones (mass of the sorbent phase 200 mg and 45.8 cm^2 as sampler surface area). PES membranes (0.1 μm pore size) and Oasis HLB sorbent were purchased from Pall Italia (Buccinasco, Italy) and Waters (Vimodrone, Italy), respectively. Before use, PES membranes were washed in a H$_2$O/CH$_3$OH solution (80:20 v/v) for 24 h at room temperature (RT) and 1 h at 40 °C. Membranes were allowed to dry and subsequently washed with CH$_3$OH for 24 h at RT and 1 h at 40 °C. After drying under a fume hood, 200 mg of HLB phase was weighted and placed between two membranes. Then, the membrane-sorbent-membrane layer was compressed between two stainless-steel support rings and held together with three thumbscrews. The prepared POCIS were stored at -20 °C.

After the exposure, samplers were collected, rinsed with Milli-Q water, wrapped in aluminum foil, transported to the laboratory under cooled conditions (4 °C) and kept at -20 °C until analysis.

Prior to processing, the samplers were thawed and rinsed with Milli-Q water to remove fouling present on POCIS membranes, although during all the sampling experiments, no significant biofouling was observed. This is probably due to the daily replacement of a part of the water in the shark tank. POCISs were dismantled and the sorbent phase was transferred using Milli-Q water into a 1 cm diameter glass syringe cartridge fitted with a polytetrafluoroethylene (PTFE) frit and glass wool and dried under a gentle nitrogen stream. Elution was performed with 10 mL of dichloromethane: isopropanol: trifluoroacetic acid (80:19.9:0.1 *v/v/v*) in a flask and the extract was reduced back down to a small volume [9]. Then, 20 mL of dichloromethane was added, and the solution was evaporated to dryness and redissolved in 1 mL of methanol for GC-MS analysis. The extracts in methanol were diluted 1:1 using Milli-Q water for LC-MS/MS. Samples were filtered through 0.2 μm PTFE filters before chromatographic analysis.

2.5. Analytical Methods

The untargeted analyses of the volatile compounds were performed on an Agilent 7890A gas chromatograph connected to a 5975C mass spectrometer (Agilent Technologies, Palo Alto, CA, USA) with an electron ionization source (EI) set at 70 eV. Chromatographic separation was achieved on a Phenomenex ZB5 capillary column (30 m $\times$ 0.25 mm $\times$ I.D. 0.25 mm) coated with 5% phenylpolysiloxane at a constant He flow rate of 1.2 mL min^{-1}. The injector was maintained at 250 °C. The temperature program was set as follows: 50 °C kept for 1 min, ramp of 10 °C min^{-1} up to 300 °C and kept constant for 10 min. The transfer line, ion source and quadrupole were held at 280 °C, 230 °C and 150 °C, respectively. The mass spectrometer was set in full-scan (SCAN) mode for non-target screening, and full-scan spectra were collected from 50 to 550 *m/z*. The mass spectra obtained in full scan for each chromatographic signal were compared with those contained in the NIST2 library (Version 2.2, June 2014). The selected ion monitoring (SIM) mode was used for the semi-quantitative analysis of chosen compounds. For each analyte, two ions were monitored in SIM mode, using the most abundant one for quantitation.

The presence of target emerging contaminants was evaluated by using liquid chromatography coupled to tandem mass spectrometry [20]. Chromatographic separations were performed by an Agilent Liquid Chromatograph Series 1200SL. HPLC was equipped with a biphenyl Raptor column (2.1 mm $\times$ 50 mm, 1.8 μm by Restek, Milan, Italy) held at 60 °C. An Agilent 6430 MSD triple-quadrupole mass spectrometer (Agilent Technologies, Santa Clara, CA, USA) equipped with an electrospray ion source (ESI) was employed for detection and used both in positive and negative ionization modes. Multiple reaction monitoring (MRM) was set to enhance sensitivity by choosing two transitions for each analyte (Table S1), using the most abundant one for quantification.

Positive acquisition involved a chromatographic run in isocratic conditions with a 0.2 mL min^{-1} flow rate and mobile phase composition of Milli-Q water:ACN = 60:40 (*v/v*). As for the analytes acquired in the negative mode, separation was performed in isocratic conditions as well. The mobile phase was Milli-Q water:ACN = 50:50 (*v/v*) containing 0.1% of acetic acid, and the flow rate was of 0.3 mL min^{-1}. Three technical replicates were performed for each sample and the injection volume was of 10 μL.

Three points calibration curves were drawn for each analyte in the 0.01–10 μg L^{-1} range. The performances of the instrumental method are reported in Table 1. The LOD and LOQ were calculated by considering a signal to noise ratio of 3 and 10, respectively; dilution factor and final volume of the extracts were considered to obtain LOD and LOQ values in the POCIS sampler.

Table 1. Limits of detection and quantitation of the monitored emerging pollutants in solution (μg L^{-1}) and in the POCIS sampler (ng/POCIS), intra-day precision and accuracy.

Compound	LOD (μg L^{-1})	LOQ (μg L^{-1})	LOD (ng/POCIS)	LOQ (ng/POCIS)	Intra-Day Precision (RSD%)	Accuracy (%)
EE2	1	5	2	10	2.9	100.7 ± 6.7
MEF	0.05	0.5	0.1	1	3.8	102.1 ± 6.7
PFOA	0.01	0.1	0.02	0.2	1.8	105.6 ± 12.8
PFOS	0.01	0.05	0.02	0.1	1.5	107.4 ± 12.2
E2	1	5	2	10	5.6	101.9 ± 10.8
BPA	0.05	0.5	0.1	1	7.5	99.2 ± 7.4
CAF	0.01	0.5	0.02	1	2.6	97.5 ± 5.3
CBZ	0.01	0.05	0.02	0.1	3.4	106.9 ± 15.6
DIC	0.05	0.5	0.1	1	2.7	101.9 ± 5.5
E1	0.5	1	1	2	4.5	98.9 ± 8.1
OD-PABA	0.01	0.05	0.02	0.1	7.0	101.2 ± 3.3
EHMC	0.01	0.5	0.02	1	2.0	105.0 ± 15.6
GBZ	0.05	0.1	0.1	0.2	2.2	99.8 ± 9.0
IBU	0.5	2	1	4	5.4	97.5 ± 6.6
BP-3	0.01	0.5	0.02	1	1.3	103.0 ± 6.8
NAP	0.05	0.1	0.1	0.2	6.0	105.7 ± 13.8
OC	0.01	1.0	0.02	2	4.4	103.5 ± 18.9
TCS	0.1	1.0	0.2	2	2.9	103.5 ± 5.6

3. Results and Discussion

3.1. Untargeted Analysis

In both monitoring campaigns, the first step was a general characterization of the most relevant contaminants present in aquarium waters. As already mentioned in the introduction, although POCISs were originally designed to sample polar analytes, the detection of lipophilic compounds in certain deployment conditions has been reported in the literature [9,10].

Therefore, an untargeted analysis by GC-MS was chosen by also considering the probable presence of volatile and semi-volatile contaminants coming from bunker fuel. GC-MS in EI mode allows the tentative identification of several compounds thanks to the match with a spectrum database. Non-targeted analysis highlighted the presence of hydrocarbons and two other chemicals as the most abundant: DPS and 2,4-DTBP. While hydrocarbons and DPS were already detected in Ligurian seawater [9], 2,4-DTBP was identified for the first time using POCIS.

3.1.1. Hydrocarbons

After a first screening in full scan mode, a more sensitive and selective analysis was performed in SIM mode. The ions monitored were at 57, 71 and 85 *m/z*, which are characteristics of linear hydrocarbons. Procedural blanks were analyzed, and their contribution in hydrocarbons was subtracted from the samples. For each sequence of analyses, an alkanes standard mixture (C10–C40, concentration 1 mg L^{-1}) was employed to verify retention times and normalize chromatographic signals in order to perform a comparative analysis of the extracts obtained from different samplers.

Results obtained in the first monitoring, reported in Figure 1, did not show significant differences among the hydrocarbons sampled in the tank (P1 and P2; P3 and P4) and in the treated water (T1 and T2; T3 and T4). Furthermore, results obtained for the samplings I and II showed a comparable profile regarding the presence of the homologous series of even n-alkanes of petrogenic origin [21]; therefore, no seasonal trend appears evident.

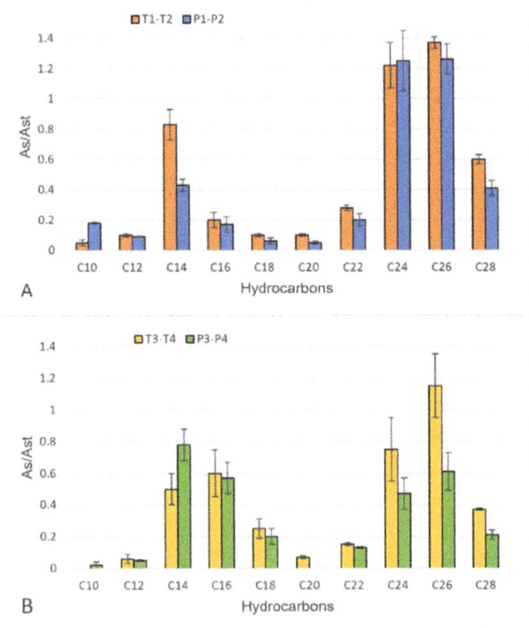

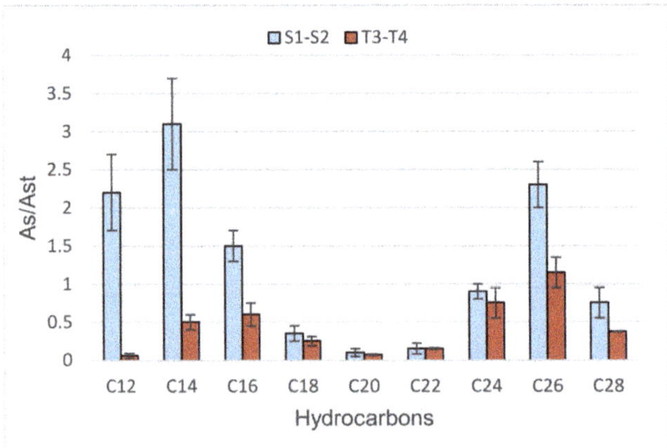

Figure 1. Results obtained from the semi-quantitative analysis of hydrocarbons (chain length: 10–28 C atoms) in treated water (T1 and T2 in orange) and shark tank water (P1 and P2 in blue) in sampling I (**A**). Results obtained from the semi-quantitative analysis of hydrocarbons (chain length: 10–28 C atoms) in treated water (T3 and T4 in yellow) and shark tank water (P3 and P4 in green) in sampling II (**B**). As/Ast represents the ratio between the hydrocarbon peak areas in the sample and in the reference standard.

In sampling II of the first monitoring, an evaluation of the treatment process was carried out by analyzing the POCIS deployed in supply water and treated water (Figure 2).

Figure 2. Results obtained from the semi-quantitative analysis of hydrocarbons (chain length: 12–28 C atoms) in supply water (S1 and S2 in light blue) and treated water (T3 and T4 in red) in sampling II. As/Ast represents the ratio between the hydrocarbon peak areas in the sample and in the reference standard.

T3 and T4 extracts showed a good decrease in hydrocarbon concentration, especially for short chain alkanes (between C12 and C16). This result suggests that sand filtration has a lower impact on long chain analytes or that shorter hydrocarbons tend to undergo degradation onto the filter [22].

Water treatment efficiency was further investigated in the second monitoring. Linear alkanes from C10 to C40 were found in S3, S4 and S5 and in T5, T6 and T7 water extracts. Analogously to the first monitoring, a general decrease was observed for each analyte, especially for short-chain compounds: C12 and C14 hydrocarbons presented the most abundant decrease (84% and 69%, respectively).

As a final remark for hydrocarbons, it is worth highlighting that the profile of supply water presented the highest peaks in the diesel range (C8–C23) [23,24] both in the first and in the second monitoring.

3.1.2. Industrial Compounds

In sampling I and II, the screening in full-scan mode of the extracts allowed the detection of DPS and 2,4-DTBP. This untargeted procedure was employed also in sampling III, where, in addition to hydrocarbons, only 2,4-DTBP was found. Subsequently, semi-quantitative analysis was carried out in SIM mode. The ions at 125 m/z and 191 m/z were monitored for DPS and 2,4-DTBP, respectively. Analogously to hydrocarbons, the ratio between the analytes peak area in the sample and in a reference standard (concentration 1 mg L^{-1}) was used to compare different POCIS extracts.

DPS is a herbicide employed as ovicide and acaricide in the past. Today, it is mainly used as a monomer in the polymer industry and is a transformation product of phenols synthesis. It enters aquatic ecosystems through various sources and Wick et al. affirmed that photochemical and chemical degradations are not significant in surficial water [25].

DPS concentrations estimated in the sampling I extracts were approximately 1 mg L^{-1}, corresponding to 1 µg/POCIS. This value was comparable to that reported in one of our previous studies, where POCISs were deployed in Santa Margherita Harbor (Liguria) [9]. Furthermore, the detection of DPS at low concentrations in seawater was reported during a micro-pollutant monitoring campaign in a fish farm [17]. On the other hand, in sampling II and III, DPS was neither detected in the supply water nor in the treated water. This difference could be related to the sampling period of the last two deployments (winter).

2,4-DTBP is a volatile organic compound used as intermediate for the preparation of UV stabilizers and antioxidants. Migration tests performed on high-density polyethylene pipes (HDPE) used in drinking water distribution networks showed 2,4-DTBP as the major migrating component from pipes into water [26]. It has fungicidal activity [27] and has shown potential endocrine disrupting effects and aquatic toxicity [28].

Regarding first monitoring, higher levels of 2,4-DTBP in treated and shark tank waters were observed in sampling I compared to sampling II (mean value approximately 3 µg/POCIS and 1 µg/POCIS, respectively). Moreover, the concentration of 2,4-DTBP was approximately 6 µg/POCIS for supply water extracts obtained from sampling II, indicating strong abatement during water treatment (6 µg/POCIS and 1 µg/POCIS detected in supply and treated water, respectively).

The second monitoring (sampling III) confirmed the good performances of the treatment system in pollutant removal. 2,4-DTBP was found in the sole supply water at a concentration around 4 µg/POCIS, while there was no evidence of the analyte presence in the treated water. The good performance of aquarium filtration and disinfection systems in 2,4-DTBP removal was an important result due to its endocrine disruptor effect and its aquatic toxicity.

It is worth noting that a lower amount of 2,4-DTBP was sampled in winter (samplings II and III) compared to spring (sampling I), confirming the trend already observed for DPS, which was only detected in POCIS deployed in spring (sampling I).

3.2. Targeted Analysis of Emerging Contaminants

In addition to the evaluation of major volatile and semi-volatile compounds, the second monitoring (sampling III) involved a further investigation on the presence of emerging contaminants (ECs) in aquarium waters (supply and treated). The selected target analytes belong to different classes (pharmaceuticals, UV filters, hormones and perfluorinated compounds); therefore, an LC-MS/MS multi-residual analysis method was used for their determination. The concentration of the analytes was expressed as the amount of analyte per POCIS (ng/POCIS) to obtain a quantitative comparison between supply and treated waters.

Thirteen analytes were identified in POCIS deployed in supply and treated waters (results in Table 2), although many of them were simply detected but not quantifiable. Sampling rates (Rs) of some of the considered chemicals are present in the literature, thus theoretically allowing the estimation of the time-weighted average (TWA) concentration of the contaminants in supply and treated waters. However, the Rs from the literature are generally affected by high variability; thus, analogously to the detected volatile compounds, the TWA concentrations of most ECs were not calculated.

Table 2. Measured mean concentrations (±standard deviation) of the emerging pollutants detected in supply (S3, S4 and S5) and treated (T5, T6 and T7) water (ng/POCIS).

Compound	Supply Water (ng/POCIS)	Treated Water (ng/POCIS)
CAF	<LOQ	<LOD
CBZ	0.51 ± 0.12	<LOQ
BP-3	<LOQ	<LOQ
DIC	<LOQ	<LOQ
OD-PABA	<LOQ	<LOQ
EHMC	15.28 ± 0.13	7.31 ± 0.32
OC	17.92 ± 1.05	5.44 ± 0.41
PFOA	<LOQ	<LOQ
PFOS	<LOQ	<LOQ
BPA	5.97 ± 0.86	<LOQ
MEF	<LOQ	<LOQ
GBZ	1.58 ± 0.01	0.322 ± 0.004
TCS	<LOQ	<LOQ

The data obtained exhibit a general decrease in concentration after treatments, confirming the good performance of filtration and disinfection systems.

The presence of different contaminants in the samples, even though at low concentrations, confirms that these compounds may be considered as semi-persistent in the environment. In fact, they are continuously released from wastewater treatment plants due to incomplete removal [1,15].

The most concentrated chemicals in the supply water were BPA and the UV-filters OC and EHMC.

BPA is an intermediate in the production of epoxy resins and polycarbonate plastics [29]. It is used in many plastic products [30] and is ubiquitous due to its frequent release [31]. BPA represents a potential concern for aquatic organisms, with reproductive and developmental effects on aquatic species [32]. In order to assess the risk for marine organisms, the Risk Characterization Ratio is used, which is the ratio between the Predicted Environmental Concentration (PEC) in water and the provisional predicted no-effect concentration (PNEC). A PNEC value of 0.15 µg L^{-1} for BPA has been proposed for marine waters by the European Union [33]. Considering the literature sampling rate for BPA obtained in seawater of 0.288 L day^{-1} [16], the TWA concentration in water results in 0.0015 µg L^{-1}. By assuming TWA as an acceptable approximation of PEC, the Risk Characterization Ratio (PEC/PNEC) should be below 1 in the analyzed waters. This result suggests that adverse effects are not expected on seawater organisms for this substance [34].

UV filter compounds are integrated in many cosmetic formulations and can reach surface waters through the release from the skin during recreational activity or through wastewater [35]. Relatively high levels of EHMC and OC and the presence of BP-3 and OD-PABA were detected in POCIS extracts. Due to hormonal activity shown by several UV filters at very low concentrations [10], these results underline the importance of their monitoring. No Rs are present in the literature regarding standard POCIS sampling of UV-filters [36]. Hence, it was not possible to estimate TWA concentrations. Nonetheless, the detection of two UV-filters at considerable concentrations in the samplers represent the first report regarding POCIS uptake of these substances from seawater. This result highlights the usefulness of this approach in pre-concentrating contaminants from large water volumes where some contaminants are subjected to huge dilution factors.

4. Conclusions

In this study, an assessment of the contamination of Genoa Aquarium supply, treated waters and shark tank waters was performed.

The untargeted analysis resulted in the identification of hydrocarbons in all samples, and chromatogram profiles showed their petrogenic origins. The industrial pollutants DPS and 2,4-DTBP were found in some samples using the non-targeted approach. Moreover, the presence of different emerging contaminants was investigated by using target analysis: the UV filters OC and EHMC were the most concentrated among ECs both in supply and treated waters. BPA, Gemfibrozil and Carbamazepine were also detected in quantifiable amounts in the supply water. Water treatment system efficacy was evaluated by comparing supply and treated waters: results generally showed a decrease in the concentration of contaminants after treatment.

The passive sampling approach permitted the enhancement of sensitivity and allowed comparisons of water quality based on time-integrative sampling. The data obtained with POCIS are mediated over the deployment period, considering fluctuations and episodic events. On the contrary, the classical sampling approach provides only a snapshot of the pollution levels at the time of the sampling. In addition, a much larger number of spot samples should be collected during the two weeks of exposure, while also considering the shark tank water volume.

The use of POCIS to monitor Genoa Aquarium water quality allowed the detection of the low concentration presence of some endocrine disruptor compounds and substances with aquatic toxicity. A more in-depth study will be necessary to establish contaminant concentrations in water based on sampling rates. This investigation would hopefully allow a risk assessment related to the health of aquarium fishes.

Supplementary Materials: The following supporting information can be downloaded at: https://www.mdpi.com/article/10.3390/app12062951/s1, Figure S1: Deployment of POCIS in beakers during sampling II: on the right the samplers S1 and S2 and on the left T3 and T4; Figure S2: Exposure tank with deployed POCIS during sampling III; Table S1: MRM transitions of the emerging pollutants monitored.

Author Contributions: Conceptualization, E.M., M.D.C., N.P. and E.C.; methodology, E.M. and M.D.C.; data curation, C.S. and B.B.; writing—original draft preparation, C.S. and B.B.; writing—review and editing, C.S., E.M. and M.D.C. All authors have read and agreed to the published version of the manuscript.

Funding: The authors declare that no funds, grants or other support were received during the preparation of this manuscript.

Institutional Review Board Statement: Not applicable.

Informed Consent Statement: Not applicable.

Data Availability Statement: The datasets generated during the current study are available from the corresponding author upon request.

Acknowledgments: Tomas Ocelka from E&H Service is gratefully acknowledged for kindly providing commercial POCIS.

Conflicts of Interest: The authors have no relevant financial or nonfinancial interests to disclose.

References

1. Comeau, F.; Surette, C.; Brun, G.L.; Losier, R. The Occurrence of Acidic Drugs and Caffeine in Sewage Effluents and Receiving Waters from Three Coastal Watersheds in Atlantic Canada. *Sci. Total Environ.* **2008**, *396*, 132–146. [CrossRef] [PubMed]
2. Alygizakis, N.A.; Gago-Ferrero, P.; Borova, V.L.; Pavlidou, A.; Hatzianestis, I.; Thomaidis, N.S. Occurrence and Spatial Distribution of 158 Pharmaceuticals, Drugs of Abuse and Related Metabolites in Offshore Seawater. *Sci. Total Environ.* **2016**, *541*, 1097–1105. [CrossRef] [PubMed]
3. Arpin-Pont, L.; Martínez-Bueno, M.J.; Gomez, E.; Fenet, H. Occurrence of PPCPs in the Marine Environment: A Review. *Environ. Sci. Pollut. Res.* **2016**, *23*, 4978–4991. [CrossRef] [PubMed]
4. Alvarez, D.; Perkins, S.; Nilsen, E.; Morace, J. Spatial and Temporal Trends in Occurrence of Emerging and Legacy Contaminants in the Lower Columbia River 2008–2010. *Sci. Total Environ.* **2014**, *484*, 322–330. [CrossRef]
5. Alvarez, D.A.; Petty, J.D.; Huckins, J.N.; Jones-Lepp, T.L.; Getting, D.T.; Goddard, J.P.; Manahan, S.E. Development of a Passive, in Situ, Integrative Sampler for Hydrophilic Organic Contaminants in Aquatic Environments. *Environ. Toxicol. Chem.* **2004**, *23*, 1640–1648. [CrossRef]
6. Godlewska, K.; Stepnowski, P.; Paszkiewicz, M. Pollutant Analysis Using Passive Samplers: Principles, Sorbents, Calibration and Applications. A Review. *Environ. Chem. Lett.* **2021**, *19*, 465–520. [CrossRef]
7. Mijangos, L.; Ziarrusta, H.; Prieto, A.; Zugazua, O.; Zuloaga, O.; Olivares, M.; Usobiaga, A.; Paschke, A.; Etxebarria, N. Evaluation of Polar Organic Chemical Integrative and Hollow Fibre Samplers for the Determination of a Wide Variety of Organic Polar Compounds in Seawater. *Talanta* **2018**, *185*, 469–476. [CrossRef]
8. Vermeirssen, E.L.M.; Dietschweiler, C.; Escher, B.I.; van der Voet, J.; Hollender, J. Transfer Kinetics of Polar Organic Compounds over Polyethersulfone Membranes in the Passive Samplers Pocis and Chemcatcher. *Environ. Sci. Technol.* **2012**, *46*, 6759–6766. [CrossRef]
9. Di Carro, M.; Magi, E.; Massa, F.; Castellano, M.; Mirasole, C.; Tanwar, S.; Olivari, E.; Povero, P. Untargeted Approach for the Evaluation of Anthropic Impact on the Sheltered Marine Area of Portofino (Italy). *Mar. Pollut. Bull.* **2018**, *131*, 87–94. [CrossRef]
10. Fent, K.; Zenker, A.; Rapp, M. Widespread Occurrence of Estrogenic UV-Filters in Aquatic Ecosystems in Switzerland. *Environ. Pollut.* **2010**, *158*, 1817–1824. [CrossRef]
11. Tanwar, S.; Di Carro, M.; Magi, E. Innovative Sampling and Extraction Methods for the Determination of Nonsteroidal Anti-Inflammatory Drugs in Water. *J. Pharm. Biomed. Anal.* **2015**, *106*, 100–106. [CrossRef]
12. Mirasole, C.; Di Carro, M.; Tanwar, S.; Magi, E. Liquid Chromatography–Tandem Mass Spectrometry and Passive Sampling: Powerful Tools for the Determination of Emerging Pollutants in Water for Human Consumption. *J. Mass Spectrom.* **2016**, *51*, 814–820. [CrossRef]
13. Jeong, Y.; Schäffer, A.; Smith, K. A Comparison of Equilibrium and Kinetic Passive Sampling for the Monitoring of Aquatic Organic Contaminants in German Rivers. *Water Res.* **2018**, *145*, 248–258. [CrossRef]
14. Jones, L.; Ronan, J.; McHugh, B.; Regan, F. Passive Sampling of Polar Emerging Contaminants in Irish Catchments. *Water Sci. Technol.* **2019**, *79*, 218–230. [CrossRef]
15. Martínez Bueno, M.J.; Herrera, S.; Munaron, D.; Boillot, C.; Fenet, H.; Chiron, S.; Gómez, E. POCIS Passive Samplers as a Monitoring Tool for Pharmaceutical Residues and Their Transformation Products in Marine Environment. *Environ. Sci. Pollut. Res.* **2016**, *23*, 5019–5029. [CrossRef]
16. Shi, X.; Zhou, J.L.; Zhao, H.; Hou, L.; Yang, Y. Application of Passive Sampling in Assessing the Occurrence and Risk of Antibiotics and Endocrine Disrupting Chemicals in the Yangtze Estuary, China. *Chemosphere* **2014**, *111*, 344–351. [CrossRef]
17. Martínez Bueno, M.J.; Hernando, M.D.; Agüera, A.; Fernández-Alba, A.R. Application of Passive Sampling Devices for Screening of Micro-Pollutants in Marine Aquaculture Using LC-MS/MS. *Talanta* **2009**, *77*, 1518–1527. [CrossRef]
18. Harman, C.; Allan, I.J.; Vermeirssen, E.L.M. Calibration and Use of the Polar Organic Chemical Integrative Sampler—A Critical Review. *Environ. Toxicol. Chem.* **2012**, *31*, 2724–2738. [CrossRef]
19. Di Carro, M.; Bono, L.; Magi, E. A Simple Recirculating Flow System for the Calibration of Polar Organic Chemical Integrative Samplers (POCIS): Effect of Flow Rate on Different Water Pollutants. *Talanta* **2014**, *120*, 30–33. [CrossRef]
20. Di Carro, M.; Lluveras-Tenorio, A.; Benedetti, B.; Magi, E. An Innovative Sampling Approach Combined with Liquid Chromatography–Tandem Mass Spectrometry for the Analysis of Emerging Pollutants in Drinking Water. *J. Mass Spectrom.* **2020**, *55*, e4608. [CrossRef]
21. Commendatore, M.G.; Esteves, J.L. Natural and Anthropogenic Hydrocarbons in Sediments from the Chubut River (Patagonia, Argentina). *Mar. Pollut. Bull.* **2004**, *48*, 910–918. [CrossRef] [PubMed]
22. Kalmykova, Y.; Moona, N.; Strömvall, A.M.; Björklund, K. Sorption and Degradation of Petroleum Hydrocarbons, Polycyclic Aromatic Hydrocarbons, Alkylphenols, Bisphenol A and Phthalates in Landfill Leachate Using Sand, Activated Carbon and Peat Filters. *Water Res.* **2014**, *56*, 246–257. [CrossRef]

23. Gaines, R.B.; Frysinger, G.S.; Hendrick-Smith, M.S.; Stuart, J.D. Oil Spill Source Identification by Comprehensive Two-Dimensional Gas Chromatography. *Environ. Sci. Technol.* **1999**, *33*, 2106–2112. [CrossRef]

24. Ou, S.; Zheng, J.; Zheng, J.; Richardson, B.J.; Lam, P.K.S. Petroleum Hydrocarbons and Polycyclic Aromatic Hydrocarbons in the Surficial Sediments of Xiamen Harbour and Yuan Dan Lake, China. *Chemosphere* **2004**, *56*, 107–112. [CrossRef]

25. Wick, L.Y.; Gschwend, P.M. By-Products of a Former Phenol Manufacturing Site in a Small Lake Adjacent to a Superfund Site in the Aberjona Watershed. *Environ. Health Perspect.* **1998**, *106*, 1069–1074. [CrossRef] [PubMed]

26. Skjevrak, I.; Due, A.; Gjerstad, K.O.; Herikstad, H. Volatile Organic Components Migrating from Plastic Pipes (HDPE, PEX and PVC) into Drinking Water. *Water Res.* **2003**, *37*, 1912–1920. [CrossRef]

27. Varsha, K.K.; Devendra, L.; Shilpa, G.; Priya, S.; Pandey, A.; Nampoothiri, K.M. 2,4-Di-Tert-Butyl Phenol as the Antifungal, Antioxidant Bioactive Purified from a Newly Isolated *Lactococcus* sp. *Int. J. Food Microbiol.* **2015**, *211*, 44–50. [CrossRef]

28. Wang, J.; Wang, J.; Liu, J.; Li, J.; Zhou, L.; Zhang, H.; Sun, J.; Zhuang, S. The Evaluation of Endocrine Disrupting Effects of Tert-Butylphenols towards Estrogenic Receptor A Androgen Receptor and Thyroid Hormone Receptor B and Aquatic Toxicities towards Freshwater Organisms. *Environ. Pollut.* **2018**, *240*, 396–402. [CrossRef]

29. Ademollo, N.; Patrolecco, L.; Rauseo, J.; Nielsen, J.; Corsolini, S. Bioaccumulation of Nonylphenols and Bisphenol A in the Greenland Shark Somniosus Microcephalus from the Greenland Seawaters. *Microchem. J.* **2018**, *136*, 106–112. [CrossRef]

30. Barboza, L.G.A.; Cunha, S.C.; Monteiro, C.; Fernandes, J.O.; Guilhermino, L. Bisphenol A and Its Analogs in Muscle and Liver of Fish from the North East Atlantic Ocean in Relation to Microplastic Contamination. Exposure and Risk to Human Consumers. *J. Hazard. Mater.* **2020**, *393*, 122419. [CrossRef]

31. Flint, S.; Markle, T.; Thompson, S.; Wallace, E. Bisphenol A Exposure, Effects, and Policy: A Wildlife Perspective. *J. Environ. Manag.* **2012**, *104*, 19–34. [CrossRef] [PubMed]

32. Miglioli, A.; Balbi, T.; Besnardeau, L.; Dumollard, R.; Canesi, L. Bisphenol A Interferes with First Shell Formation and Development of the Serotoninergic System in Early Larval Stages of *Mytilus galloprovincialis*. *Sci. Total Environ.* **2021**, *758*, 144003. [CrossRef] [PubMed]

33. Aschberger, K.; Munn, S.; Olsson, H.; Pakalin, S.; Pellegrini, G.; Vegro, S.; Paya Perez, B. *Updated European Union Risk Assessment Report: 4,4'-Isopropylidenediphenol (Bisphenol-A): Environment Addendum of February 2008*; Publications Office: Luxembourg, 2010. [CrossRef]

34. Muñoz, I.; Martínez Bueno, M.J.; Agüera, A.; Fernández-Alba, A.R. Environmental and Human Health Risk Assessment of Organic Micro-Pollutants Occurring in a Spanish Marine Fish Farm. *Environ. Pollut.* **2010**, *158*, 1809–1816. [CrossRef] [PubMed]

35. Nguyen, K.T.N.; Scapolla, C.; Di Carro, M.; Magi, E. Rapid and Selective Determination of UV Filters in Seawater by Liquid Chromatography–Tandem Mass Spectrometry Combined with Stir Bar Sorptive Extraction. *Talanta* **2011**, *85*, 2375–2384. [CrossRef] [PubMed]

36. MacKeown, H.; Magi, E.; Di Carro, M.; Benedetti, B. Unravelling the role of membrane pore size in polar organic chemical integrative samplers (POCIS) to broaden the polarity range of sampled analytes. *Anal. Bioanal. Chem.* **2022**, *414*, 1963–1972. [CrossRef]

applied sciences

Article

Assessment of PCBs in Surface Waters at Ultratrace Levels: Traditional Approaches and Biomonitoring (Lake Baikal, Russia)

Alexander G. Gorshkov *, Olga V. Kustova and Yurij S. Bukin

Limnological Institute, Siberian Branch of the Russian Academy of Sciences, Ulan-Batorskaya St. 3, 664033 Irkutsk, Russia; delete_21@mail.ru (O.V.K.); bukinyura@mail.ru (Y.S.B.)
* Correspondence: gorchkov_ag@mail.ru

Abstract: This article presents the results of the assessment of PCB concentrations in surface waters at ultratrace level of concentrations. The assessment of PCB concentrations is based on data from monitoring PCBs in Baikal water within the conventional approach as well as from biomonitoring of PCBs using Baikal omul, *Coregonus migratorius*, Georgi, 1775 (*C. migratorius*), as a bioindicator. The time cycle of the monitoring covered the period from 2014 to 2021. The concentrations of PCBs in the water were estimated from the concentrations of seven indicator congeners: 28, 52, 101, 118, 138, 153, and 180, and from congeners of dioxin-like (dl) PCBs in the tissues of *C. migratorius*. The average value and the statistically significant range of the detected total concentrations (Σ_7PCBs) in Baikal water were 0.30 and 0.26–0.34 ng/L, respectively. In the tissues of *C. migratorius*, the average value and the range of Σ_7PCB concentrations were 5.6 and 4.9–6.3 ng/g (ww), respectively, and for dl-PCBs, 1.5 and 1.3–1.7 ng/g (ww), respectively. The total toxicity equivalent of the detected dl-PCBs was in the WHO-TEQ $_{(2005)}$ range from 0.03 to 0.06 pg/g (ww). The concentrations of Σ_7PCBs in Baikal water and dl-PCBs in the tissues of *C. migratorius* corresponded to the concentration levels in the European alpine lakes and the tissues of *S. trutta* fish inhabiting these lakes.

Keywords: PCBs; monitoring; biomonitoring; statistical analysis; Lake Baikal

Citation: Gorshkov, A.G.; Kustova, O.V.; Bukin, Y.S. Assessment of PCBs in Surface Waters at Ultratrace Levels: Traditional Approaches and Biomonitoring (Lake Baikal, Russia). *Appl. Sci.* **2022**, *12*, 2145. https://doi.org/10.3390/app12042145

Academic Editor: Elida Nora Ferri

Received: 24 January 2022
Accepted: 14 February 2022
Published: 18 February 2022

Publisher's Note: MDPI stays neutral with regard to jurisdictional claims in published maps and institutional affiliations.

1. Introduction

Accumulation of persistent organic pollutants (POPs) in surface waters is the most important problem of the international community within the framework of its Sustainable Development Goal. The development of reliable methods for monitoring and assessing the concentrations of POPs at the background level of concentrations is one of the tasks, the solution of which requires priority measures [1,2]. Among POPs, polychlorinated biphenyls (PCBs) stand out for their exceptional chemical and biological stability and global distribution: pollutants of this class are found in air, water, soil, and biota. PCBs have many unique physical and chemical properties that have determined their widespread application in various fields of technology. The total world production of PCBs is 1.5 million tons, of which approximately 10% remain in the environment [3–6]. When their toxic effect on living organisms was determined in extremely low doses, together with the ability to accumulate in biological objects, the Stockholm Convention prohibited the production and use of PCBs. They were included in the list of POPs and the OSPAR Convention included them in the list of priority pollutants [7,8].

In the surface water of background areas, PCBs are present at ultratrace concentration levels. For example, in the Arctic water, PCBs were detected in the concentration range from 1.3 to 21 pg/L [9,10]. It should be noted that data on the concentrations of PCBs in the water is scarce. Methodological problems may complicate the detection of PCBs at trace levels of concentrations, in particular, due to hydrophobicity of PCBs, i.e., sorption on the surface of suspended particles in the water column and accumulation in biota lipids. Increasing the volume of water samples up to 100 L or more [11,12] to reduce the lower limit of the range of detected PCB concentrations increases expended time and resources. A decrease in the number of samples (during the multiple high-volume sampling) increases the likelihood of missing local areas with high concentrations of PCBs. Otherwise, estimation based on the extreme short-term periods of water pollution in individual areas is possible. Passive sampling techniques [13] can estimate the concentrations of dissolved PCBs, but the assessment of the total concentration of PCBs in two phases of water matrix (aqueous solution and solid phase) as a biota habitat is more significant at the background level of pollution. In turn, the detection of PCBs at trace levels of concentration with the required reliability and accuracy is possible using modern gas chromatography–mass spectrometry techniques that provide high sensitivity and specificity [14,15].

Biological monitoring, a determination of pollutants in aquatic bioindicator organisms directly exposed to these substances, is an alternative method to control POPs in surface water. Bioindicators must have certain relevant characteristics, in particular, accumulate PCBs in the amounts sufficient for their detection; their habitat must correspond to the boundaries of the studied water body, and it is possible to collect a sufficient number of specimens of these organisms [16]. Fish is the most widespread bioindicator of PCBs in water bodies; PCBs reflecting water pollution have been detected over a wide range of concentrations for both marine and freshwater species [17–27]. The dependence of the levels of PCB accumulation in fish on the concentrations of these substances in the water was used to assess temporal and spatial variations in surface waters, investigate vertical transport of pollutants from the surface to deep water layers [19,28,29], and estimate the content of pollutants at the background level of concentrations [30–34].

When the industrial production of PCBs was ceased, their concentrations in fish, bottom sediments, and the atmosphere of the Arctic regions decreased [35–37]. At the present stage, the sources of PCBs include dismantling and disposal of equipment containing these substances as a structural material as well as various chemical and thermal processes, during which PCBs are formed as byproducts from landfills, waste disposal sites, or illegal stockpiles [38–41]. As shown in [3,42], the number of PCBs in the environment is currently not decreasing but even increasing in some regions of the world. For instance, as the climate warms, many POPs accumulated in the Arctic environment re-enter the atmosphere as a result of sea ice retreat and temperature rise [43].

We chose Lake Baikal as a model for estimating PCBs in surface water at ultratrace levels of concentration. Lake Baikal covers an area of 31,700 km^2 and contains up to 20% of the world reserves of surface freshwater, ~23,000 km^3. Water in the lake is clean and has a low mineralization degree, and a minimal content of suspended organic matter and POPs [44–47]. The presence of PCBs in the aquatic ecosystem of Lake Baikal was first reported in [48]. At that time, the level of their concentrations in the water corresponded to the world background level [11,49–51]. According to independent studies, there was up to a 4.5-fold increase in the PCB concentrations in 2014 and 2015 [11,52] compared to the levels recorded in 1992 and 1993.

The increase in the PCB concentrations in Baikal water is of special concern because the lake is the most important water resource on the planet. This has highlighted the scientific problem of PCB control in the ecosystem of Lake Baikal, which is associated with the detection of PCBs in the surface water at the ultratrace level of concentrations. This study evaluates the results of PCB monitoring, which were obtained during the observation period from 2014 to 2021 at Lake Baikal. Monitoring of PCBs was carried out within the framework of the conventional approach, including water sampling according to a standardised procedure followed by the determination of the required pollutants in them by the GC–MS/MS method, as well as within the framework of biomonitoring using Baikal omul, *Coregonus migratorius*, Georgi, 1775 (*C. migratorius*), as a bioindicator. To identify the relationship between the concentrations of indicator PCB congeners and their total concentrations in the water and *C. migratorius*, key factors influencing the concentrations of PCBs in the water and bioindicator, we also carried out a comparative statistical analysis.

2. Materials and Methods

2.1. Sampling

Water samples from the pelagic zone of Lake Baikal were collected by using an SBE-32 cassette sampler (Carousel Water Sampler, Sea-Bird Electronics, Bellevue, WA, USA) at 21 stations (Figure 1) in May 2015, May and September 2016, May 2018, May and September 2019, and June and September 2020 and 2021 (Table S1, supplementary materials). At each station, two samples were taken in 1 L glass bottles, to which 0.5 mL of a 1 M aqueous solution of sodium azide (Merck, Darmstadt, Germany) was added as a preserving agent. Water bottles were closed using a lid with an aluminium foil gasket and stored at 5 °C until laboratory analysis.

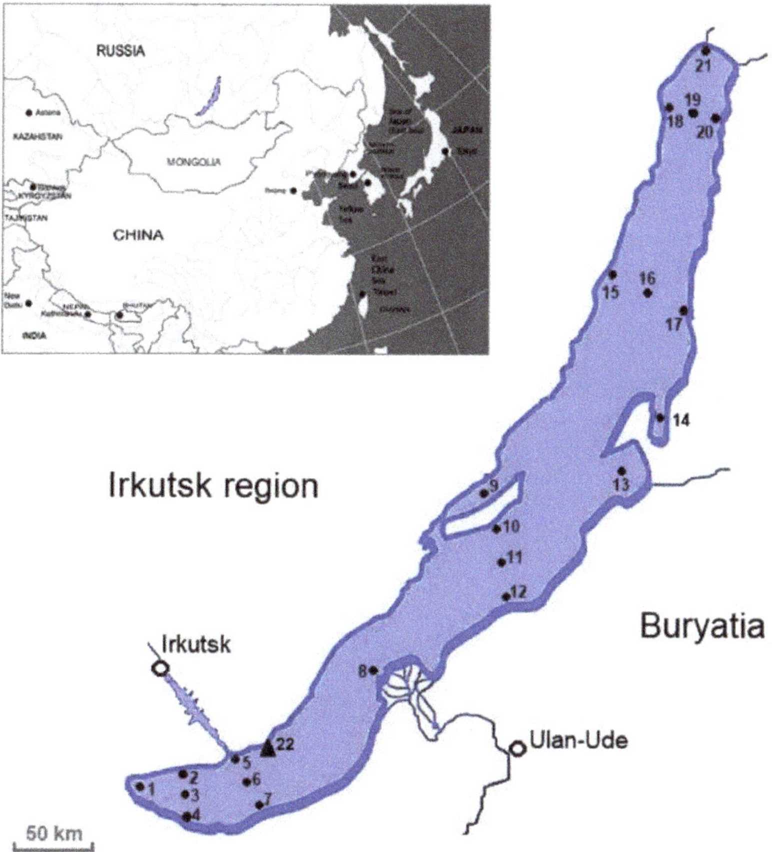

Figure 1. Map of Lake Baikal and sampling sites. At stations 1, 2, 4, 5, 7, 10, 12, 17, 18, 20, and 21, sampling was carried out from the upper water layer (5 m) at a distance of up to 3 km from the coast; at station 3, 6, 8, 11, 16, and 19 at reference central stations; at station 6—from the water column at depths of 5 to 1200 m; at station 11—from the water column at depths of 5 to 1600 m; at station 16—from the water column at depths of 5 to 800 m; and at station 22, *C. migratorius* was caught.

Baikal omul, *Coregonus migratorius*, Georgi, 1775 (*C. migratorius*), is fish species meeting the criteria for biological objects to be selected as bioindicators [16]: (i) *C. migratorius* migrates throughout the entire water area of Lake Baikal in a depth range up to 350 m, including the zones of river runoff; (ii) it is a commercial fish species, ensuring economic efficiency of the sampling stage during monitoring; (iii) individuals at the age of three to seven years accumulate PCBs in an amount sufficient for their measurement with the required accuracy [53,54]. *Coregonus migratorius* was caught near station 22 (Figure 1) in March to July 2014, February 2016, March 2018–2020, and September 2017–2018. Collected fish were packed in aluminium foil, frozen, and stored at a temperature of ≤minus 18–20 °C. Identification characteristics of the selected Baikal omul (sex and age, Table S2) were determined according to the methods in [55,56]. The group of fish selected as bioindicators of PCBs included *C. migratorius* aged from three to seven years. The choice of this species of Baikal fish as bioindicators ensured the efficiency of the sampling because Baikal omul belongs to a commercial species.

The following reasons determined the choice of sampling seasons for water and *C. migratorius*. In the spring, after the ice melts, PCBs enter the lake water due to the melting of ice on the water surface of the lake and snow on its coast, in which PCBs have accumulated during the winter. Water samples collected in autumn reflect the accumulation of PCBs in water, which have come from the atmosphere in summer.

2.2. Sample Processing

In water samples and omul specimens, indicator congeners 28, 52, 101, 118, 138, 153, and 180, and dioxin-like congeners (dl-PCBs) 105, 114, 118, 123, and 126 (dl-PCBs 118 is one of the indicator congeners) were determined according to the technique shown in [55,56] (Tables S3–S5). This technique included the following steps: extraction by n-hexane from 1 L of unfiltered water, the concentration of the extract to the volume of ~0.1 mL, and direct analysis of the extracts by GC–MS/MS. Before the determination of PCBs in *C. migratorius*, the fish selected for the analysis were thawed; the head, fins, and entrails of the fish were removed. The fish was then homogenised, discarding the skin, and large bones. PCBs were extracted from the mince via ultrasound extraction with a mixture of organic solvents: n-hexane/acetone (1:1). The extracts were purified with concentrated H_2SO_4.

2.3. Instrumental Analysis

Prepared extracts of the samples were analysed using an Agilent Technologies 7890B GC System 7000C GC–MS Triple Quad chromatography–mass spectrometer with an HT-8, SCE Analytical Science capillary column (30 m × 0.25 mm × 0.25 μm). PCB peaks were recorded using the MRM mode and identified by relative retention times. PCB congeners were quantitated by the method of internal standards using a Marker-7 PCB Mixture ^{13}C (Cambridge Isotope Laboratories, Inc., Shirley, NY, USA) as a surrogate internal standard (Tables S3 and S4). The limits of determination (LOD) of PCB congeners in the water and fish were 0.01–0.02 ng/L and 0.10–15 ng/g (ww), respectively. The LOD method was assessed based on the peak-to-peak signal-to-noise response of each of the PCB peaks (N/S) and at the lowest standard concentration. Values for S/N > 10 were employed to determine the limit of quantitation (LOQ) for each target compound. Relative standard deviation (RSD_{Rl}) for the determination procedure was 35% for individual indicator congeners in the water: 28, 52, 101, 118, 138, 153, and 180; 25% for individual indicator congeners and dl-PCBs in the fish: 28, 52, 101, 105, 114, 118, 123, 126, 138, and 153; and 45% for congener 180.

2.4. Toxicity Assessment

The dioxin toxicity equivalents (TEQ) for dl-PCBs were assessed as a product of the concentration of dl-PCB congeners (C) and its corresponding toxicity equivalency factor (TEF) established by the International Program on Chemical Safety (IPCS) and World Health Organisation (WHO):

$$TEQ = C \times TEF,$$

where C (pg/L) and TEF_i are the concentrations and toxic equivalent factors (TEFs) of the dl-PCB congeners [57].

2.5. Statistical Methods

Confidence intervals for the mean values of the concentrations of PCB congeners and the total concentrations of seven indicator PCB congeners (Σ_7PCBs) were estimated using the bootstrap method in the «boot» package and the R programming language. Because of technical problems, a small part of the primary data on some PCB congeners for *C. migratorius* was unspecified. For further analysis, these missing data were replaced with averages for these congeners according to the recommendations [58].

PERMANOVA testing in the R package 'vegan' [59] was used for analysing an impact of the explanatory parameter: the year, month, the basin of Lake Baikal, and sampling depth for water, along with the year, month, age, and sex for *C. migratorius* (Tables S6 and S7). In the range of concentrations of the studied PCB *congeners* and Σ_7PCBs in the samples of Baikal water and *C. migratorius*, *p* values of the PERMANOVA analysis were corrected for false discovery rate in multiple comparisons using the Benjamini–Hochberg equation [60].

To determine the similarity and difference between water samples and separately between the specimens of *C. migratorius* in the concentrations of various PCB *congeners* and Σ_7PCBs, principal component analysis (PCA) was applied by using the «factoextra» package [61] for the R programming language. For analysis, data on water and *C. migratorius* were previously converted into 'integral samples', which were average values for the concentrations of PCB congeners over a set of explanatory parameters, indicating, according to the results of PERMANOVA analysis, a significant effect on PCB concentrations in the samples. Before PCA, all PCB concentrations were transformed to eliminate the physical dimensions by ranging from 0 to 1.

Pairwise correlations between all concentrations of the studied PCB *congeners* and the total concentration of PCB in samples of Baikal water and Baikal omul (*C. migratorius*) were estimated with Spearman's *r* correlation coefficient. Only reliable correlation coefficients ($p < 0.05$) were included in analysis. *p* values for the correlation coefficients were calculated using Spearman's «W» statistics and corrected for false discovery rate in multiple comparisons using the Benjamini–Hochberg equation [60]. Unreliable values of correlation coefficients were replaced with 0 values. Pairwise correlations were visualized with a heat map generated using «gplots» [62] in R. Lines and columns in the correlation matrix were clustered and grouped in order of similarity (i.e., Euclidean distance metric and the complete-link clustering method).

To determine the relationship between the concentrations of the studied PCB congeners and Σ_7PCBs for the samples of Baikal water and *C. migratorius* averaged over the year of sampling, the Mantel test [63,64] was used based on the Pearson correlation coefficient; the significance of the correlation was determined by the permutation test (10,000 permutations). Distance matrices for the Mantel test that determine the differences for concentration of PCB congeners between the water samples and between *C. migratorius* samples were calculated using the Euclidean metric. The Mantel test was performed using the R package 'vegan' [59].

Relative error statistics for repeated measurements of PCB concentrations for all samples from Lake Baikal were estimated as a relative difference between the maximum and minimum concentration of analytical repetitions of one measurement (relative error = $100 \times (\text{max} - \text{min})/\text{max}$, %) (Tables S8 and S9).

All primary data are available in tables table_Sup_1.xlsx—water data, table_Sup_2.xlsx— *C. migratorius* data in supplementary material.

3. Results and Discussion

3.1. PCBs in the Upper Water Layer of the Pelagic Zone and Deep Layers of the Water Column of Lake Baikal

The concentrations of PCBs in the water of Lake Baikal were characterised by seasonal and interannual variability (Figure 2). In May 2015, we detected the maximum total concentrations of PCB congeners (ΣPCBs) ranging from 1.4 to 6.6 ng/L. There were 24 to 34 congeners of PCBs dominated by homologues of penta-dichlorobiphnyls as well as 7 indicator congeners 28, 52, 101, 118, 153, 138, and 180. The total concentrations of indicator PCB congeners ($\sum_7$PCBs) ranged from 0.43 to 2.0 ng/L, and the contribution from $\sum_7$PCBs to the total concentration of all detected congeners ranged from 28 to 35%. In September 2019, the concentration of $\sum_7$PCBs also increased at some stations (19 and 21) to 0.24–2.8 ng/L with an increase in the number of the detected congeners to 27. Indicator congeners were detected in the concentration range from 0.17 to 1.3 ng/L with a contribution to $\sum$PCBs of up to 40–70%. The minimum concentrations of $\sum$PCBs in Baikal water ranged from 0.11 to 0.82 ng/L, and the number of the detected congeners were recorded in September 2016, May 2018 and 2019, and in the spring and summer of 2020 and 2021. The concentrations of $\sum_7$PCBs in these seasons ranged from $\leq$0.11 to 0.47 ng/L. The average concentration and statistically significant range of detected Σ_7PCB concentrations in Baikal water were 0.30 and 0.26–0.34 ng/L, respectively. Notably, the indicated level of the detected $\sum$PCB concentrations was comparable with the concentrations of PCBs in Lake Baikal identified previously in 1992 and 1993 (0.08–1.9 ng/L) and estimated as background [49,51].

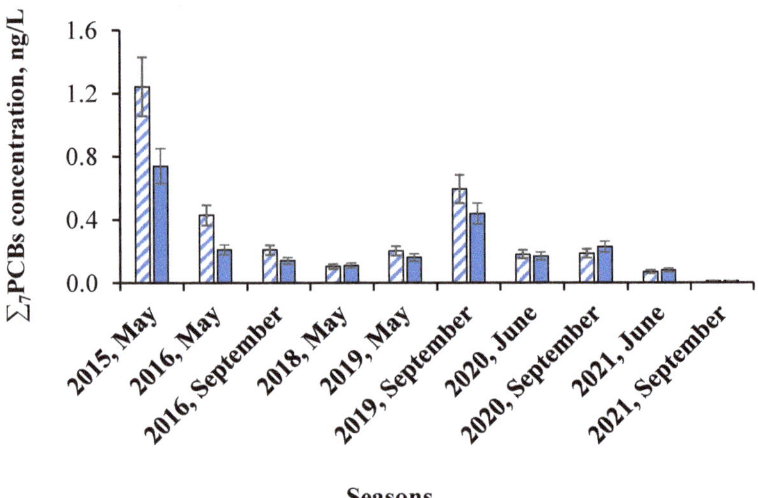

Figure 2. Average concentrations of $\sum_7$PCBs in the pelagic zone and deep layers of Lake Baikal determined during the monitoring of PCBs in Baikal water.

The global atmospheric transport is the dominant source of PCBs in Baikal water [11,49–51]. An increase in the concentrations of $\sum$PCBs in May 2015 and September 2019 was likely due to the influx of PCBs from additional sources, which a change in the ratio of indicator congeners indicated. In spring 2015, the proportion of indicator congeners with a high degree of chlorination (138, 153, and 180) increased in water samples up to 30–37%. In autumn 2019, the proportion of pentachlorinated congeners 101 and 118 increased, and the proportion of 'light' tri- and tetrachlorinated congeners 28 and 52 decreased compared to the monitoring data during spring 2015. During the minimum concentrations of PCBs in the water of the pelagic zone of Lake Baikal, tri-, tetra-, and pentachlorinated homologues (28, 52, 101, and 118) predominated in the composition of indicator congeners, whose total concentrations reached 80–90% in the fraction of indicator PCBs.

Seasonal changes in the PCB concentrations in the upper water layer (5 m) of the pelagic zone in May 2015 and September 2019 were not tracked throughout the water column to depths of 800–1600 m (depending on the depth of the lake's basin). In May 2015, a high level of Σ_7PCBs in the upper water layer (5 m) of the southern basin of the lake, up to 3.1 ng/L, was recorded at some stations. However, in deep water layers of the water column (50–1200 m), the Σ_7PCB concentrations remained in a narrow range from 0.20 to

0.51 ng/L. An increase in the Σ_7PCB concentrations in the upper water layer at stations 19 and 21 to 1.6–2.2 ng/L in autumn 2019 was also not accompanied by an increase in the Σ_7PCB concentrations at depths of 50 to 800 m, which remained at a minimum level from 0.31 to 0.69 ng/L.

3.2. PCBs in Coregonus migratorius

We studied Baikal omul, *C. migratorius*, as a bioindicator of PCBs in Baikal water. In the course of biomonitoring, we collected and analysed 46 individuals of the *C. migratorius* population of Lake Baikal. The analysis revealed that the concentration of PCBs in fish specimens depends on the season and year of sampling (Figure 3). In the investigated specimens, we detected from 30 to 47 PCB congeners, including seven indicator congeners. The average concentrations and statistically significant range of the detected Σ_7PCBs in *C. migratorius* were 5.6 and 4.9–6.3 ng/g (ww), respectively.

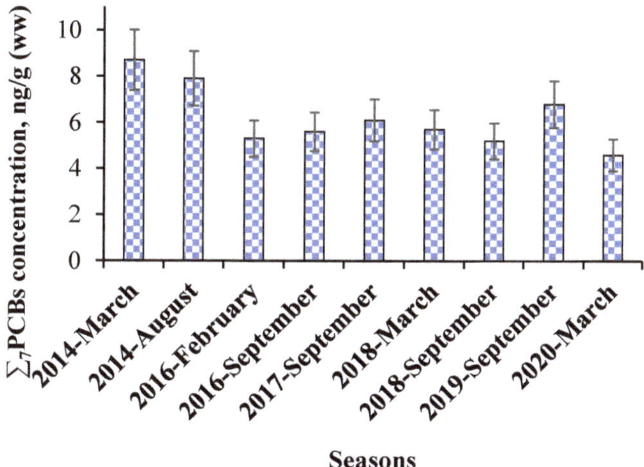

Seasons

Figure 3. Average concentrations of Σ_7PCBs in specimens of *C. migratorius* (per season) collected during the monitoring of PCBs in Baikal water.

Biomonitoring of PCBs in the lake water revealed five dl-congeners: 105, 114, 123, 126, and 156, which were present in the water below the LOQ of the method used to analyze the water. The total concentrations of dl-congeners in *C. migratorius* ranged from 1.3 to 1.7 ng/g (ww). The most toxic congener, 126 (the toxicity equivalence factor, TEF$_{2005}$ is 0.1), was found in single individuals of the older age group (4 out of the 46 investigated individuals), reaching 0.5 ng/g (ww).

The age of fish is a critical biomonitoring factor, since the concentration of PCBs in individuals of the older age group increases [31]. During the biomonitoring, we chose individuals of *C. migratorius* aged from three to seven years. In this group of fish, the level of concentration of Σ_7PCBs changes insignificantly: Σ_7PCBs were 3.9 (3 year), 4.9 (5 year) and 5.4–5.6 (6–7 years) ng/g (ww), and the proportion of congeners with a high degree of chlorination (153, 138, 180) increased. In the muscle tissues of the individuals of the older age group (8 years, 2 specimens), the number of PCBs reached extreme values: Σ_7PCBs to 32 ng/g (ww) and Σ_{dl}PCBs to 9.0 ng/g (ww).

3.3. Assessment of the PCB Concentration in the Water of Lake Baikal

Global atmospheric transport is the dominant source of PCBs in the aquatic ecosystem of Lake Baikal. Therefore, the assessment of the concentrations of pollutants of this class in Baikal water by the concentrations of indicator congeners is reasonable and provides adequate results [65]. We carried out a comparative statistical analysis to study the influence of the year and season of sampling, locations of sampling stations in the pelagic zone of the lake and layers of the water column, and the age and sex of omul on the ratio and concentrations of indicator congeners of PCBs detected in the water samples and tissues of *C. migratorius*. To identify relationships between the concentrations of PCBs in Baikal water and *C. migratorius*, selected as a bioindicator of PCBs, a correlation analysis of the results of determining PCBs in water and *C. migratorius* was carried out.

3.3.1. Statistical Analysis of the Results of 'Conventional' Monitoring

The results of the PERMANOVA analysis (Table S6) indicated that among all studied factors only year and month of sampling significantly influenced the concentrations of the investigated PCB congeners, including Σ_7PCBs, in Baikal water (year factor—$R^2 = 0.163$; month factor—$R^2 = 0.074$; p-value < 0.05). At the same time, based on the coefficients of the R^2 covariation, the sampling year factor had the maximum effect on the concentrations of PCBs. Further data analysis by PCA based on the PERMANOVA results was carried out for the values averaged by the factors of the pelagic zone of the lake's basin and sampling depth.

The analysis of the results of the determination of PCB indicator congeners by PCA revealed that all data were divided into two groups (Figure 4a). The first group in the second quarter of the coordinate plane included the average data obtained in September 2016, May 2018–2019, and June and September 2020–2021. All points of this group were characterised by minimum concentrations of all indicator congeners and Σ_7PCB concentration ranging from ≤ 0.11 to 0.47 ng/L. The second group (the first quarter of the coordinate plane) includes the data obtained in May 2016 and September 2019. These data referred to extreme concentrations of all indicatory congeners and Σ_7PCBs (1.3–2.0 ng/L) relative to the minimum level determined during the monitoring between 2015 and 2021.

On the PCA biplot (Figure 4a), the PCB concentration gradient vectors form a codirectional group that includes all investigated indicator congeners of PCBs and Σ_7PCBs, except for congener 28 (a congener group with a positive correlation). The vector of congener 28 is directed almost perpendicular to the main group, and the concentration of this pollutant depends (correlates) to a lesser extent on the concentrations of other indicator PCBs. Perhaps, the sources of congener 28, its further fate in the aquatic ecosystem, and (or) congener properties (the minimum hydrophobicity and the maximum water solubility among indicator congeners) affect its concentration in Baikal water.

The PCA data on the concentrations of PCBs in Baikal water allowed us to draw the following conclusions: (i) an increase or a decrease in all indicator congeners, except for congener 28 and the concentration of Σ_6PCBs, occurred synchronously; (ii) within the framework of conventional monitoring, the assessment of the PCB concentrations in Baikal water based on the total concentration of six indicator congeners (Σ_6PCBs, excluding congener 28) would meet more stringent criteria.

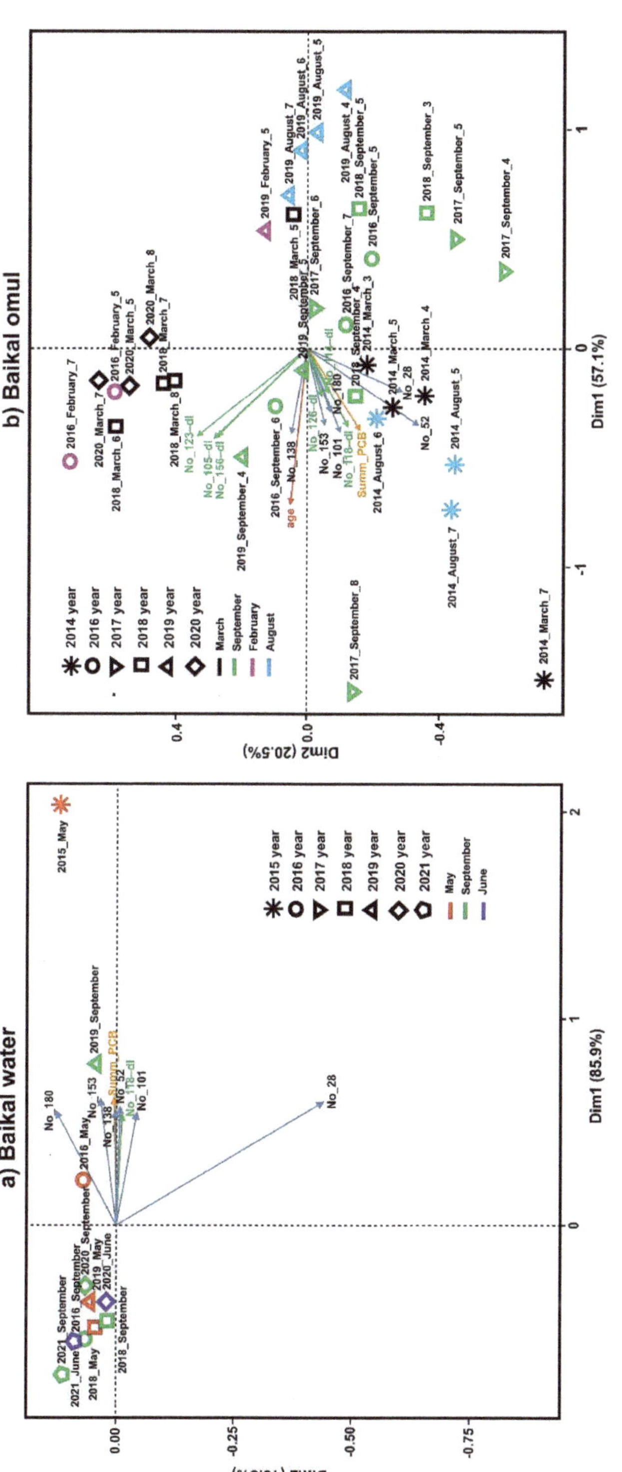

Figure 4. Results of the concentration distribution analysis of the investigated PCB types and the total PCB concentration based on PCA. (**a**) Analysis of water samples: point on the plane characterise the average PCB concentrations in Baikal water (entire water area of the lake) by the year and month of sampling (a point is the average PCB concentration in water samples taken in a certain month of one year); (**b**) analysis of the distribution of PCB concentrations in omul specimens: points on the plane characterise the average PCB concentrations in omul by the year, month, and age of the individuals (a point is the average PCB concentration in individuals of the same age collected in the same year and the same month). Vectors on the planes show the directions of the concentration gradients of the investigated PCB types (blue vectors and green vectors are dioxin-like PCBs), the total PCB concentration (orange vectors), and the age of omul individuals (red vector).

3.3.2. Statistical Analysis of Biomonitoring Results

Based on the PERMANOVA data (Table S7), among all studied factors, the year and month of sampling and the age of individuals (year factor—R^2 = 0.193; month factor—R^2 = 0.107; age factor—R^2 = 0.075; p-value < 0.05) significantly influenced the concentrations of PCBs in the muscle tissue of *C. migratorius*. The R^2 covariation coefficients indicated that the year of sampling was the main factor of influence followed by the less significant month of sampling and the age of fish (Figure 4b). As the sex of the *C. migratorius* individuals was not a critical factor (R^2 = 0.002), PCA for *C. migratorius* was based on the sex-averaged data.

PCA of the concentrations of PCB congeners in *C. migratorius* distinguished three groups of averaged data (Figure 4b). The first group is located in the upper part of the coordinate plane at the border of the first and second quarters and included the data on *C. migratorius* caught in February 2016 and March 2018 and 2020. The age of the individuals in this group ranged from five to eight years.

All fish from this group were characterised by high dl-PCB concentrations ranging from 1.0 to 2.2 ng/g (ww), whose proportion in Σ_7PCBs reached 33%, and by relatively low total concentrations of congeners 28 and 52, ranging from 1.1 to 2.6 ng/g (ww). Notably, this group of the *C. migratorius* fish was caught in the spring during the freezing period on the lake after the winter lifecycle with the corresponding feeding structure [66].

The second group of data located at the border of the first and fourth quarters of the coordinate plane included the results of the determination of PCB concentrations in *C. migratorius* that were caught at the end of the feeding period in August and September of 2016, 2017, 2018, and 2019. In this group of specimens, the concentration of PCBs was in the Σ_7PCB range from 5.2 to 6.8 ng/g (ww) with the concentrations of Σ_{dl}PCBs ranging from 0.7 to 1.2 ng/g (ww). The age of the individuals was from four to seven years and predominated by five-year-olds. The group of points in the fourth coordinate plane includes the data on the *C. migratorius* specimens that differed in high concentrations of 'light' congeners 28 and 52, with up to 33% of the Σ_7PCB concentration.

In the third quarter of the coordinate plane, there was the third group of the data on *C. migratorius* caught in March and August 2014 and September 2018. The presence of PCBs in this group was characterised by the maximum concentrations of Σ_7PCBs ranging from 6.8 to 8.7 ng/g (ww) with the dominance of congeners 28 and 52 in the total proportion, with 50% of Σ_7PCBs. Individual points in this part of the coordinate plane characterised the *C. migratorius* individuals caught from March to August 2014 and in September 2017. PCBs detected in these fish specimens had high concentrations of Σ_7PCBs, from 10 to 13 ng/g (ww). Noteworthy was the result of the determination of PCBs in an eight-year-old *C. migratorius* individual. The PCB concentrations in this individual reached extreme values: Σ_7PCBs 32 ng/g (ww) and Σ_{dl}PCBs 9.0 ng/g (ww), with minimum total concentrations of 'light' congeners 28 and 52 at less than 7.0% of Σ_7PCBs.

On the PCA biplot (Figure 4b), the gradient vectors of average data on the PCB concentrations in *C. migratorius* formed three codirectional groups. The first group included the vectors of indicator congeners 138, 153, and 101, and dl-congeners 126, 118, and 114, together with the vector of the age of individuals. All these characteristics, including the age of individuals, were positively correlated with each other. The second *group* included the vectors of congeners 28 and 52, whose concentrations were associated with a strong positive correlation with each other and a weaker positive correlation with the vectors of the first group. The third group of vectors included dl-congeners 156, 105, and 123, whose concentrations were associated with each other by a strong positive correlation and a weaker positive correlation with the vectors of the first group.

The vectors of the third and second groups were almost perpendicular to each other. Therefore, the concentrations of PCBs in the specimens from these groups were not correlated and interconnected with each other.

The evaluation of *C. migratorius* as a bioindicator of PCBs in Baikal water by PCA based on the analysis of the results of the detection of PCBs in its tissues revealed that (i) the age of the fish determined the concentrations of indicator congeners with a high degree of chlorination (101, 138, and 153) and dioxin-like congeners 114 and 118, including the most toxic congener 126; (ii) the concentrations of 'light' congeners 28 and 52 did not depend on the age of the fish and were not associated with the accumulation of other congeners.

3.3.3. Analysis of Correlations

The heatmap in Figure 5a shows an analysis of correlations between the concentrations of various PCB congeners and the Σ_7PCB concentrations in Baikal water. Between all concentrations of PCBs and Σ_7PCBs, there were significant positive ($p < 0.05$) pairwise correlations with correlation coefficients ranging from 0.6 to 0.9. Among all correlation clusters, a group of indicators stood out, including the concentrations of congeners 52, 101, and 153, dl-congener 118 and Σ_7PCBs in the water with a high pairwise correlation (correlation coefficient from 0.85 to 0.98). Additionally, there was the correlation group of congeners 138 and 180. The correlation coefficients of the concentrations of these congeners with each other, with the concentrations of other congeners, and with Σ_7PCBs had lower values (from 0.75 to 0.85). The concentration of congener 28 was distinguished by the minimum correlation coefficient with the concentration of congener 180 (0.6). The results of the correlation analysis of the indicator PCBs in Baikal water indicated a strong agreement in the change in the concentrations of chlorobiphenyls (28, 52, 101, and 118) with each other and with the concentration of Σ_7PCBs. An increase or a decrease in the concentrations of one of these congeners was accompanied by an increase or a decrease in the concentrations of all congeners from the abovementioned group. The concentrations of congeners 138 and 180 were less correlated with the concentrations of other PCBs. Obviously, various factors determined the qualitative composition of PCB congeners in Baikal water: PCB sources (local and long-range atmospheric transport), water solubility and hydrophobicity, sorption on the particles of the solid phase of the water column, and transition to the bottom sediments.

PCBs identified in the tissues of *C. migratorius* were characterised by more complicated pairwise correlations between the concentrations of different congeners compared to the results of the corresponding analysis of PCBs in Baikal water (Figure 5b). Positive values of the correlation coefficients above 0.5 ($p < 0.05$) were classified as significant. Based on the range of pairwise correlation coefficients and the clusterisation results, all PCB congeners were divided into three groups.

The first group included the concentrations of congeners 153, 180, 101, and 118 together with the concentration of Σ_7PCBs. In this group, the concentrations of the components were positively correlated with each other and with the rest of the PCB congeners. The second group included congener 138 and dl-congeners 105, 123, and 156. The concentrations of congeners from this group were associated by high correlations with each other (r > 0.7) and a small correlation with congeners from the first group (0.3 < r < 0.6). There was no significant correlation between the concentrations of congeners from the second group and congeners 28 and 52, and dl-congeners 114 and 126. The third group included congeners 114 and 126 and dl-congeners 114 and 126 that had high correlations with each other (r > 0.7) and weak correlations with congeners from the first group (0.3 < r < 0.6). There was no significant correlation between the concentrations of congeners from the third and the second groups. The ages of the *C. migratorius* individuals were significantly correlated with the concentrations of congeners 138, 101, and 153 and dl-congeners 123, 105, and 118. At the same time, the correlation coefficients were relatively small, from 0.3 to 0.5.

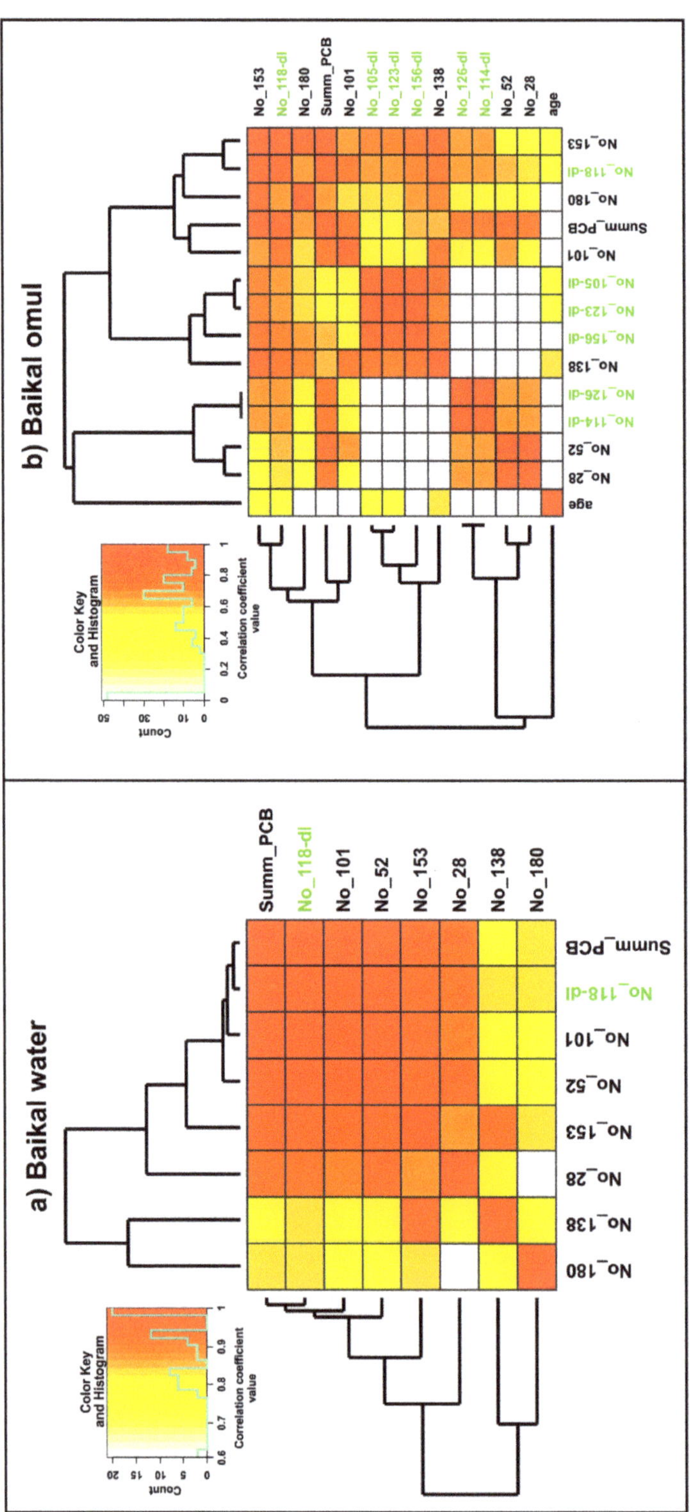

Figure 5. Heatmaps reflecting the calculations of pairwise correlation coefficients (**a**) between the concentrations of indicator congeners and Σ_7 PCBs in Baikal water samples and (**b**) between the concentrations of indicator congeners and Σ_7 PCBs in the tissues of *C. migratorius* and the age of the investigated omul individuals from Lake Baikal. Rows and columns in the heatmaps cluster by the proximity of the spectra of the correlation coefficients. Insignificant values of the correlation coefficients are replaced with zero values; dl-PCBs are marked in green.

The division of the investigated congeners in *C. migratorius* into three groups was likely related to the mechanisms of their accumulation and metabolism in fish bodies. Some indicator congeners of PCBs had a wide range of hydrophobicity properties (log K_{ow} = 5.67–7.36 [67]), and the level of their accumulation in the tissues of fish corresponded to their hydrophobicity. For this reason, congeners 28 and 52 predominated in the water, having the minimum log K_{ow} values among indicator congeners. The concentrations of congeners with the maximum numbers of chlorine atoms in the structure increased with the age of individuals. Consequently, the ratio of congeners in fish tissues did not reflect their ratio in water samples. The change in the ratio of congeners in fish tissues was likely owing to not only hydrophobicity of chlorobiphenyls but also to the faster metabolism of congeners with a low degree of chlorination [68]. It was also due to the barriers that occurred during their passage through lipid membranes of cells and (or) to the adsorption caused by the low water solubility [69,70].

Analysis of the relationship between the levels of PCB concentrations in Baikal water and the tissues of *C. migratorius* averaged over years using the Mantel test (Figure 6a) revealed the low and insignificant ($p > 0.05$) Spearman correlation coefficient. This indicated that the change in the concentrations of PCBs in Baikal water did not lead to similar changes in the level of the PCB concentrations in the tissues of *C. migratorius*. The relationship between average annual concentrations of PCBs in the water and their accumulation in the tissues of *C. migratorius* was not statistically confirmed.

Pairwise correlation analysis between average annual concentrations of PCBs in Baikal water and *C. migratorius* (Figure 6b) also did not revealed significant correlations. All pairwise correlation coefficients between the concentration of the investigated PCB congeners in the water and their accumulation in *C. migratorius* did not significantly differ from zero ($p > 0.05$), confirming the conclusions of the Mantel test.

For the Mantel test and correlation analysis between the pairs of dependent and insignificant variables, we took the same-year data on both the water and *C. migratorius*. Perhaps, due to this fact, there was no relationship between the concentrations of PCBs in the water and their accumulation in the tissues of *C. migratorius*. Omul accumulates PCBs gradually during several years, and not only the current concentration of PCBs but also the impact of PCBs in previous years determines the concentrations of the investigated congeners in the tissues of *C. migratorius*. Identification of this relationship requires long series of observations (ten years or more), taking into account other factors that are difficult to consider. These factors affect the processes of PCB accumulation in omul during the observation period, particularly, the change in the qualitative composition of dominant phytoplankton and zooplankton species that are part of *C. migratorius* food web. They also lead to interannual changes in temperature and ice regimes of the lake, which change the speed and direction of migrations and feeding behaviour in fish. Moreover, various factors also affect the qualitative composition of PCB congeners in the water of the lake: PCB sources (local and long-range atmospheric transport), hydrophobicity and water solubility of PCBs, their sorption on particles of the solid phase of the water column, and transition to the bottom sediments.

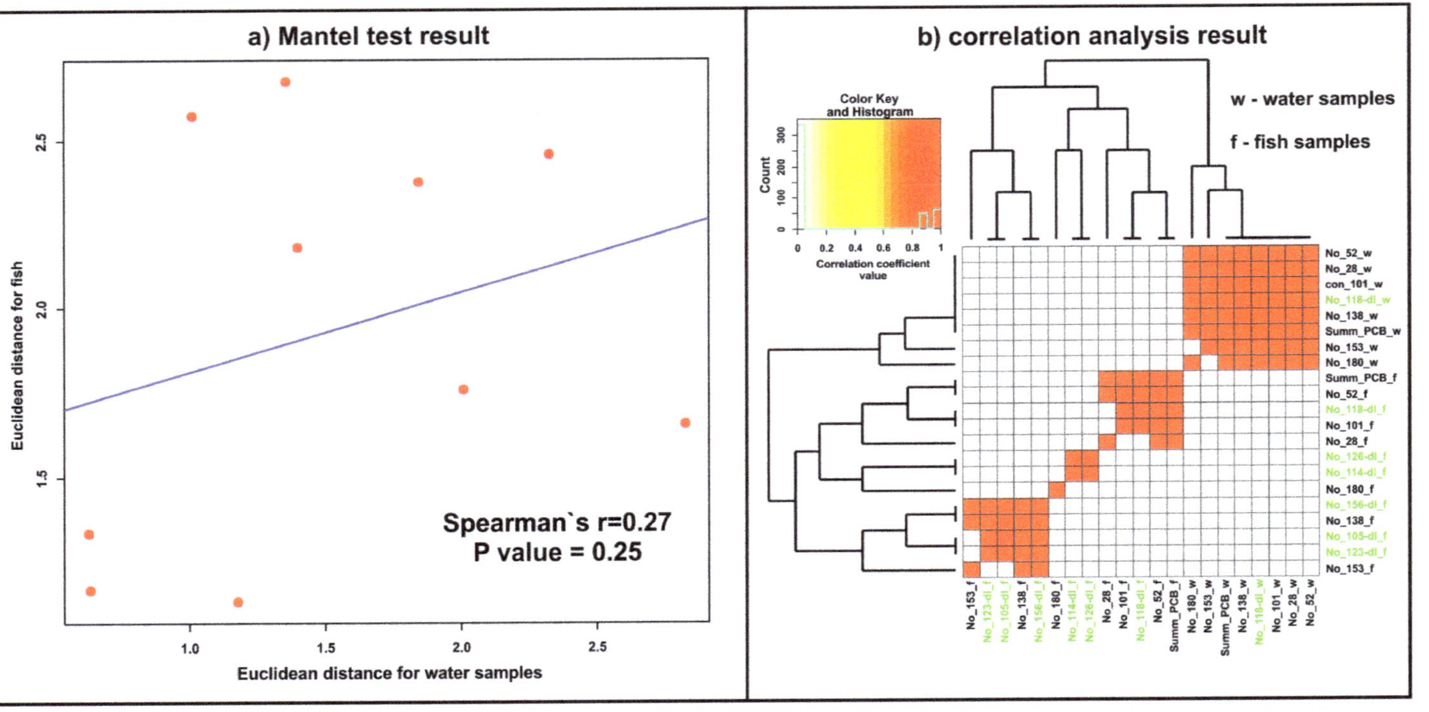

Figure 6. (**a**) Results of the analysis of the relationship between the concentrations of the investigated congeners and Σ_7PCBs in the water and *C. migratorius* using the Mantel test (the analysis was carried out for values averaged by years). (**b**) Heatmaps reflect the calculations of pairwise correlation coefficients between the concentrations of indicator congeners and Σ_7PCBs in the water and *C. migratorius* (the analysis was carried out for values averaged over years). Rows and lines on the heatmap are clustered based on the proximity of spectra of correlation coefficients; insignificant values of the correlation coefficients were replaced with zero values; dl-PCBs are marked in green.

3.3.4. Toxicity Equivalence for Dioxin-like PCBs

In Baikal water, we did not identify congeners of polychlorinated dibenzo-p-dioxins and dibenzofurans exceeding the detection limit of 0.01 pg/L [11]. Only one dl-PCB congener, 118, was identified during the monitoring of PCBs between 2015 and 2021. Its concentrations ranged from ≤ 10 to 110 pg/L. In seasons with the extreme concentrations of PCBs, its concentration in the water increased to 350 pg/L. The toxicity of this congener was assessed by the dioxin toxicity equivalents (TEQs) for dl-PCB within the WHO-TEQ $_{(2005)}$ pg/L range from 3×10^{-4} to 3.3×10^{-3}. In seasons with the extreme concentrations, the WHO-TEQ $_{(2005)}$ pg/L value reached 0.01 in May 2015 in the southern basin and 7×10^{-3} in September 2019 in the northern basin of Lake Baikal. In the muscle tissues of *C. migratorius*, the total toxicity equivalent of the detected dl-congeners (except for congener 126) was within the WHO-TEQ $_{(2005)}$ pg/g (ww) range from 0.03 to 0.06, and for congener 126, the value was two orders of magnitude higher, up to 4.7.

3.3.5. Comparative Analysis of the PCB Concentrations in the Surface Water of the Background Sites and in Bioindicator Fish

Currently, there is no common criterion for assessing the pollution of water by PCBs. According to the interim environmental quality standard (EQS), the PCB concentration in the surface waters of the European Union should not exceed 1 ng/L [71]. According to the standards established in Russia, the PCB concentration in water bodies for drinking and domestic water use should not exceed 1000 ng/L, but the PCB concentrations are not allowed in the water bodies for fishery purposes [72]. According to the standards of the US Environmental Protection Agency [73], the PCB concentration in water less than 14 ng/L does not pose a danger to aquatic organisms and human health.

The level of PCBs in Baikal water was higher than in the water from the Antarctic and Arctic regions but comparable with or much lower than the concentrations of PCBs in waters of the continental lakes such as Como (Italy), the Caspian Sea (northern part, Russia), and Lake Baiyangdian (China) (Table 1). The use of fish to assess the pollution of different water bodies by PCB pollutants based on the level of their accumulation in fish was a rather difficult task due to the discrepancy between the characteristics of fish (age and the number of lipids) as well as to the methods used for the detection of PCBs (number of congeners).

We compared the accumulation level of $\sum_7$PCBs in the muscle tissue of *C. migratorius* with the accumulation of $\sum_7$PCBs in *S. trutta fario* inhabiting alpine lakes of Southern and Central Europe. The species *S. trutta fario* was the closest to *C. migratorius* in terms of taxonomic characters. Individuals corresponding in age and lipid content were taken for comparison. Comparable levels of $\sum_7$PCBs in *C. migratorius* and *Salmo trutta* indicated the same concentration level of bioavailable PCBs in Baikal water and alpine lakes [30–33]. In the tissues of *Salmo trutta* from Lake Iseo and Lake Endine situated in the industrial regions of Northern Italy [22], the $\sum_7$PCB value was three to seven times higher than in *C. migratorius*: 2600, 1100, and 330 ng/g (lw), respectively.

Table 1. The concentrations of PCB congeners in the water from background sites and in the tissues of fish chosen as PCB bioindicators (ng/g, ww).

Area	Year	PCBs in the Water, ng/L	References
Arctic	2011	$\sum_{13}$PCBs 0.0013–0.021	[9]
Arctic Norway	1999	$\sum_{12}$PCBs 0.023–0.129	[74]
Canadian Archipelago	2018	$\sum$PCBs 0.65–46	[42]
Tropical Atlantic Ocean	2021	$\sum$PCBs 0.0041 ng/L	[12]
Antarctic, Victoria Land	2011–2012	$\sum_{127}$PCBs 0.046–0.143	[74]
European alpine lakes	2000–2001	$\sum_7$PCBs 0.048–0.123	[75]
Himalayas	2007	$\sum_5$PCBs 0.02–0.45	[76]
Lake Baikal, Russia	2015–2021	$\sum_{24-34}$PCBs 1.4–9.7 $\sum_7$PCBs 0.30; 0.26–0.34	This study
Lake Como, Italy	2007	$\sum_7$PCBs 0.30	[77]
Caspian Sea, Russia	2002	$\sum_6$PCBs: 0.10–7.3	[78]
Lake Baiyangdian, China	2008	$\sum_{39}$PCBs 19–132	[79]

Table 1. *Cont.*

Area	Year	PCBs in the tissues of fish, ng/g ww	
Lake Baikal, Russia	2014–2020	*C. migratorius*, $\sum_7$ PCBs 5.6; 4.9–6.3	This study
Alpine lakes of Western Europe	2003–2004	*Salmo trutta*, $\sum_7$ PCBs 6.3–13	[30–33]
Southern Moravia, Czech Republic	2013	Carp, $\sum_7$PCBs 2.4–14, Bream, $\sum_7$PCBs 8.0–210 Pike perch, $\sum_7$PCBs 2.4–4	[27]
Baltic Sea, Polish fishing area	2012	*Salmo salar*, $\sum_6$PCBs 7.7–61 *Salmo trutta*, $\sum_6$PCBs: 31–56 *Gadus morhua callarisa*, $\sum_6$PCBs: 0.07–2.8	[80]
Legally binding documents, EU	2019	Food products, muscle meat of fish, $\sum_6$PCBs 75	[81]
Area	Year	PCBs in the tissues of fish, pg WHO-TEQ/g lw	
Lake Baikal, Russia	2014–2020	*C. migratorius* 0.03–0.06	This study
Baltic Sea, Polish fishing area	2003	*Salmo salar*, 1.0–7.4 *Salmo trutta*, 4.2–9.0 *Gadus morhua callarisa*, 0.62–0.67	[80]
Legally binding documents, EU	2019	Food products, muscle meat of fish, 3.0	[81]

In the muscle tissues of *C. migratorius*, the total toxicity of four dl-congeners 118, 105, 123, and 156 did not exceed the WHO-TEQ $_{(2005)}$ pg/g (ww) value of 0.06. The higher toxicity level of the detected dl-congeners, up to the WHO-TEQ $_{(2005)}$ pg/g (ww) value of 4.7, was observed in the individuals of the older age group (eight years and older) due to the accumulation of congener 126. In the fish from the Baltic Sea, the total toxicity equivalent of dl-congeners was by two orders of magnitude higher than in *C. migratorius* (WHO-TEQ $_{(1997)}$ pg/g (ww) from 0.64 to 6.0) [80].

3.4. Opportunities for Monitoring

The conventional monitoring of PCBs in surface waters with pollutant concentrations close to the background level suggests two stages. The first stage includes water sampling in different seasons, from different depths, and at different stations, taking into account the presence of potential pollution sources, and determining the PCB concentrations in the samples The second stage includes analysis of the relationship between the detected concentrations of PCBs and sampling factors (year, season, station location, and depth of water layer). Based on the correlation analysis of the data, the congener(s) is (are) selected as indicators (indices) of the group with high positive correlations between the concentrations of congeners in the group. After that, during systematic monitoring, the concentrations of PCBs in a water body are estimated from the concentration of a congener selected as an indicator.

Biomonitoring provides detection of congeners in a water body, whose concentrations are below the LOQ level of the method used, in particular, dl-PCBs. A biological object chosen as a bioindicator must meet all the criteria that bioindicator objects should have [16]. Based on the correlation analysis of the data, a congener is also selected in the groups with high positive correlations between the concentrations of dl-PCBs characterised by higher concentrations and, hence, higher reliability and accuracy of detection. Thereafter, the current monitoring of dl-PCBs is possible for one (several) selected congener(s), a sharp increase in the concentration of which may trigger a more detailed analysis of the monitoring results.

4. Conclusions

The results of assessing concentrations of PCBs in the water of Lake Baikal obtained within the framework of conventional monitoring and biomonitoring confirmed the background level of PCB concentrations. The average concentration and statistically significant range of the detected Σ_7PCB concentrations in Baikal water were 0.30 and 0.26–0.34 ng/L, respectively. This concentration level was higher than in the waters from the Arctic and Antarctic areas but is comparable with or much lower than the PCB concentrations in the waters of continental lakes. Biomonitoring of PCBs using *C. migratorius* as a bioindicator in the lake water revealed five dl-congeners that were present in the water below LOQ of the method used. The most toxic congener 126 (the toxicity equivalence factor, TEF_{2005} is 0.1) was found in single individuals of the older age group. The total concentrations of dl-congeners in *C. migratorius* ranged from 1.3 to 1.7 ng/g (ww). The total toxicity equivalent of the detected dl-congeners (except for congener 126) was within the $WHO\text{-}TEQ_{(2005)}$ pg/g (ww) range from 0.03 to 0.06. This level of dl-PCBs corresponded to their concentrations in fish inhabiting European alpine lakes but was much lower than in fish from the Baltic Sea.

In the course of choosing a system for monitoring PCBs in Baikal water, the preliminary stage of the study included water sampling at 21 stations (water surface area 31,700 km^2). Based on the analysis of the monitoring results, five reference stations were selected for water sampling: 3, 6, 11, 16, and 19. The reduction in the number of stations did not distort the characteristics for the assessment of the PCB concentrations in the lake water. The control of PCBs in the surface water was possible by determining the concentrations of two of the seven indicator congeners, 52 and 101, because the concentrations of these congeners were highly correlated with each other, all indicator congeners, and Σ_7PCBs. They also have the maximum level of concentrations in water samples, and homogenous peaks of PCBs were recorded in chromatograms. According to the data on the correlation analysis, the congeners identified in *C. migratorius* were divided into three groups, and we proposed indicator congeners for these groups: congener 101 (correlation group from congeners 118-dl, 153, and 180), 138 (correlation group of congeners 105-dl, 123-dl, and 156-dl) and 52 (correlation group of congeners 28, 114-dl, and 126-dl).

Supplementary Materials: The following supporting information can be downloaded at: https://www.mdpi.com/article/10.3390/app12042145/s1, Table S1: Sampling stations, Table S2: Characteristics of the *C. migratorius* individuals caught in the course of biomonitoring, Table S3: Concentrations of indicator congeners in the water, ng/L, Table S4: Concentrations of indicator congeners in *C. migratorius*, ng/g (ww), Table S5: Concentrations of dl-congeners in *C. migratorius*. ng/g (ww), Table S6: Results of the analysis of the data from conventional monitoring by the PERMANOVA method, Table S7: Results of the analysis of the data from biomonitoring by the PERMANOVA method, Table S8: Relative error statistics for repeated measurements of PCB concentrations in water and *C. migratorius*, Table S9: Statistical parameters of the average values of PCB concentrations in water and *C. migratorius*.

Author Contributions: A.G.G.: concept of the study, methodology, writing, and preparation of the original draft; O.V.K.: water and fish sampling, chromatography–mass spectrometric analysis; Y.S.B.: statistical analysis. All authors have read and agreed to the published version of the manuscript.

Funding: This study was carried out within the framework of the State Task, project No. 0279-2021-0005 (121032300224-8).

Institutional Review Board Statement: Not applicable.

Informed Consent Statement: Not applicable.

Data Availability Statement: All primary data are available in tables Tables S3–S5.

Acknowledgments: Chromatography mass spectrometry analysis was carried out in the Shared Research Facilities for Physical and Chemical Ultramicroanalysis LIN SB RAS. The authors thank Vitaly A. Khutoryanskiy for the sampling of *C. migratorius* and Anatoly M. Mamontov for determining the age of fish carried out within the framework of the project by the Russian Foundation for Basic Research and the Government of the Irkutsk Region, No. 17-45-388077 r_a, and Elena V. Dzyuba for advice on preparing fish for analysis.

Conflicts of Interest: The authors declare no conflict of interest.

References

1. World Economic Forum. *Global Risks*, 10th ed.; World Economic Forum: Geneva, Switzerland, 2015; Available online: https://www.weforum.org/reports/global-competitiveness-report-2015 (accessed on 24 September 2021).
2. Kirschke, S.; Avellán, I.; Bärlund, T.; Bogardi, J.J.; Carvalho, L.; Chapman, D.; Dickens, C.W.S.; Irvine, K.; Lee, S.B.; Mehner, T.; et al. Capacity challenges in water quality monitoring: Understanding the role of human developmet. *Environ. Monit. Assess.* **2020**, *192*, 298. [CrossRef] [PubMed]
3. Wong, F.; Hung, H.; Dryfhout-Clark, H.; Aas, W.; Bohlin-Nizzetto, P.; Breivik, K.; Mastromonaco, M.N.; Lundén, E.B.; Ólafsdóttir, K.; Sigurðsson, Á.; et al. Time trends of persistent organic pollutants (POPs) and Chemicals of Emerging Arctic Concern (CEAC) in Arctic air from 25 years of monitoring. *Sci. Total Environ.* **2021**, *775*, 145109. [CrossRef] [PubMed]
4. Wolska, L.; Mechlinska, A.; Rogowska, J.; Namiesnik, J. Sources and fate of PAHs and PCBs in the marine environment. *Crit. Rev. Environ. Sci. Technol.* **2012**, *42*, 1172–1189. [CrossRef]
5. Cipro, C.V.Z.; Colabuono, F.I.; Taniguchi, S.; Montone, R.C. Persistent organic pollutants in bird, fish and invertebrate samples from King George Island. *Antarct. Antarct. Sci.* **2013**, *25*, 545–552. [CrossRef]
6. Gakuba, E.; Moodley, B.; Ndungu, P.; Birungi, G. Occurrence and significance of polychlorinated biphenyls in water, sediment pore water and surface sediments of Umgeni River, KwaZulu-Natal, South Africa. *Environ. Monit. Assess.* **2015**, *187*, 568. [CrossRef]
7. Stockholm Convention. About Persistent Organic Pollutants (CO3). As Amended in 2017. Available online: http://Downloads/UNEP-POPS-COP-CONVTEXT-2017 (accessed on 21 December 2021).
8. OSPAR Commission. Protecting and Conserving the North-East Atlantic and Its Resources. Available online: http://www.ospar.org/convention (accessed on 21 December 2021).
9. Carrizo, D.; Gustafsson, Ö. Distribution and Inventories of Polychlorinated Biphenyls in the Polar Mixed Layer of Seven Pan-Arctic Shelf Seas and the Interior Basins. *Sci. Technol.* **2011**, *45*, 1420–1427. [CrossRef] [PubMed]
10. Lohmann, R.; Dachs, J. *World Seas: An Environmental Evaluation*, 2nd ed.; Volume III: Ecological Issues and Environmental Impacts; Academic Press: Cambridge, MA, USA, 2019; pp. 269–282. [CrossRef]
11. Samsonov, D.P.; Kochetkov, A.I.; Pasynkova, E.M.; Zapevalov, M.A. Levels of Persistent Organic Pollutants in the Components of the Lake Baikal Unique Ecosystem. *Russ. Meteorol. Hydrol.* **2017**, *42*, 345–352. [CrossRef]
12. Lohmann, R.; Markham, E.; Khalova, I.; Sunderland, E.M.; Kukucka, P.; Pribylova, P.; Gong, X.; Pockalny, R.; Yanishevsky, T.; Wagner, C. Trends of Diverse POPs in Air and Water Across the Western Atlantic Ocean: Strong Gradients in the Ocean but Not in the Air. *Environ. Sci. Technol.* **2021**, *55*, 9498–9507. [CrossRef]
13. Lohmann, R.; Muir, D. Global Aquatic Passive Sampling (AQUA-GAPS): Using passive samplers to monitor POPs in the waters of the world. *Environ. Sci. Technol.* **2010**, *44*, 860–864. [CrossRef]
14. Reddy, A.V.B.; Moniruzzaman, M.; Aminabhavi, T.M. Polychlorinated biphenyls (PCBs) in the environment: Recent updates on sampling, pretreatment, cleanup technologies and their analysis. *Chem. Eng. J.* **2019**, *358*, 1186–1207. [CrossRef]
15. Muir, D.; Sverko, E. Analytical methods for PCBs and organochlorine pesticides in environmental monitoring and surveillance: A critical appraisal. *Anal. Bioanal. Chem.* **2006**, *386*, 769–789. [CrossRef] [PubMed]
16. Tanabe, S.; Subramanian, A. *Bioindicators of POPs*; University Press and Trans Pacific Press: Kyoto, Japan, 2006; 172p.
17. Manirakiza, P.; Covaci, A.; Nizigiymana, L.; Ntakimazi, G.; Schepens, P. Persistent chlorinated pesticides and polychlorinated biphenyls in selected fish species from Lake Tanganyika, Burundi, Africa. *Environ. Pollut.* **2002**, *117*, 447–455. [CrossRef]
18. Szlinder-Richert, J.; Barska, I.; Mazerski, J.; Usydus, Z. PCBs in fish from the southern Baltic Sea: Levels, bioaccumulation features, and temporal trends during the period from 1997 to 2006. *Mar. Pollut. Bull.* **2009**, *58*, 85–92. [CrossRef]
19. Szlinder-Richert, J.; Barska, I.; Usydus, Z.; Ruczynska, W.; Grabic, R. Investigation of PCDD/Fs and dl-PCBs in fish from the southern Baltic Sea during the 2002–2006 period. *Chemosphere* **2009**, *74*, 1509–1515. [CrossRef] [PubMed]
20. Jaikanlaya, C.; Settachan, D.; Denison, M.S.; Ruchirawat, M.; van den Berg, M. PCBs contamination in seafood species at the Eastern Coast of Thailand. *Chemosphere* **2009**, *76*, 239–249. [CrossRef]
21. Blocksom, K.A.; Walters, D.M.; Jicha, T.M.; Lazorchak, J.M.; Angradi, T.R.; Bolgrien, D.W. Persistent organic pollutants in fish tissue in the mid-continental great rivers of the United States. *Sci. Total Environ.* **2010**, *408*, 1180–1189. [CrossRef] [PubMed]
22. Piersanti, A.; Amorena, M.; Manera, M.; Tavoloni, T.; Lestingi, C.; Perugini, M. PCB concentrations in freshwater wild brown trouts (*Salmo trutta trutta* L) from Marche rivers, Central Italy. *Ecotoxicol. Environ. Saf.* **2012**, *84*, 355–359. [CrossRef]
23. Baptista, J.; Pato, P.; Pereira, E.; Duarte, A.; Pardal, M.A. PCBs in the fish assemblage of a southern European estuary. *J. Sea Res.* **2013**, *76*, 22–30. [CrossRef]
24. Cerveny, D.; Zlabek, V.; Velisek, J.; Turek, J.; Grabic, R.; Grabicova, K.; Fedorova, G.; Rosmus, J.; Lepic, P.; Randak, T. Contamination of fish in important fishing grounds of the Czech Republic. *Ecotoxicol. Environ. Saf.* **2014**, *109*, 101–109. [CrossRef] [PubMed]
25. Moraleda-Cibrián, N.; Carrassón, M.; Rosell-Melé, A. Polycyclic aromatic hydrocarbons, polychlorinated biphenyls and organochlorine pesticides in European hake (*Merluccius merluccius*) muscle from the Western Mediterranean Sea. *Mar. Pollut. Bull.* **2015**, *95*, 513–519. [CrossRef] [PubMed]
26. Net, S.; Henry, F.; Rabodonirina, S.; Diop, M.; Merhaby, D.; Mahfouz, C.; Amara, R.; Ouddane, B. Accumulation of PAHs, Me-PAHs, PCBs and total Mercury in sediments and Marine Species in Coastal Areas of Dakar, Senegal: Contamination level and impact. *Inter. J. Environ. Res.* **2015**, *9*, 419–432.

27. Zelníčková, L.; Svobodová, Z.; Maršálek, P.; Dobšíková, R. Persistent organic pollutants in muscle of fish collected from the Nové Mlýny reservoir in Southern Moravia, Czech Republic. *Environ. Monit. Assess.* **2015**, *187*, 448–454. [CrossRef] [PubMed]
28. Froescheis, O.; Looser, R.; Cailliet, G.M.; Jarman, W.M.; Ballschmiter, K. The deep-sea as a regional global sink of semivolatile persistent organic pollutants? Part I: PCBs in surface and deep-sea dwelling fish of the North and South Atlantic and the Monterey Bay Canyon (California). *Chemosphere* **2000**, *40*, 651–660. [CrossRef]
29. Flanagan Pritz, C.M.; Schrlau, J.E.; Massey Simonich, S.L.; Blett, T.F. Contaminants of emerging concern in fish from western U.S. and Alaskan national parks—Spatial distribution and health thresholds. *J. Am. Water Resour. Assoc.* **2014**, *50*, 309–323. [CrossRef]
30. Vives, I.; Grimalt, J.O.; Catalan, J.; Rosseland, B.O.; Battarbee, R.W. Influence of altitude and age in the accumulation of arganochlorine compounds in fish from high mountain lakes. *Environ. Sci. Technol.* **2004**, *38*, 690–698. [CrossRef] [PubMed]
31. Vives, I.; Grimalt, J.O.; Venture, M.; Catalan, J.; Rosseland, B.O. Age dependence of the accumulation of organochlorine pollutants in brown trout (*Salmo trutta*) from a remote high mountain lake (Redo, Pyrenees). *Environ. Pollut.* **2005**, *133*, 343–350. [CrossRef]
32. Demers, M.J.; Kelly, E.N.; Blais, J.M.; Pick, F.R.; St Louis, V.L.; Schindler, D.W. Organochlorine Compounds in Trout from Lakes over a 1600 Meter Elevation Gradient in the Canadian Rocky Mountains. *Environ. Sci. Technol.* **2007**, *41*, 2723–2729. [CrossRef] [PubMed]
33. Schmid, P.; Kohler, M.; Gujer, E.; Zennegg, M.; Lanfranchi, M. Persistent organic pollutants, brominated flame retardants and synthetic musks in fish from remote alpine lakes in Switzerland. *Chemosphere* **2007**, *67*, 16–21. [CrossRef]
34. Yang, R.; Wang, Y.; Li, A.; Zhang, Q.; Jing, C.; Wang, T.; Wang, P.; Li, Y.; Jiang, G. Organochlorine pesticides and PCBs infish from lakes of the Tibetan plateau and the implications. *Environ. Pollut.* **2010**, *158*, 2310–2316. [CrossRef] [PubMed]
35. Hung, H.; Katsoyiannis, A.A.; Brorström-Lundén, E.; Olafsdottir, K.; Aas, W.; Breivik, K.; Bohlin-Nizzetto, P.; Sigurdsson, A.; Hakola, H.; Bossi, R.; et al. Temporal trends of Persistent Organic Pollutants (POPs) in arctic air: 20 years of monitoring under the Arctic Monitoring and Assessment Programme (AMAP). *Environ. Pollut.* **2016**, *217*, 52–61. [CrossRef] [PubMed]
36. Anttila, P.; Brorström-Lundén, E.; Hansson, K.; Hakola, H.; Vestenius, M. Assessment of the spatial and temporal distribution of persistent organic pollutants (POPs) in the Nordic atmosphere Author links open overlay panel. *Atmos. Environ.* **2016**, *140*, 22–23. [CrossRef]
37. Bossi, R.; Vorkamp, K.; Skov, H. Concentrations of organochlorine pesticides, polybrominated diphenyl ethers and perfluorinated compounds in the atmosphere of North Greenland. *Environ. Pollut.* **2016**, *217*, 4–10. [CrossRef]
38. Hu, D.; Hornbuckle, K.C. Inadvertent polychlorinated biphenyls in commercial paint pigments. *Environ. Sci. Technol.* **2010**, *44*, 2822–2827. [CrossRef]
39. Lohmann, R.; Northcott, G.; Jones, K. Assessing the contribution of diffuse domestic burning as a source of PCDD/Fs, PCBs, and PAHs to the UK atmosphere. *Environ. Sci. Technol.* **2000**, *34*, 2892–2899. [CrossRef]
40. Ikonomou, M.; Sather, P.; Oh, J.; Choi, W.; Chang, Y. PCB levels and congener patterns from Korean municipal waste incinerator stack emissions. *Chemosphere* **2002**, *49*, 205–216. [CrossRef]
41. Vijgen, L.; Weber, R.; Lichtensteiger, W.; Schlumpf, M. The legacy of pesticides and POPs stockpiles—A threat to health and the environment. *Environ. Sci. Pollut. Res. Int.* **2018**, *32*, 31793–31798. [CrossRef]
42. Fernández, P.; Grimalt, J.O. On the global distribution of persistent organic pollutants. *Environ. Chem.* **2003**, *57*, 514–521. [CrossRef]
43. Ma, Y.; Adelman, D.A.; Bauerfeind, E.; Cabrerizo, A.; Carrie, A.; McDonough, C.A.; Derek Muir, D.; Soltwedel, T.; Sun, C.; Wagner, C.C.; et al. Concentrations and Water Mass Transport of Legacy POPs in the Arctic Ocean. *Geophys. Res. Lett.* **2018**, *45*, 12972–12981. [CrossRef]
44. Rusenek, O.T. (Ed.) *Baicalogy*; Nauka: Novosibirsk, Russia, 2012; 466p, Available online: https://search.rsl.ru/ru/record/010067 19741 (accessed on 23 January 2022). (In Russian)
45. Domysheva, V.M.; Sorokovikova, L.M.; Sinyukovich, V.N.; Onishchuk, N.A.; Sakirko, M.V.; Tomberg, I.V.; Zhuchenko, N.A.; Golobokova, L.P.; Khodzher, T.V. Ionic Composition of Water in Lake Baikal, Its Tributaries, and the Angara River Source during the Modern Period. *Russ. Meteorol. Hydrol.* **2019**, *44*, 687–694. [CrossRef]
46. Yoshioka, N.; Ueda, S.; Khodzher, T.; Bashenkaeva, N.; Korovyakova, I.; Sorokovikova, L.; Gorbunova, L. Distribution of dissolved organic carbon in Lake Baikal and its watershed. *Limnology* **2002**, *3*, 159–168. [CrossRef]
47. Gorshkov, A.G.; Kustova, O.V.; Izosimova, O.N.; Babenko, T.A. POPs Monitoring System in Lake Baikal—Impact of Time or the First Need? *Limnol. Freshwater Biol.* **2018**, *1*, 43–48. [CrossRef]
48. Bobovnikova, T.I.; Virchenko, U.P.; Dibtsova, F.V.; Yablokov, A.V.; Solntseva, G.N.; Pastukhov, D.V. Aquatic mammals—Indicators of the presence of organochlorine pesticides and polychlorinated biphenyls in the aquatic environment. *J. Hydrobiol.* **1986**, *22*, 63–68. (In Russian)
49. Kucklik, J.R.; Bidleman, T.F.; McConnell, L.L.; Walla, M.D.; Ivanov, G.P. Organochlorines in the water and biota of Lake Baikal, Siberia. *Environ. Sci. Technol.* **1994**, *28*, 31–37. [CrossRef]
50. Kucklik, J.R.; Harvey, H.R.; Ostrom, P.H.; Ostrom, H.E.; Baker, J.E. Organochlorine dynamics in the pelagic food web of Lake Baikal. *Environ. Toxicol. Chem.* **1996**, *15*, 1388–1400. [CrossRef]
51. Iwata, H.; Tanabe, S.; Ueda, K.; Tatsukawa, R. Persistent Orgnochlorine Residues in Air, Water, Sediments, and Soils from the Lake Baikal Region, Russia. *Environ. Sci. Technol.* **1995**, *29*, 792–801. [CrossRef]
52. Gorshkov, A.G.; Kustova, O.V.; Dzyuba, E.V.; Zakharova, Y.R.; Shishlyannikov, S.M.; Khutoryanskiy, V.A. Polychlorinated biphenyls in Lake Baikal ecosystem. *Chem. Sustain. Dev.* **2017**, *25*, 269–278. [CrossRef]

53. Kustova, O.V.; Stepanov, A.S.; Gorshkov, A.G. Determining the Indicator Congeners of Polychlorinated Biphenyls in Water at Ultratrace Concentration Level Using Gas Chromatography-Tandem Mass-Spectrometry. *J. Anal. Chem.* **2021**, *76*, 1028–1037. [CrossRef]

54. FR.1.31.2021.40284. Method for Measuring the Mass Fractions of Polychlorinated Biphenyl Congeners in Fish (omul) Samples from Lake Baikal by Chromatography-Mass Spectrometry with Detection in the Monitoring Mode of Specified Reactions. Certificate No. 222.0156/RA.RU.311866/2021. Available online: https://docs.cntd.ru/document/437249035 (accessed on 23 January 2022).

55. Pravdin, I.F. *Guide to the Study of Fish (Mainly Freshwater)*; Food Industry: Moscow, Russia, 1966; 376p, Available online: https://search.rsl.ru/ru/record/01006163511 (accessed on 23 January 2022). (In Russian)

56. Chugunova, N.I. *Guidelines for the Study of the Age and Growth of Fish*; RAS: Moscow, Russia, 1959; 168p, Available online: http://kpabg.ru/sites/default/files/chugunova1957.pdf (accessed on 23 January 2022). (In Russian)

57. Van den Berg, M.; Birnbaum, L.S.; Denison, M.; De Vito, M.; Farland, W.; Feeley, M.; Fiedler, H.; Hakansson, H.; Hanberg, A.; Haws, L.; et al. The 2005 World Health Organization Reevaluation of Human and Mammalian Toxic Equivalency Factors for Dioxins and Dioxin-Like Compounds. *Toxicol. Sci.* **2006**, *93*, 223–241. [CrossRef] [PubMed]

58. Grzymala-Busse, J.W.; Goodwin, L.K.; Grzymala-Busse, W.J.; Zheng, X. Handling missing attribute values in preterm birth data sets. In *International Workshop on Rough Sets, Fuzzy Sets, Data Mining, and Granular-Soft Computing, Regina, SK, Canada, 31 August–3 September 2005*; Springer: Berlin/Heidelberg, Germany, 2005; pp. 342–351. Available online: https://sci2s.ugr.es/keel/pdf/specific/congreso/grzymala_busse_goodwin05.pdf (accessed on 23 January 2022).

59. Oksanen, J.; Blanchet, G.F.; Friendly, M.; Kindt, R.; Legendre, P.; McGlinn, D.; Minchin, P.R.; O'Hara, R.B.; Simpson, G.L.; Solymos, P.; et al. Vegan: Community Ecology Package. R Package Version 2.5-6. 2019. Available online: https://CRAN.R-project.org/package=vegan (accessed on 24 June 2020).

60. Benjamini, Y.; Hochberg, Y. Controlling the false discovery rate: A practical and powerful approach to multiple testing. *J. R. Statist. Soc. B* **1995**, *57*, 289–300. [CrossRef]

61. Kassambara, A.; Mundt, F. *Factoextra: Extract and Visualize the Results of Multivariate Data Analyses*, R Package Version; R Project for Statistical Computing: Vienna, Austria, 2017.

62. Warnes, G.R.; Bolker, B.; Bonebakker, L.; Gentleman, R.; Liaw, W.H.A.; Lumley, T.; Maechler, M.; Magnusson, A.; Moeller, S.; Schwartz, M.; et al. *Package "gplots": Various R Programming Tools for Plotting Data*, R Package Version 2.17.0; Science Open: Berlin, Germany, 2015. Available online: https://cran.r-project.org/web/packages/gplots/index.html(accessed on 24 June 2020).

63. Mantel, N. The detection of disease clustering and a generalized regression approach. *Cancer Res.* **1967**, *27*, 209–220. [PubMed]

64. Legendre, P.; Legendre, L. *Numerical Ecology*; Elsevier: Amsterdam, The Netherlands, 2012; 990p.

65. Megson, D.; Benoit, N.B.; Sandau, C.D.; Chaudhuri, S.R.; Long, T.; Coulthard, E.; Johnson, G.W. Evaluation of the effectiveness of different indicator PCBs to estimating total PCB concentrations in environmental investigations. *Chemosphere* **2019**, *237*, 124429. [CrossRef] [PubMed]

66. Smirnov, V.V.; Smirnova-Zalumi, L.V.; Sukhanova, L.V. *Microevolution of the Baikal Omul Corogonus Autumnalis Migratorius (Georgi)*; SB RAS: Novosibirsk, Russia, 2009; 246p. (In Russian)

67. IARC. *Polychlorinated Biphenyls and Polybrominated Biphenyls*; Monographs on the Evaluation of Carcinogenic Risks to Humans; IARC: Lyon, France, 2015; Volume 107.

68. Fox, K.; Zauke, G.P.; Butte, W. Kinetics of Bioconcentration and Clearance of 28 Polychlorinated Biphenyl Congeners in Zebrafish (*Brachydanio rerio*). *Ecotoxicol. Environ. Safety.* **1994**, *28*, 99–109. [CrossRef]

69. Gobas, F.A.P.C.; Opperhuizen, A.; Hutzinger, O. Bioconcentration of hydrophobic chemicals in fish: Relationship with membrane permeation. *Environ. Toxicol. Chem.* **1986**, *5*, 637–646. [CrossRef]

70. Gobas, F.A.P.C.; Zhang, X.; Wells, R. Environ. Gastrointestinal magnification: The mechanism of biomagnification and food chain accumulation of organic chemicals. *Environ. Sci. Technol.* **1993**, *27*, 2855–2863. [CrossRef]

71. INERIS. Données Technico-Economiques sur les Substances Chimiques en France: Les polyChloroBiphenyles (PCB). DRC-11-118962-11081A. 2011; p. 89. Available online: http://www.ineris.fr/substances/fr (accessed on 18 June 2016).

72. State Committee of the Russian Federation for Environmental Protection. 1999. Order of 13 April, No 165. Available online: http://www.dioxin.ru/doc/prikaz165.htm (accessed on 23 March 2020). (In Russian)

73. US EPA 440.5-88.006. 1998. Water Quality Standards Criteria Summaries: A Compilation of State/Federation Criteria. Available online: http://www.epa.gov/waterscience/criteria/wqcriteria.html (accessed on 23 March 2020).

74. Vecchiato, M.; Zambon, S.; Argiriadis, E.; Barbante, C.; Gambaro, A.; Piazza, R. Polychlorinated biphenyls (PCBs) and polybrominated diphenyl ethers (PBDEs) in Antarctic ice-free areas: Influence of local sources on lakes and soils. *Microchem. J.* **2015**, *120*, 26–33. [CrossRef]

75. Fernandez, P.; Carrera, G.; Grimalt, J.O. Persistent organic pollutants in remote freshwater ecosystems. *Aquat. Sci.* **2005**, *67*, 263–273. [CrossRef]

76. Guzzella, L.; Poma, G.; De Paolis, A.; Roscioli, C.; Viviano, G. Organic persistent toxic substances in soils, waters and sediments along an altitudinal gradient at Mt. Sagarmatha, Himalayas, Nepal. *Environ. Pollut.* **2011**, *159*, 2552–2564. [CrossRef]

77. Villa, S.; Bizzotto, E.C.; Vighi, M. Persistent organic pollutant in a fish community on a sub-alpine lake. *Environ. Pollut.* **2011**, *159*, 932–939. [CrossRef]

78. ESIM. Unified State System of Information on the Situation in the World Ocean ЕСИМ. 2004. Available online: http://esimo.oceanography.ru/esp2/index/index/esp_id/2/section_id/8/menu_id/4218 (accessed on 24 December 2021). (In Russian)

79. Dai, G.; Liu, X.; Liang, G.; Han, X.; Shi, L.; Cheng, D.; Gong, W. Distribution of organochlorine pesticides (OCPs) and poly-chlorinated biphenyls (PCBs) in surface water and sediments from Baiyangdian Lake in North China. *J. Environ. Sci.* **2011**, *23*, 1640–1649. [CrossRef]

80. Piskorska-Pliszczynska, J.; Maszewski, S.; Warenik-Bany, M.; Mikolajczyk, S.; Goraj, L. Survey of persistent organochlorine contaminants PCDD, PCDF, and PCB in fish collected from the Polish Baltic fishing areas. *Sci. World J.* **2012**, *2012*, 973292. [CrossRef]

81. Eurofins. *Dioxins and PCBs in Food Products: Maximum Levels and Action Levels*; Eurofins: Nantes, France, 2019.

applied sciences

Article

Effects of Virgin Microplastics on Growth, Intestinal Morphology and Microbiota on Largemouth Bass (*Micropterus salmoides*)

Chaonan Zhang [1,2], Qiujie Wang [2], Shaodan Wang [2], Zhengkun Pan [2], Di Sun [1,2], Yanbo Cheng [1], Jixing Zou [2,3,*] and Guohuan Xu [1,*]

[1] State Key Laboratory of Applied Microbiology Southern China, Guangdong Provincial Key Laboratory of Microbial Culture Collection and Application, Institute of Microbiology, Guangdong Academy of Sciences, Guangzhou 510070, China; zhangchaonan@stu.scau.edu.cn (C.Z.); sundi@stu.scau.edu.cn (D.S.); huaqingzq@163.com (Y.C.)

[2] Joint Laboratory of Guangdong Province and Hong Kong Region on Marine Bioresource Conservation and Exploitation, College of Marine Sciences, South China Agricultural University, Guangzhou 510642, China; wongchaukit@stu.scau.edu.cn (Q.W.); 20182048005@stu.scau.edu.cn (S.W.); 20192150008@stu.scau.edu.cn (Z.P.)

[3] Guangdong Laboratory for Lingnan Modern Agriculture, South China Agricultural University, Guangzhou 510642, China

[*] Correspondence: zoujixing@scau.edu.cn (J.Z.); wswsl@gdim.cn (G.X.)

Abstract: Microplastics (MPs), classified as plastic debris less than 5 mm in size, are widely found in various aquatic environments. However, there have been few studies regarding their potential threat under aquaculture conditions. The aim of this study was to investigate the general health, intestinal morphology and microbiota of virgin polypropylene MPs (3–4 mm) on largemouth bass (*Micropterus salmoides*) over a 28-d exposure period. Four groups were divided according to whether the MPs were added in water or in food. The results disproved the hypothesis that MPs expose may adversely affect the growth of fish. Largemouth bass expelled MPs with minimal harm to the organism. MPs exposure had no significant effect on the community composition or diversity of intestinal microbial, although it could partly influence intestinal morphology, and the recombination process of the intestinal microbial community. Fish may be more sensitive to answer MPs exposure in water than in feed. Proteobacteria could potentially be pathogenic bacteria phylum in fish gut when affected by MPs. This research represents an innovative attempt to investigate the impact of virgin MPs on largemouth bass using a manipulative feeding experiment. The results could provide new insight on commercial fish health when challenged with MPs pollution.

Keywords: microplastics; largemouth bass; growth; intestinal morphology; intestinal microbiota

Citation: Zhang, C.; Wang, Q.; Wang, S.; Pan, Z.; Sun, D.; Cheng, Y.; Zou, J.; Xu, G. Effects of Virgin Microplastics on Growth, Intestinal Morphology and Microbiota on Largemouth Bass (*Micropterus salmoides*). *Appl. Sci.* **2021**, *11*, 11921. https://doi.org/10.3390/app112411921

Academic Editor: Elida Nora Ferri

Received: 17 November 2021
Accepted: 7 December 2021
Published: 15 December 2021

Publisher's Note: MDPI stays neutral with regard to jurisdictional claims in published maps and institutional affiliations.

1. Introduction

Microplastics (MPs) are identified as plastic particles smaller than 5 mm in size as referenced by the standards setting of the US National Oceanic and Atmospheric Administration (NOAA). MPs are generally categorized in four shapes: fibers, films, fragments and spheres [1–3]. Fibers are mainly derived from textiles and worn fishing nets [4,5]. Spheres come from microbeads in beauty and health products such as cleansers [6]. Fragments and films are irregularly shaped microparticles originated from the degradation of larger plastics in the environment due to physical, chemical and biological processes [7,8]. As a potential threat to nature, MPs have appeared in nearly all environments, thus they are becoming a growing concern worldwide [9–12].

Several reviews have described how adverse effects of MPs in aquatic organisms point to important factors that lead to bioaccumulation [13–15]. Due to the attractive color, buoyancy and food-like properties, fish are particularly prone to ingesting MPs [16]. Numerous animal studies have shown that MPs can lead to physical damage and histopathological alterations in the intestines and can even change swimming behavior and the lipid metabolism [17–20]. We suspected that these adverse effects are generally observed in acute toxicity studies with high concentrations of MPs that may not be present in the real environment. Moreover, most MPs exposure test used regular shape microspheres as materials, while spheres were not the common shape found in MPs field investigations. The chosen concentration and shape of MPs should be reliable in the experimental design as for the caution of potential harm in aquaculture.

The definition of virgin MPs is opposite to commercial MPs. Virgin MPs mean the materials of MPs from human daily life, and generally with no uniform shape. Fibers, fragments and films, are common shapes of virgin MPs, which also commonly found in water and organisms. Few studies have reported harm to the organism coming from virgin MPs. It seems unclear whether virgin MPs cause imminent harm to fish or not, as different researchers have shown discrepant results. For example, there was no significant difference in fish mass between control group and harbor collected plastics treatment, which fed to small juvenile glassfish (*Ambassis dussumieri*) daily for 95 d [21]. Similarly, there were no significant differences in growth performance of gilthead seabream (*Sparus aurata*) with an experimental diet containing 500 mg kg^{-1} of PVC-MPs for 30 d [22]. In a related study, exposure to six common types of virgin MPs in the diet of gilthead seabream caused no observable pathology or change in growth rate [23]. Furthermore, virgin industrial MPs extracts showed no toxic effect on early life stages of Japanese medaka (*Oryziaz latipes*) [24]. After 45 d exposure period, variations in body weight and survival rate of African catfish (*Clarias gariepinus*) were not significantly different among the PVC-treated groups relative to the control, although fish were fed PVC MPs spiked diets at maximum 3% [25]. However, virgin MPs altered blood biochemical parameters and changed the transcription of the reproductive-axis genes of African catfish [26]. Furthermore, researchers have also shown that marine-derived fishmeal might be a source of MPs, which could be transmitted to farmed fish, thus causing a concern in the aquaculture industry [27]. Still, the effect of virgin MPs on aquaculture is not clear.

Recently, severe MPs pollution was found in the aquaculture water of fish ponds from the Pearl River Estuary [14], which raised concerns about the safety of aquatic production. To address these concerns, this study aimed to investigate the effect of two kinds of virgin MPs on commercial fish health. We chose largemouth bass (*Micropterus salmoides*) as a biological model, given its key role in the Chinese fishery market as well as other important parameters, including physiological characteristics, high predation rate, and upper-middle dwelling [28,29]. The amount of Chinese production of bass was 477,808 tons in 2019 (China fishery Statistical Yearbook, 2020). The shapes of MPs in this study were restricted to fragments and fibers, since these are the most predominant shapes of MPs found in aquatic environments and in fish [3,30]. The size of MPs was 3–4 mm, and material of MPs used in this study was polypropylene, which were related to actual contamination. It was also reported that the contamination of seafood-based feeds by MPs threaten the fish health [31]. In reference to the sources of MPs that farmed fish may be exposed to, we set up two exposure ways (water and feed). The exposure concentrations of virgin MPs in water and feed were set according to previous field results [3] and study [23].

The objectives of this study were to expose largemouth bass to two shapes of virgin MPs in water and feed, finally assess their growth, intestinal morphology and microbial after four weeks. Four groups were divided according to whether the MPs were added in water or in food. We also tried to elucidate the effects of shape and exposure ways of virgin MPs on aquaculture. This research represents an innovative attempt to investigate the impact of virgin MPs on largemouth bass using a manipulative feeding experiment.

2. Materials and Methods

2.1. Experimental Fish and Culture Conditions

Largemouth bass juvenile (5.64 ± 0.6 in length, 4.05 ± 0.6 in weight) were purchased from a local fishery in Shunde (Guangdong, China), and acclimatized in a circulating breeding system in South China Agricultural University (Guangzhou, China). For the acclimatization period, fish were kept in tanks (50 cm × 60 cm × 60 cm) with air-pumped circulating diluted water at least two weeks prior to the experiment. Fish were fed twice per day using commercial fish feeds (Fenghua Feed Industrial Co., Ltd., Dongguan, China) with a ratio of 5% fish body weight. Formal experiment was carried out in the same tanks with 90 L water and 40 largemouth bass per tank. Three experimental groups and one control group were set up with three replicates in each group, giving a total of 12 tanks and 480 fish. Weight and length information of largemouth bass are shown in Supplementary Table S1. The culture conditions during fish adaptation period were: 12 h light, 12 h dark photoperiod, dissolved oxygen concentration from 5.5 to 6.5 mg/L, temperature 26 ± 1.5 °C, pH 7.1 and ammonia and nitrite levels of 0.01–0.02 mg/L.

2.2. Experimental Design and Sampling

The materials of virgin MPs were fibers and disposable plastic cups, which were purchased from a local store (Guangzhou, China). Materials were identified through a Fourier transform infrared spectrometer (FTIR Spectrometer, Vertex 70, Bruker) and found to be polypropylene (PP) (Supplementary Figure S1). Plastic cups were cut into pieces by surgical scissors and then filtered by a 5 mm steel screen. These fragments made from plastic cups were 3–4 mm. The virgin fibers themselves were also 3–4 mm in length. The majority of microplastics (more than ninety percent) used in this study were 3–4 mm through measured statistically.

After, we ensured the materials, processed fragments and virgin fibers were added into fish feed; Supplementary Figure S2 shows the process of making the feed contained MPs. Specifically, the commercial feed was soaked in water to make a dough, then 0.3% weight MPs were added into the dough according to a previous protocol [21]. The feed pellets were 2–5 mm in size and 3.33 g/kg of MPs in concentration, which mean there were 3.33 g fibers and fragments in 1 kg feed. Therein, 3.33 g MPs contained 1.665 g fibers and 1.665 g fragments. After being dried in air for 48 h, the pellets were stored in airtight bags until use. The control group of feed was treated in the same way above, but without the addition of MPs.

The concentration of MPs set in the water was 1 mg/L at environmentally relevant level [2,3]. Fibers and fragments represented 50%. Specifically, 90 L water contained 45 mg fibers and 45 mg fragments. MPs were thoroughly mixed in the water before fish put in. The water was refreshed every week, and MPs in water were supplemented at the same time. Fifty percent of water was changed.

Four groups were divided according to whether the MPs were added in water or in feed. In particular, MPs were found in both feed and water in Group A. MPs were found only in water and not in feed in Group B, while in Group D, MPs were found only in feed not in water. Group C is the control group, and there was no MPs in feed or water.

The fish were cultured for 28 d in such setting groups, starting from 28 June 2020. The culture conditions were: 12 h light, 12 h dark photoperiod, dissolved oxygen concentration from 5.3 to 6.8 mg/L, temperature 26.2 ± 1.3 °C, pH 6.8–7.2, and ammonia and nitrite levels of 0.01–0.02 mg/L. Fish were fed feed of 5% body mass twice per day in the regular feeding time (9:00, 17:00). During cultivation, the health condition of fish in each group was observed and recorded every day. Dead fish was removed from the tank as soon as possible. At the end of the experiment, 10 fish were randomly selected from each tank for euthanasia with 3-aminobenzoic acid ethyl ester methane sulfonate (MS-222) (50 mg/L) (Adamas-beta, Shanghai, China). The middle part of intestinal specimens was dissected from anaesthetized fish, placed in 1.5 mL sterile centrifuge tubes and stored in −80 °C for the microbial assay. The process was repeated five times with fifteen samples in each group.

The other five intestinal specimens were removed, and fixed in paraformaldehyde fixing solution (Servicebio, Wuhan, China) for histopathology analyses. This study was carried out in strict accordance with the recommendations in the Guide for the Care and Use of Laboratory Animals of the National Institutes of Health. The protocol was according to the National Institute of Health Guide for the Care and Use of Laboratory Animals of China. All experiments were approved by the Animal Care and Use Committee of South China Agricultural University (identification code: 20200127; date of approval: 20 May 2020). All surgery was performed under anesthesia with MS-222, and all efforts were made to minimize suffering.

2.3. Growth Performance

Body weight and length of five fish from the different groups were measured before and after the trial. Growth was measured by obtaining the weight gain (WG%) and specific growth rate (SGR) according to [22].

$$WG\% = (\text{final weight} - \text{initial weight}) \times 100/\text{initial weight}.$$

$$SGR = [(\text{Ln (final weight)} - \text{Ln (initial weight)}) \times 100]/\text{days}.$$

2.4. Intestinal Histopathology

Fixed intestinal samples were processed by standard methods by placing into paraffin blocks, cutting at 5 μm, and staining with hematoxylin and eosin (H and E). Tissue sections were examined with a microscope (Nikon, Tokyo, Japan) and photographed using the Mshot Image Analysis System. The intestinal folds length was measured from basement to top of the intestinal folds in each group.

2.5. Intestinal Microbial Assay

The five intestinal specimens in each group were mixed for the microbial assay. Each group had three repeats. Microbial DNA was extracted from intestine samples using the E.Z.N.A.® soil DNA Kit (Omega Bio-tek, Norcross, GA, USA) according to the manufacturer's protocols. Specific steps of PCR reactions are presented in Supporting Text S1. Purified amplicons were reacted on the Illumina MiSeq platform (Illumina, San Diego, CA, USA) according to the standard protocols provided by Majorbio Bio-Pharm Technology Co. Ltd. (Shanghai, China). Processing of raw fastq files is presented in Supporting Text S2. The Kruskal–Wallis test was used to compare the difference among groups.

2.6. Statistical Analysis

All data were quantified as the mean ± standard deviation (SD). One-way analysis of variance (ANOVA) and Duncan's multiple range test was performed via the R program (V5.1.3) (Development Core Team, 2018), SPSS 17.0 statistical package (SPSS Inc., Chicago, IL, USA). GraphPad Prism 7 (GraphPad Software, San Diego, CA, USA) were used to assess the differences among different groups. Differences were considered significant at $p < 0.05$ and highly significant at $p < 0.01$. Intestinal microbial analyses are shown in Supporting Text S3.

3. Results

3.1. Growth Performance

During the manipulative feeding experiments, largemouth bass readily consumed the different feed and showed the similar feeding behavior. The survival rate of each group was above 90%, and there was no significant difference (Group A: 92.5%; Group B: 95%; Group C: 90%; and Group D: 92.5%). At the end of the trial, no significant differences were detected in WG% or SGR among the different groups of largemouth bass (Figure 1). Weight and length information of largemouth bass is presented in Supplementary Table S1.

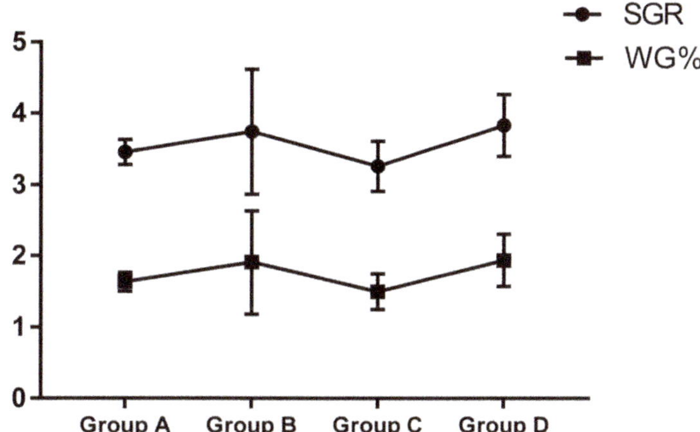

Figure 1. SGR and WG% of largemouth bass among each group after the trial.

3.2. Intestinal Histopathology

The intestinal histomorphology of largemouth bass among each group was observed by microscopy (Figure 2). Fish in the control group had normal intestinal wall morphology and a smooth epithelial layer (Figure 2C). However, a variety of intestinal abnormalities were found in the MPs treatment group. The folds of the intestine were irregularly arranged, and exfoliation of tissue to varying degrees were found in the intestinal folds (Figure 2A,B,D). The degree of intestinal damage was similar in all MPs treatment groups. However, there was no statistically significant difference in the intestinal folds length among the groups (Supplementary Figure S3).

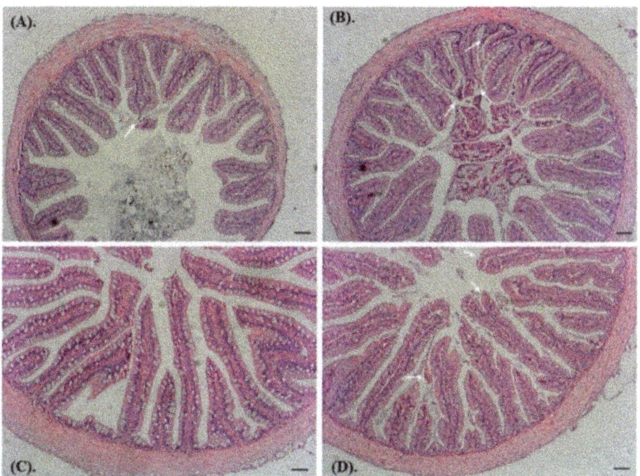

Figure 2. The intestine morphology of largemouth bass in Group A (**A**), Group B (**B**), Group C (**C**) and Control group D (**D**). Scale bar = 50 μm. White arrow indicates the location of the lesion.

3.3. Intestinal Microbiota Diversity

The intestinal microbial compositions in largemouth bass in four groups were investigated in this study. A total of 726,114 high quality sequences were retrieved from

12 samples (three repetitions in each group) belonging to four study groups. The saturation plateau of the rarefaction curves was well established (Supplementary Figure S4), indicating that this study covered nearly all the microbial diversity. According to the results of Good's coverage estimator, the microorganisms obtained from the 12 samples reached 100%. All matching tags were divided into 1428 OTUs, with 97% tagging similarity values. Venn of OTUs in each group showed Group A and B had independent 201 and 211 OTUs, respectively, while Group C and D only had 64 and 72 OTUs (Figure 3). MPs in water may be the major factor in restructuring the intestinal microbial composition.

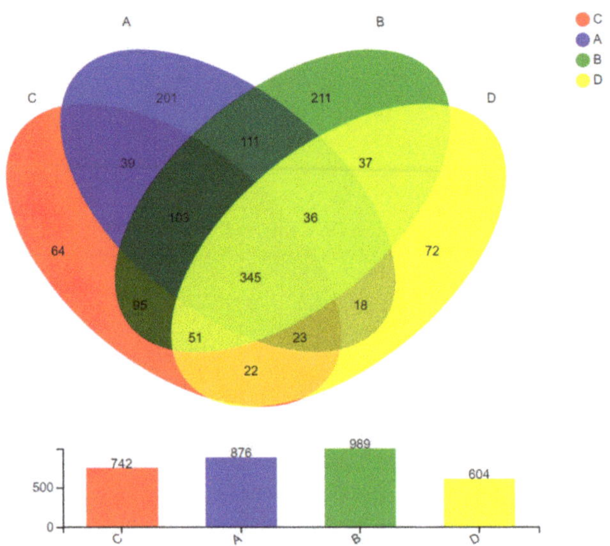

Figure 3. Venn of OTUs in each group.

The α-diversities of the intestinal microbiota sequencing completeness in each group were determined by the community richness index (Chao, Ace) and community diversity index (Simpson, Shannon), that are shown in Supplementary Table S2. There was no significant difference between Group A, B, C or D analyzed with a *t*-test (Figure 4), which demonstrated that the diversity of intestinal microbes in largemouth bass was not changed when exposure to MPs occurred in different ways.

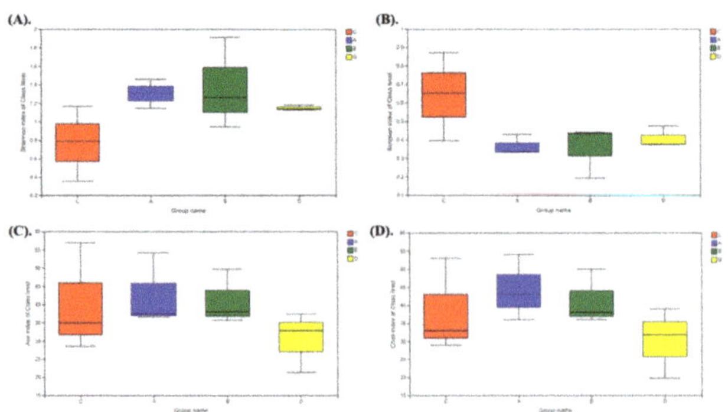

Figure 4. Box plots of intestinal microbiota diversity and richness in largemouth bass among each group after trial. (**A**) Shannon index of class level; (**B**) Simpson index of class level; (**C**) Ace index of class level; (**D**) Chao index of class level.

3.4. Intestinal Microbial Community Structures at Phylum and Genus Taxonomic Level

Bacterial compositions of the different communities and results of the Kruskal–Wallis test for differences in these microbial taxa among each group at the phylum level are shown in Figure 5. The dominant Phyla among the four groups were Firmicutes, Proteobacteria and Fusobacteriota (Figure 5A). The relative abundance of Proteobacteria and Actinobacteria gradually increased compared with the control group, but there was no significant difference among each group based on the Kruskal–Wallis test (Figure 5B). Intestinal microbial compositions at the genus level were also detected (Figure 6). The dominant genera in the gut of largemouth bass were Mycoplasma, Cetobacterium and Plesiomonas (Figure 6A). Therein, the percentage of Mycoplasma in intestinal microbial composition was the highest in healthy largemouth bass (70%), while the proportion of Mycoplasma in virgin MPs treatment groups was under 40%. The relative abundance of Cetobacterium in Group A gradually decreased, while Blautia increased compared with the control group. However, there was no significant effect on the microbial community composition at the genus level among each group statistically (Figure 6B).

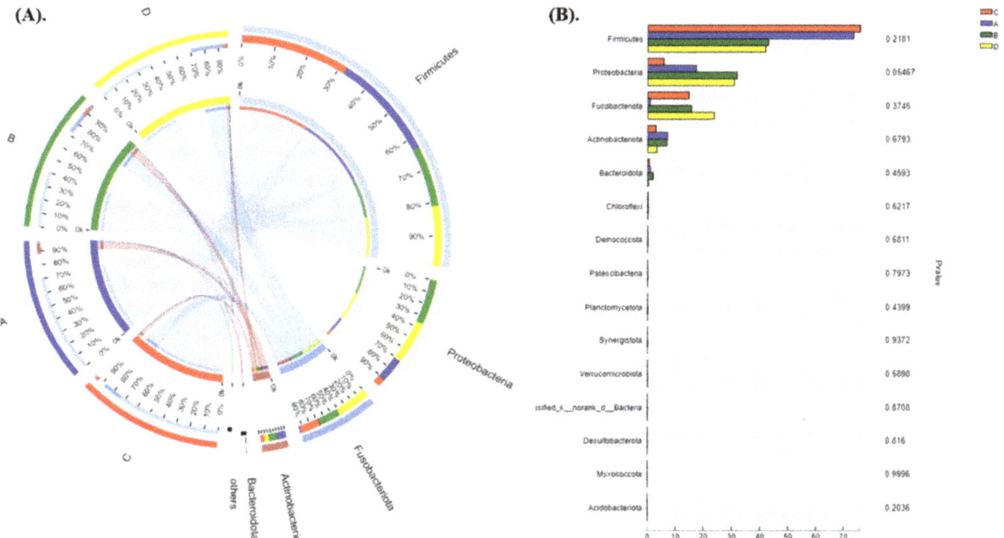

Figure 5. Bacterial compositions of the different communities (**A**) and results of the Kruskal–Wallis test for differences in these microbial taxa (**B**) among each group at the phylum level.

Principal coordinate analysis (PCA) based on the unweighted UniFrac scores of the microbial communities was performed to reflect the differences and distances between microbes among each group. The results showed that the largemouth bass in three MPs treatment groups (A, B and D) possessed similar compositions of microbes, while the fish in the control group was different from the experimental groups (Figure 7). Virgin MPs exposure could partly influence the recombination process of the intestinal microbial community.

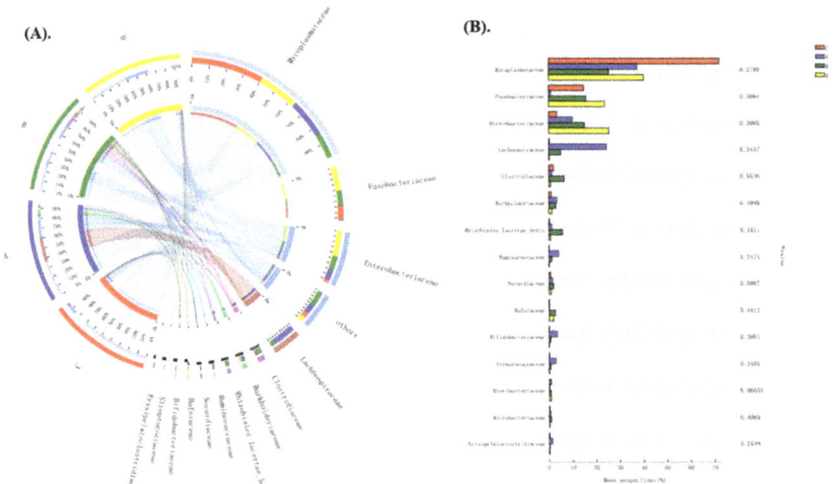

Figure 6. Bacterial compositions of the different communities (**A**) and results of the Kruskal–Wallis test for differences in these microbial taxa (**B**) among each group at the genus level.

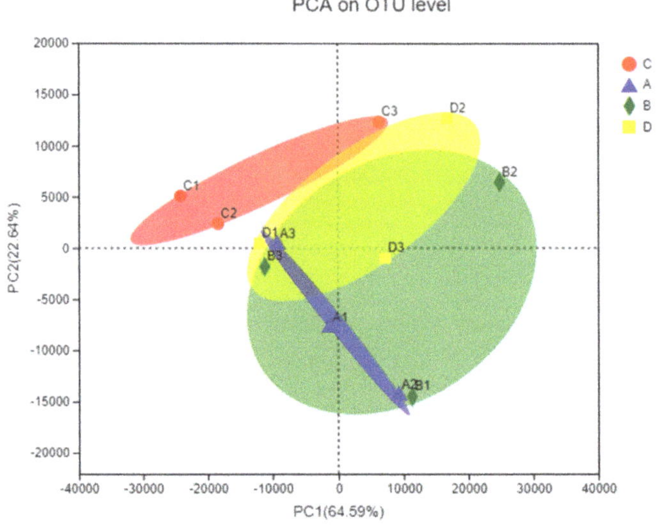

Figure 7. Principal coordinate analysis (PCA) on OTU level among each group.

4. Discussion

4.1. The Exposure Condition of MPs

Taking the shape and size of MPs into consideration, there are several problems in the existing literature that report the effects of MPs on aquatic organisms. For example, there is a wide range of polymer sizes reported in the literature, but more of a homogeneous shape used (i.e., microbeads) [32]. It is possible that the millimeter-sized MPs might be excreted more quickly by the organism [20]. The middle part of the gut of largemouth bass was examined after the feeding experiment. Nearly no MPs residue was found in any group. Our results were similar with Jabeen et al., (2018), which they found little MPs residue in goldfish intestine since the fish chewed fragments and pellets and then expelled them [33].

Exposure to MPs has been proven as having negative effects on various aspects of aquatic organisms, such as growth, behavior, development, reproduction and even death [13–15]. However, the experimental setup in which organisms were exposed to excessive concentration of MPs, far exceeds the native concentration in aquatic environments. However, in the present study, 1 mg/L was set as the water MPs concentration, and 0.3% of the total daily feed as the feed MPs concentration [2,3,21]. The ingested MPs content by fish per day could not exceed this setting, even in aquatic environments with a high MPs concentration. In addition, juvenile fish are more susceptible to environmental disturbances than adult fish [34]; therefore, juvenile fish may be more vulnerable to MPs treatment [21]. However, virgin MPs were not causing imminent harm to largemouth bass juveniles after dietary or water exposure.

4.2. Effect of MPs on Fish Growth

It is generally accepted that MPs would decrease fish body length or weight, since the ingestion of MPs might increase the energy burden on organisms and decrease energy reserves through the catalysis of lipids [35]. Specifically, the MPs-treated fish possibly have to redirect energy used for growth toward other important functions, such as systemic removal and detoxification of the MPs in the body [21]. Energy is also required to cope with other stresses, such as inflammation and a compromised endocrine system, burdened liver detox function and aberrant food absorption.

The hypothesis stating that exposure to virgin MPs adversely effects fish growth was not supported by the data in our study. After four weeks, the treatment fish did not show lower growth in length or body weight compared to controls. Growth index is one of the most important parameters that is measured in aquaculture. While the results presented here may provide an optimistic outlook for the aquaculture industry, laboratory results may not reflect the true situation in aquaculture where there are many other environmental factors (i.e., antibiotics, heavy metals and hormones). Although there are papers where microplastics did not show any effect on fish growth [36,37], and there are also a lot of studies showing retarded growth of fish [21,38]. This topic very much depends on fish species, microplastic shape and size, behavior and feeding habits of fish, etc. Further research is needed on the combined toxicity of MPs and other environmental contaminants.

4.3. Effect of MPs on Fish Intestinal Morphology

In this study, virgin MPs caused structural damage to intestines of largemouth bass, although we didn't find residual MPs in the microphotographs or statistically significance in the intestinal folds length among the groups. It was reported that morphological measurements of the intestinal folds length could help predict the absorption and digestion mechanisms of the fish intestinal tract [39]. The shortening of intestinal folds may be related to the toxin stress [40]. Fish intestinal abnormalities were found in many MPs toxic studies; however, statistically significant differences were rarely found in the data. For example, tissue morphology of rainbow trout (*Oncorhynchus mykiss*) did not change after dietary exposure to MPs for 4 weeks through qualitative histological examination of different intestinal segments [41]. The histopathology scores for leukocyte infiltration in the intestine of gilthead seabream with six virgin MPs for 45 d were not significantly different [23]. No tissue damage was found in silver barb (*Barbodes gonionotus*) fry internal organs or gills when exposed to polyvinyl chloride (PVC) fragments because of their relatively smooth surface [42]. Differences in pathological statistics might require longer experimental period and observation of specific tissue parts. Pedà et al., (2016) found the distal part of intestine had mostly been affected by pathological alterations after 90 d of exposure [43]. Pathological changes are generally descriptive, reflecting characteristics through comparison with the control group, but there are individual differences that require more research.

4.4. Effect of MPs on Fish Intestinal Microbiota

In this study, diversity or composition of intestinal microbiota was not significantly changed by virgin MPs exposure. However, PCA analysis revealed that MPs in water shaped the bacterial community composition by separating Group A and B from the control group. The relative abundance of Proteobacteria and Actinobacteria gradually increased compared with the control group at the phylum level. The dominant genera in the gut of largemouth bass were Mycoplasma in healthy largemouth bass, while they were decreased after MPs treatment. Huang et al., (2020) also found the gut microbiota in juvenile guppy (*Poecilia reticulata*) were dominated by Proteobacteria and Actinobacteria, and exposure to MPs resulted in an increase of Proteobacteria at phylum level [44]. Similar results were found in zebrafish gut, Proteobacteria significantly decreased ($p < 0.05$) after exposure to 5 µm microbeads for 21 d [45]. Proteobacteria were more prevalent in contaminated environments with higher MPs concentrations [46]. The growth of Proteobacteria and other bacteria were encouraged by the inflammation induced oxidation that connected with microbiota dysbiosis [47]. Collectively, Proteobacteria could potentially be pathogenic bacteria phylum in fish gut when affected by MPs.

Moreover, the general resilience of fish intestinal microbiota was found in a study of common carp (*Cyprinus carpio*) exposed to environmentally relevant concentrations of MPs for 30 d and followed by MPs excretion for another 30 d [48]. We tried to use experiments to simulate real aquacultural situations and found that the fish themselves can cope with the pressure of MPs more than our expectation. Nevertheless, we should also acknowledge that the absence of biofilm can be advantageous in reducing the variables that can affect the gut microbiological diversity, thus simplifying the interpretation of the results.

5. Conclusions

There are growing concerns about the potential toxic effects of MPs on aquatic organisms. These toxic pollutants may have far-reaching disadvantages on the aquaculture industry and on human health. This study offered new insight about the effect of virgin polypropylene MPs (3–4 mm) on commercial fish health. Firstly, the results presented here do not support the hypothesis that the addition of MPs has an adverse effect on fish growth. After a four-week trial, the body length and weight of the treatment largemouth bass were indistinguishable from the control group. Secondly, MPs exposure had no significant effect on the community composition or diversity of intestinal microbial, although it could partly influence intestinal morphology, and the recombination process of the intestinal microbial community. Virgin MPs did not cause imminent harm to largemouth bass. However, the results only offer temporary relief, since pollutants in the natural environment are more complex. We used MPs that were brand new. There were no weathering effects or biofilm on their smooth surface. This is unrealistic for natural pond MPs pollution. Furthermore, the choice of size and polymer of MPs affects the results. More attention should be addressed on combined toxicity of MPs and other contaminants in further research.

Supplementary Materials: The following are available online at https://www.mdpi.com/article/10.3390/app112411921/s1, Table S1: Weight and length information of largemouth bass before and after the experiment; Table S2: Alpha diversity of intestinal microbiota; Figure S1: Identification of materials; Figure S2: The process of making fodder contained MPs; Figure S3: Villi length of largemouth bass among each group after the trial; Figure S4: Rarefaction curves; Supporting Text S1: Specific steps of PCR reactions; Supporting Text S2: Processing of raw fastq files; Supporting Text S3: Intestinal microbial analyses.

Author Contributions: Conceptualization, C.Z. and G.X.; methodology, S.W. and Z.P.; software, Q.W.; validation, J.Z., Q.W. and S.W.; formal analysis, C.Z.; investigation, D.S. and Y.C.; resources, J.Z.; data curation, Q.W.; writing—original draft preparation, C.Z.; writing—review and editing, G.X.; visualization, J.Z.; supervision, G.X.; project administration, J.Z.; funding acquisition, G.X. All authors have read and agreed to the published version of the manuscript.

Funding: This research was funded by the Key Realm R&D Program of Guangdong Province (No. 2020B0202080005), and China Agriculture Research System of MOF and MARA (CARS-45-50).

Institutional Review Board Statement: The study was conducted according to the Animal Care and Use Committee of South China Agricultural University (identification code: 20200127; date of approval: 20 May 2020).

Informed Consent Statement: Informed consent was obtained from all subjects involved in the study. Written informed consent has been obtained from the patient(s) to publish this paper.

Data Availability Statement: The data presented in this study are available on request from the corresponding author.

Conflicts of Interest: The authors report no conflict of interest.

References

1. Hidalgo-Ruz, V.; Gutow, L.; Thompson, R.C.; Thiel, M. Microplastics in the marine environment: A review of the methods used for identification and quantification. *Environ. Sci. Technol.* **2012**, *46*, 3060–3075. [CrossRef]
2. Jabeen, K.; Su, L.; Li, J.; Yang, D.; Tong, C.; Mu, J.; Shi, H. Microplastics and mesoplastics in fish from coastal and fresh waters of China. *Environ. Pollut.* **2017**, *221*, 141–149. [CrossRef]
3. Zhang, C.; Wang, S.; Sun, D.; Pan, Z.; Zhou, A.; Xie, S.; Wang, J.; Zou, J. Microplastic pollution in surface water from east coastal areas of Guangdong, South China and preliminary study on microplastics biomonitoring using two marine fish. *Chemosphere* **2020**, *256*, 127202. [CrossRef] [PubMed]
4. Hernandez, E.; Nowack, B.; Mitrano, D.M. Polyester Textiles as a Source of Microplastics from Households: A Mechanistic Study to Understand Microfiber Release During Washing. *Environ. Sci. Technol.* **2017**, *51*, 7036–7046. [CrossRef]
5. Feng, Z.; Zhang, T.; Li, Y.; He, X.; Wang, R.; Xu, J.; Gao, G. The accumulation of microplastics in fish from an important fish farm and mariculture area, Haizhou Bay, China. *Sci. Total Environ.* **2019**, *696*, 133948. [CrossRef]
6. Mcdevitt, J.P.; Criddle, C.S.; Morse, M.; Hale, R.C.; Bott, C.B.; Rochman, C.M. Addressing the Issue of Microplastics in the Wake of the Microbead-Free Waters Act-A New Standard Can Facilitate Improved Policy. *Environ. Sci. Technol.* **2017**, *51*, 6611–6617. [CrossRef] [PubMed]
7. Zbyszewski, M.; Corcoran, P.L. Distribution and Degradation of Fresh Water Plastic Particles Along the Beaches of Lake Huron, Canada. *Water Air Soil Pollut.* **2011**, *220*, 365–372. [CrossRef]
8. Alimi, O.S.; Farner, B.J.; Hernandez, L.M.; Tufenkji, N. Microplastics and Nanoplastics in Aquatic Environments: Aggregation, Deposition, and Enhanced Contaminant Transport. *Environ. Sci. Technol.* **2018**, *52*, 1704–1724. [CrossRef]
9. Wright, S.L.; Kelly, F.J. Plastic and Human Health: A Micro Issue? *Environ. Sci. Technol.* **2017**, *51*, 6634–6647. [CrossRef] [PubMed]
10. Prata, J.C.; Da, C.J.; Lopes, I.; Duarte, A.C.; Rocha-Santos, T. Environmental exposure to microplastics: An overview on possible human health effects. *Sci. Total Environ.* **2020**, *702*, 134455. [CrossRef]
11. Wang, W.; Ge, J.; Yu, X.; Li, H. Environmental fate and impacts of microplastics in soil ecosystems: Progress and perspective. *Sci. Total Environ.* **2020**, *708*, 134841. [CrossRef] [PubMed]
12. Zhou, A.; Zhang, Y.; Xie, S.; Chen, Y.; Li, X.; Wang, J.; Zou, J. Microplastics and their potential effects on the aquaculture systems: A critical review. *Rev. Aquacult.* **2021**, *13*, 719–733. [CrossRef]
13. Jacob, H.; Besson, M.; Swarzenski, P.W.; Lecchini, D.; Metian, M. Effects of Virgin Micro- and Nanoplastics on Fish: Trends, Meta-Analysis, and Perspectives. *Environ. Sci. Technol.* **2020**, *54*, 4733–4745. [CrossRef]
14. Ma, J.; Niu, X.; Zhang, D.; Lu, L.; Ye, X.; Deng, W.; Li, Y.; Lin, Z. High levels of microplastic pollution in aquaculture water of fish ponds in the Pearl River Estuary of Guangzhou, China. *Sci. Total Environ.* **2020**, *744*, 140679. [CrossRef] [PubMed]
15. Xu, S.; Ma, J.; Ji, R.; Pan, K.; Miao, A.J. Microplastics in aquatic environments: Occurrence, accumulation, and biological effects. *Sci. Total Environ.* **2020**, *703*, 134699. [CrossRef] [PubMed]
16. Garrido, G.E.; Ryder, J.; Elvevoll, E.O.; Olsen, R.L. Microplastics in Fish and Shellfish—A Threat to Seafood Safety? *J. Aquat. Food Prod. T.* **2020**, *29*. [CrossRef]
17. Lei, L.; Wu, S.; Lu, S.; Liu, M.; Song, Y.; Fu, Z.; Shi, H.; Raley-Susman, K.M.; He, D. Microplastic particles cause intestinal damage and other adverse effects in zebrafish Danio rerio and nematode Caenorhabditis elegans. *Sci. Total Environ.* **2018**, *619–620*, 1–8. [CrossRef] [PubMed]
18. Gu, H.; Wang, S.; Wang, X.; Yu, X.; Hu, M.; Huang, W.; Wang, Y. Nanoplastics impair the intestinal health of the juvenile large yellow croaker Larimichthys crocea. *J. Hazard. Mater.* **2020**, *397*, 122773. [CrossRef]
19. Liyun, Y.; Haiyan, L.; Hongwu, C.; Bijuan, C.; Lingli, L.; Fan, W. Impacts of polystyrene microplastics on the behavior and metabolism in a marine demersal teleost, black rockfish (*Sebastes schlegelii*). *J. Hazard. Mater.* **2019**, *380*, 120861.
20. Hirt, N.; Body-Malapel, M. Immunotoxicity and intestinal effects of nano- and microplastics: A review of the literature. *Part. Fibre Toxicol.* **2020**, *17*, 57. [CrossRef]
21. Naidoo, T.; Glassom, D. Decreased growth and survival in small juvenile fish, after chronic exposure to environmentally relevant concentrations of microplastic. *Mar. Pollut. Bull.* **2019**, *145*, 254–259. [CrossRef]

22. Cristóbal, E.; Alberto, C.; María, Á.E. Effects of dietary polyvinylchloride microparticles on general health, immune status and expression of several genes related to stress in gilthead seabream (*Sparus aurata* L.). *Fish Shellfish Immunol.* **2017**, *68*, 251–259.

23. Jovanovic, B.; Gokdag, K.; Guven, O.; Emre, Y.; Whitley, E.M.; Kideys, A.E. Virgin microplastics are not causing imminent harm to fish after dietary exposure. *Mar. Pollut. Bull.* **2018**, *130*, 123–131. [CrossRef] [PubMed]

24. Pauline, P.; Bénédicte, M.; Christelle, C.; Jennifer, L.; Coline, C.; Jérôme, C. Toxicity assessment of pollutants sorbed on environmental microplastics collected on beaches: Part II-adverse effects on Japanese medaka early life stages. *Environ. Pollut.* **2019**, *248*, 1098–1107.

25. Stanley, C.I.; Gregory, E.O. Dietary exposure to polyvinyl chloride microparticles induced oxidative stress and hepatic damage in *Clarias gariepinus* (Burchell, 1822). *Environ. Sci. Pollut. Res.* **2020**, *27*, 21159–21173.

26. Karami, A.; Romano, N.; Galloway, T.; Hamzah, H. Virgin microplastics cause toxicity and modulate the impacts of phenanthrene on biomarker responses in African catfish (*Clarias gariepinus*). *Environ. Res.* **2016**, *151*, 58–70. [CrossRef] [PubMed]

27. Hanachi, P.; Karbalaei, S.; Walker, T.R.; Cole, M.; Hosseini, S.V. Abundance and properties of microplastics found in commercial fish meal and cultured common carp (Cyprinus carpio). *Environ. Sci. Pollut. Res.* **2019**, *26*, 23777–23787. [CrossRef]

28. Nicholas, R.; Nathan, E.; Herbert, Q.; Anita, K.; Amit, K.S. The effects of water hardness on the growth, metabolic indicators and stress resistance of largemouth bass *Micropterus salmoides*. *Aquaculture* **2020**, *527*, 735469.

29. Wang, Y.; Ni, J.; Nie, Z.; Gao, J.; Sun, Y.; Shao, N.; Li, Q.; Hu, J.; Xu, P.; Xu, G. Effects of stocking density on growth, serum parameters, antioxidant status, liver and intestine histology and gene expression of largemouth bass (*Micropterus salmoides*) farmed in the in-pond raceway system. *Aquac. Res.* **2020**, *51*, 5228–5240. [CrossRef]

30. Wang, S.; Zhang, C.; Pan, Z.; Sun, D.; Zhou, A.; Xie, S.; Wang, J.; Zou, J. Microplastics in wild freshwater fish of different feeding habits from Beijiang and Pearl River Delta regions, south China. *Chemosphere* **2020**, *258*, 127345. [CrossRef]

31. Yao, C.; Liu, X.; Wang, H.; Sun, X.; Qian, Q.; Zhou, J. Occurrence of Microplastics in Fish and Shrimp Feeds. *Bull. Environ. Contam. Toxicol.* **2021**, *107*, 684–692. [CrossRef]

32. Phuong, N.N.; Zalouk-Vergnoux, A.; Poirier, L.; Kamari, A.; Châtel, A.; Mouneyrac, C.; Lagarde, F. Is there any consistency between the microplastics found in the field and those used in laboratory experiments? *Environ. Pollut.* **2016**, *211*, 111–123. [CrossRef]

33. Jabeen, K.; Li, B.; Chen, Q.; Su, L.; Wu, C.; Hollert, H.; Shi, H. Effects of virgin microplastics on goldfish (*Carassius auratus*). *Chemosphere* **2018**, *213*, 323–332. [CrossRef]

34. Lima, A.R.A.; Barletta, M.; Costa, M.F.; Ramos, J.A.A.; Dantas, D.V.; Melo, P.A.M.C.; Justino, A.K.S.; Ferreira, G.V.B. Changes in the composition of ichthyoplankton assemblage and plastic debris in mangrove creeks relative to moon phases. *J. Fish Biol.* **2016**, *89*, 619–640. [CrossRef] [PubMed]

35. Ouyang, M.Y.; Liu, J.H.; Wen, B.; Huang, J.N.; Feng, X.S.; Gao, J.Z.; Chen, Z.Z. Ecological stoichiometric and stable isotopic responses to microplastics are modified by food conditions in koi carp. *J. Hazard. Mater.* **2021**, *404*, 124121. [CrossRef] [PubMed]

36. Critchell, K.; Hoogenboom, M.O. Effects of microplastic exposure on the body condition and behaviour of planktivorous reef fish (*Acanthochromis polyacanthus*). *PLoS ONE* **2018**, *13*, e193308. [CrossRef] [PubMed]

37. Malinich, T.D.; Chou, N.; Sepúlveda, M.S.; Höök, T.O. No evidence of microplastic impacts on consumption or growth of larval *Pimephales promelas*. *Environ. Toxicol. Chem.* **2018**, *37*, 2912–2918. [CrossRef] [PubMed]

38. Xia, X.; Sun, M.; Zhou, M.; Chang, Z.; Li, L. Polyvinyl chloride microplastics induce growth inhibition and oxidative stress in *Cyprinus carpio* var. larvae. *Sci. Total Environ.* **2020**, *716*, 136479. [CrossRef]

39. Dawood, M.A.O.; Eweedah, N.M.; Khalafalla, M.M.; Khalid, A.; Asely, A.E.; Fadl, S.E.; Amin, A.A.; Paray, B.A.; Ahmed, H.A. Saccharomyces cerevisiae increases the acceptability of Nile tilapia (*Oreochromis niloticus*) to date palm seed meal. *Aquac. Rep.* **2019**, *17*, 100314. [CrossRef]

40. Tan, X.; Sun, Z.; Zhou, C.; Huang, Z.; Tan, L.; Xun, P.; Huang, Q.; Lin, H.; Ye, C.; Wang, A. Effects of dietary dandelion extract on intestinal morphology, antioxidant status, immune function and physical barrier function of juvenile golden pompano Trachinotus ovatus. *Fish Shellfish Immunol.* **2018**, *73*, 197–206. [CrossRef]

41. Asmonaite, G.; Sundh, H.; Asker, N.; Almroth, B.C. Rainbow Trout Maintain Intestinal Transport and Barrier Functions Following Exposure to Polystyrene Microplastics. *Environ. Sci. Technol.* **2018**, *52*, 14392–14401. [CrossRef]

42. Romano, N.; Ashikin, M.; Teh, J.C.; Syukri, F.; Karami, A. Effects of pristine polyvinyl chloride fragments on whole body histology and protease activity in silver barb Barbodes gonionotus fry. *Environ. Pollut.* **2018**, *237*, 1106–1111. [CrossRef] [PubMed]

43. Pedà, C.; Caccamo, L.; Fossi, M.C.; Gai, F.; Andaloro, F.; Genovese, L.; Perdichizzi, A.; Romeo, T.; Maricchiolo, G. Intestinal alterations in European sea bass *Dicentrarchus labrax* (Linnaeus, 1758) exposed to microplastics: Preliminary results. *Environ. Pollut.* **2016**, *212*, 251–256. [CrossRef]

44. Huang, J.; Wen, B.; Zhu, J.; Zhang, Y.; Gao, J.; Chen, Z. Exposure to microplastics impairs digestive performance, stimulates immune response and induces microbiota dysbiosis in the gut of juvenile guppy (*Poecilia reticulata*). *Sci. Total. Environ.* **2020**, *733*, 138929. [CrossRef]

45. Qiao, R.; Sheng, C.; Lu, Y.; Zhang, Y.; Ren, H.; Lemos, B. Microplastics induce intestinal inflammation, oxidative stress, and disorders of metabolome and microbiome in zebrafish. *Sci. Total Environ.* **2019**, *662*, 246–253. [CrossRef] [PubMed]

46. Qiao, R.; Deng, Y.; Zhang, S.; Wolosker, M.B.; Zhu, Q.; Ren, H.; Zhang, Y. Accumulation of different shapes of microplastics initiates intestinal injury and gut microbiota dysbiosis in the gut of zebrafish. *Chemosphere* **2019**, *236*, 124334. [CrossRef]
47. Ni, J.; Wu, G.D.; Albenberg, L.; Tomov, V.T. Gut microbiota and IBD: Causation or correlation? *Nat. Rev. Gastroenterol. Hepatol.* **2017**, *14*, 573–584. [CrossRef]
48. Ouyang, M.; Feng, X.; Li, X.; Wen, B.; Liu, J.; Huang, J.; Gao, J.; Chen, Z. Microplastics intake and excretion: Resilience of the intestinal microbiota but residual growth inhibition in common carp. *Chemosphere* **2021**, *276*, 130144. [CrossRef]

applied
sciences

MDPI

Article

Prediction and Remediation of Groundwater Pollution in a Dynamic and Complex Hydrologic Environment of an Illegal Waste Dumping Site

Thatthep Pongritsakda, Kengo Nakamura *, Jiajie Wang, Noriaki Watanabe and Takeshi Komai

Department of Environmental Studies for Advanced Society, Graduate School of Environmental Studies, Tohoku University, Sendai 9808579, Japan; pongritsakda.thatthep.t1@dc.tohoku.ac.jp (T.P.); wang.jiajie.e4@tohoku.ac.jp (J.W.); noriaki.watanabe.e6@tohoku.ac.jp (N.W.); takeshi.komai.e7@tohoku.ac.jp (T.K.)
* Correspondence: kengo.nakamura.e8@tohoku.ac.jp

Citation: Pongritsakda, T.; Nakamura, K.; Wang, J.; Watanabe, N.; Komai, T. Prediction and Remediation of Groundwater Pollution in a Dynamic and Complex Hydrologic Environment of an Illegal Waste Dumping Site. *Appl. Sci.* **2021**, *11*, 9229. https://doi.org/10.3390/app11199229

Academic Editor: Elida Nora Ferri

Received: 10 September 2021
Accepted: 30 September 2021
Published: 4 October 2021

Publisher's Note: MDPI stays neutral with regard to jurisdictional claims in published maps and institutional affiliations.

Abstract: The characteristics of groundwater pollution caused by illegal waste dumping and methods for predicting and remediating it are still poorly understood. Serious 1,4-dioxane groundwater pollution—which has multiple sources—has been occurring at an illegal waste dumping site in the Tohoku region of Japan. So far, anti-pollution countermeasures have been taken including the installation of an impermeable wall and the excavation of soils and waste as well as the monitoring of contamination concentrations. The objective of this numerical study was to clarify the possibility of predicting pollutant transport in such dynamic and complex hydrologic environments, and to investigate the characteristics of pollutant transport under both naturally occurring and artificially induced groundwater flow (i.e., pumping for remediation). We first tried to reproduce the changes in 1,4-dioxane concentrations in groundwater observed in monitoring wells using a quasi-3D flow and transport simulation considering the multiple sources and spatiotemporal changes in hydrologic conditions. Consequently, we were able to reproduce the long-term trends of concentration changes in each monitoring well. With the predicted pollutant distribution, we conducted simulations for remediation such as pollutant removal using pumping wells. The results of the prediction and remediation simulations revealed the highly complex nature of 1,4-dioxane transport in the dumping site under both naturally occurring and artificially induced groundwater flows. The present study suggests possibilities for the prediction and remediation of pollution at illegal waste dumping sites, but further extensive studies are encouraged for better prediction and remediation.

Keywords: 1,4-dioxane; dynamic; pollution; dumping site; transport phenomena

1. Introduction

Many areas around the world still rely heavily on groundwater for daily water consumption. Thus, the maintenance of a suitable groundwater quality is crucial [1–5]. During the 2000s, however, various human activities significantly impacted groundwater quality and availability through various forms of pollution [4], with landfills, mines, and industrial plants being some of the main sources of groundwater pollutants [6,7]. The most common pollutants include heavy metals, non-aqueous phase liquids (NAPLs), and volatile organic compounds (VOCs) [8,9]. These pollutants are hazardous chemicals that can potentially negatively impact human health. Unfortunately, pollution from illegal waste dumping sites also occurs, causing serious health and environmental problems [10,11]. Previous studies have demonstrated that the respiratory exposure of VOCs via the inhalation route is associated with the risk of specific diseases [12]. However, the characteristics of pollution caused by illegal waste dumping as well as ways of predicting and addressing the problem are still poorly understood. This limited understanding is due to the dynamic and highly complex nature of pollution caused by illegal waste dumping, and the complexity originates from multiple pollution sources and the artificially induced pollution-prevention changes (e.g., excavation and pumping) in hydrologic environments.

At illegal waste dumping sites, pollutants leak from the waste to the groundwater over a long period of time. At the time of discovery, the distribution of contamination may be widespread, requiring countermeasures such as monitoring and remediation, which generally require considerable time and expense [13–15]. Consequently, it would be desirable to understand the characteristics of pollutant transport and to predict it in dynamic and complex hydrologic environments in illegal waste dumping sites [16]. For this purpose, simulations of pollutant transport in the presence of groundwater flow require consideration of the characteristic history of the illegal waste dumping site in addition to the conventionally considered subsurface flow and transport properties [17,18]. However, such simulations have not been conducted yet.

In the prefecture of the Tohoku region, Japan, an illegal dumping site was discovered in 1990, with various countermeasures having been implemented to date (2021) due to the existence of 1,4-dioxane groundwater pollution—a regulated substance (groundwater standard: ≤ 0.05 mg/L) [19–23]. At this site, multiple pollutant sources have been expected based on data from monitoring wells, and multiple countermeasures (e.g., the installation of impermeable walls) that potentially impact the hydrologic environment have been implemented. 1,4-Dioxane is highly miscible with water, and its biodegradation and adsorption to soil may be neglected [24–26], making this site suitable for fundamental studies on pollutant transport in the dynamic and complex hydrologic environments of illegal waste dumping sites. In this context, the objective of this study was to clarify the possibility of numerically predicting pollutant transport in the dynamic and complex hydrologic environments at this site in Japan, and to investigate the characteristics of pollutant transport under both naturally occurring and artificially induced groundwater flow using various countermeasures including remediation with pumping.

2. Materials and Methods

2.1. Site Information

History of Site Modification and Monitoring 1,4-dioxane

The illegal waste dumping site is located in the Tohoku region of Japan, Figure 1 shows a map of the site. The site covers an area of approximately 0.16 square kilometers and has illegally collected and buried industrial waste for decades. Approximately 270,000 m^3 of waste has been dumped at this site including incinerator ash, sludge, and refuse-derived fuel materials, which has caused the spread of serious contaminants in both the soil and groundwater.

Table 1 shows the history of the illegal waste dumping. A company started illegally dumping waste in the 1990s, and when the local government in one of Japan's prefectures discovered this, they immediately conducted a site survey. The prefectural government conducted an initial field survey in 2000, and for 10 years (from 2004 to 2014), it conducted additional surveys, observed groundwater contamination, and cleaned up the buried waste. To prevent groundwater from the waste burial site in one prefecture from contaminating the neighboring prefecture, a barrier wall (i.e., impermeable wall) was installed along the prefectural border (between 2005 and 2007). In 2009, the municipality implemented a groundwater treatment plant, installing pumping wells along the barrier wall in 2010. Moreover, the municipality monitored and documented 1,4-dioxane contamination from 2013 to 2020, as 1,4-dioxane had been detected in monitoring wells installed on the site.

This study focused on an area of 500 × 500 m for this site, as shown in Figure 1. For this area, the elevation and groundwater levels were obtained from a Geological Information System (GIS) and site investigation data [27]. There is a groundwater divide in this area, which divides groundwater flow into two major directions. The groundwater level was used to determine the flow potential in the governing equations for subsurface flows as described below.

Table 1. History of site modifications (well installation and monitoring period).

Year	Events
1990	Started illegally dumping industrial waste
2000	Started removing waste
2005–2007	Installed impermeable wall
2009	Implemented a groundwater treatment plant
2010	Installed pumping well
2013	Started monitoring of 1,4-dioxine (Under survey (2021))
2014	Finished removing waste

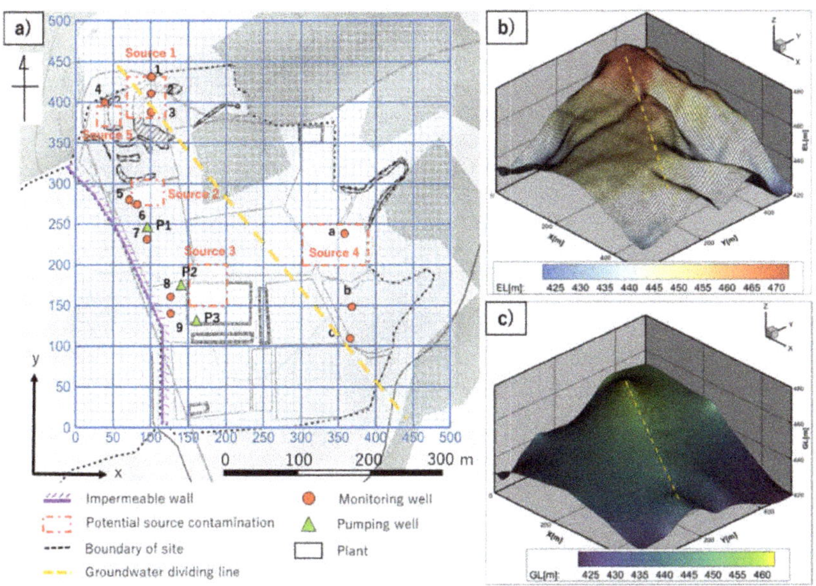

Figure 1. Map of an illegal waste dumping site. The blue grid means the calculated area. (**a**) Site map and simulation area. (**b**) Distribution of elevation level (EL) from a GIS. (**c**) Distribution of the groundwater level (GL). The EL and GL is the distance from the sea surface.

2.2. Flow and Transport Simulation

2.2.1. Governing Equation

In this study, the aquifer was assumed to be a single layer with a thickness of 5 m because it was difficult to obtain a detailed geological cross section of the aquifer. The flow in the aquifer was modeled as a 2D two-phase flow of water and non-aqueous phase liquid (NAPL) in the x-y coordinates, where NAPL is 1,4-dioxane [27]. However, the flow potential of the model takes into account the groundwater level (i.e., difference between land surface elevation and depth to water). As a result, the model is a quasi 3D model. The governing equations for the flows for NAPL and groundwater with/without dissolved NAPL and the advection-diffusion equation for NAPL in water are respectively represented by Equations (1)–(3), as follows.

$$\frac{\partial}{\partial x}\left(\frac{Kk_{ro}}{\mu_o}\rho_o\frac{\partial}{\partial x}\Phi_o\right) + \frac{\partial}{\partial y}\left(\frac{Kk_{ro}}{\mu_o}\rho_o\frac{\partial}{\partial y}\Phi_o\right) = \frac{\partial(\varnothing\rho_o S_o)}{\partial t} \tag{1}$$

$$\frac{\partial}{\partial x}\left(\frac{Kk_{rw}}{\mu_w}\rho_w\frac{\partial}{\partial x}\Phi_w\right) + \frac{\partial}{\partial y}\left(\frac{Kk_{rw}}{\mu_w}\rho_w\frac{\partial}{\partial y}\Phi_w\right) = \frac{\partial(\varnothing\rho_w S_w)}{\partial t} \tag{2}$$

$$\frac{\partial}{\partial x}\left(x_{ocw,k}\frac{Kk_{rw}}{\mu_w}\rho_w\frac{\partial}{\partial x}\Phi_w\right) \quad +\frac{\partial}{\partial y}\left(x_{ocw,k}\frac{Kk_{rw}}{\mu_w}\rho_w\frac{\partial}{\partial y}\Phi_w\right)$$
$$+\frac{\partial}{\partial x}\left[D_{ocw,k}\frac{\partial}{\partial x}\left(x_{ocw,k}\varphi\rho_w S_w\right)\right] \tag{3}$$
$$+\frac{\partial}{\partial y}\left[D_{ocw,k}\frac{\partial}{\partial y}\left(x_{ocw,k}\varphi\rho_w S_w\right)\right] = \frac{\partial\left(x_{ocw,k}\varphi\rho_w S_w\right)}{\partial t}$$

where K is the absolute permeability (m^2); k_{ro} is the relative permeability for NAPL (fraction) [28–30]; k_{rw} is the relative permeability for water (fraction) [28–30]; μ_o is the NAPL viscosity (Pa·s); μ_w is the water viscosity (Pa·s); ρ_o is the molar density of NAPL (mol/m^3); ρ_w is the molar density of water (mol/m^3); Φ_o is the flow potential of NAPL (Pa); Φ_w is the flow potential of water (Pa); S_o is the NAPL saturation (fraction); S_w is the water saturation (fraction); φ is the porosity (fraction); $D_{ocw,k}$ is the diffusion coefficient (m^2/s); $x_{ocw,k}$ is the concertation of NAPL dissolved in water (fraction); and t is time (s). The parameter values used in this study are listed in Table 2. The parameter values for the groundwater layer were set by assuming a clay soil, and the parameter values for 1,4-dioxane were acquired from a database [31–33]. Note that the biodegradation and adsorption to soil for 1,4-dioxane were neglected based on the chemical properties of 1,4-dioxane [34].

Table 2. Basic 1,4-dioxane information and geotechnical information from the site.

Symbol	Parameter	Unit	Input Values	Reference
K	Permeability	m^2	1.4×10^{-12}	[35]
φ	Porosity	-	0.3	
μ_o	Viscosity of 1,4-dioxane	Pa·s	1.31	
μ_w	Viscosity of water	Pa·s	1.138	
ρ_o	Molar density of 1,4-dioxane	kmol/m^3	18.36	
ρ_w	Molar density of water	kmol/m^3	17.83	
$D_{ocw,k}$	Molecular diffusion coefficient	m/s	1.0×10^{-9}	

These governing equations were solved using the finite difference method by applying the implicit pressure explicit saturation solution method [36] for implicit solutions for pressure and explicit solutions for saturation and concentration. To solve the governing equations, the 500×500 m area shown in Figure 1 was divided into a 5×5 m grid, and a hydrologically opened boundary condition was applied for each 500 m side and a constant flow-rate boundary for each pumping well. The water level was kept constant at the hydrologically opened boundaries. No detailed geological data were available for the sites targeted in this study. Therefore, the layer was assumed to be a single layer with the parameters used in Table 2.

2.2.2. Prediction and Remediation Simulations

The prediction simulation computed the evolution of the groundwater flow and the concentration of 1,4-dioxane from their initial conditions based on the history of the illegal waste dumping site (Table 1). The initial groundwater flow was determined based on the distribution of the groundwater level (GL) (Figure 1). The initial concentrations of 1,4-dioxane for the assumed contamination source areas 1–5, shown in Figure 1, were obtained by applying a constant annual input rate of 1,4-dioxane from the waste/contaminated soils for each grid within each source area, whereas the initial concentration was set to zero for the other locations. The input rate was set to zero to simulate the removal of waste/contaminated soils after excavation. The total input volume of 1,4-dioxane for each area is listed in Table 3, which indicates that Source 1 is where the largest amount of 1,4-dioxane was discarded. To simulate the permeability changes due to the installation of the impermeable wall, the absolute permeability at the corresponding location after the wall installation was changed to 1.00×10^{-15} m^2. Additionally, when monitoring wells were added, the corresponding grids were changed to a constant flow-rate boundary at a prescribed discharge rate (pumping rate). More specifically, the pumping rates were 10 m^3/day for pumping well 1, 31 m^3/day for pumping well 2, and 23 m^3/day for pumping well 3, respectively [37].

Table 3. Assumed sources of contamination at the site.

Name	Pollution Area [m^2]	Amount of Source [kg]
Source 1	2500	600
Source 2	1500	20
Source 3	2500	30
Source 4	4500	15
Source 5	400	30

These values were assumed based on fitting by the model.

In the remediation simulation, the 40-year evolution of groundwater flow and the concentrations of 1,4-dioxane from the final conditions in the prediction simulation were computed. We conducted two types of remediation simulations considering passive and active treatments with pumping [37,38]. The concentration of 1,4-dioxane in the studied area was expected to decrease during the passive and active treatments due to dilution and removal, respectively, by naturally occurring and pumping-induced groundwater. In the remediation simulation with active treatment, in addition to preexisting pumping wells 1–3, the monitoring wells were used as new pumping wells at a constant pumping rate of 250 m^3/day. Table 4 summarizes the pumping rates of each well.

Table 4. Pump specifications by active treatment.

Name	Symbol in Figure 1	Pumping Rate [m^3/day]	Reference
Monitoring well 1	1	250	
Monitoring well 2	2	250	
Monitoring well 3	3	250	
Monitoring well 4	4	250	
Monitoring well 5	5	250	
Monitoring well 6	6	250	
Monitoring well 7	7	250	[19]
Monitoring well 8	8	250	
Monitoring well 9	9	250	
Monitoring well a	a	250	
Monitoring well b	b	250	
Monitoring well c	c	250	
Pumping well 1	P1	10	
Pumping well 2	P2	31	[35]
Pumping well 3	P3	23	

3. Results and Discussion

3.1. Reproduction of Concentration of 1,4-Dioxane Groundwater Pollution in Monitoring Well and Distribution Prediction

A comparison between the reproduced 1,4-dioxane concentrations in each monitoring well using a model that considered groundwater flow (Figure 1) and the field monitoring data is shown in Figure 2. Between 2013 and 2017 (4 years), the reproduced data and monitoring data for all monitoring wells generally exhibited good fitting results. In 2013, both reproduced data and observed data showed relatively higher 1,4-dioxane concentrations in monitoring wells 1, 2, and 3 for Source 1, and monitoring well 6 for Source 2 than in the other wells. Significant decreases in 1,4-dioxane concentration were observed from 2014, when the excavation was completed. The good fitting results suggest that predicting the 1,4-dioxane concentrations by considering groundwater flow is possible due to its water-soluble characteristics.

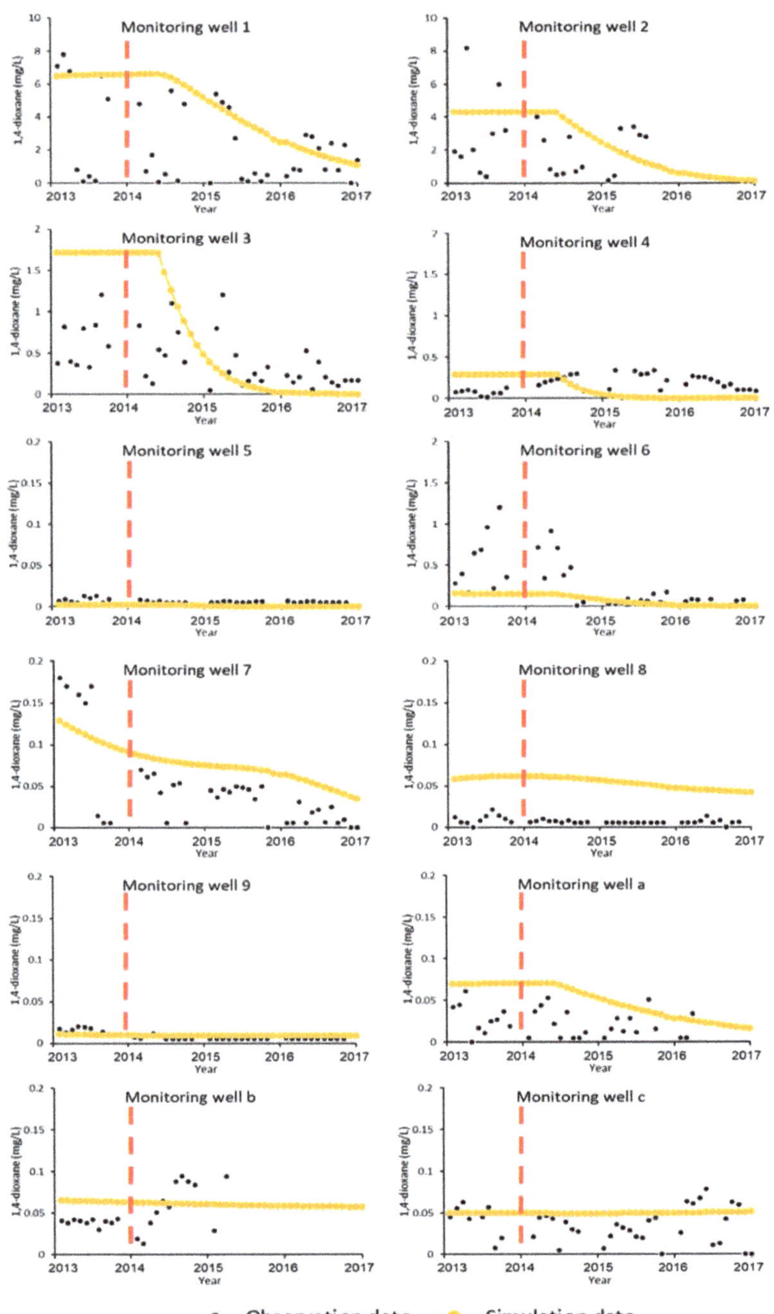

Figure 2. Comparison between the simulation results and observed data of 1,4-dioxane [26]. Red line means the excavation removal completed at the site (2014).

It should be noted that in terms of short periods of time such as each year, the reproduced 1,4-dioxane concentration did not fit well with the monitoring data. All monitoring wells, especially monitoring wells 1, 2, and 3, demonstrated highly fluctuating

1,4-dioxane concentrations each year. These fluctuations may be related to the complex geological structure of the site and dynamic environment such as seasonal rainfall or artificial activities, for example, pumping wells, impermeable walls, and excavation work, which may have increased the complexity of the groundwater environments. However, the difference between the monitoring data and reproduced data over short periods does not influence their consistency in long-term trends; therefore, this model can be used to predict 1,4-dioxane concentrations and distribution.

Using the reproduced 1,4-dioxane concentrations shown in Figure 2, the distribution of 1,4-dioxane in the groundwater of the study area over 25 years (1995–2020) is shown in Figure 3. The GL changes over time are also illustrated in the figure (light blue). In 1995 and 2000, the flow of groundwater containing pollutants was generally divided into two directions depending on the groundwater divide (northwest to southeast) of this dumping site area.

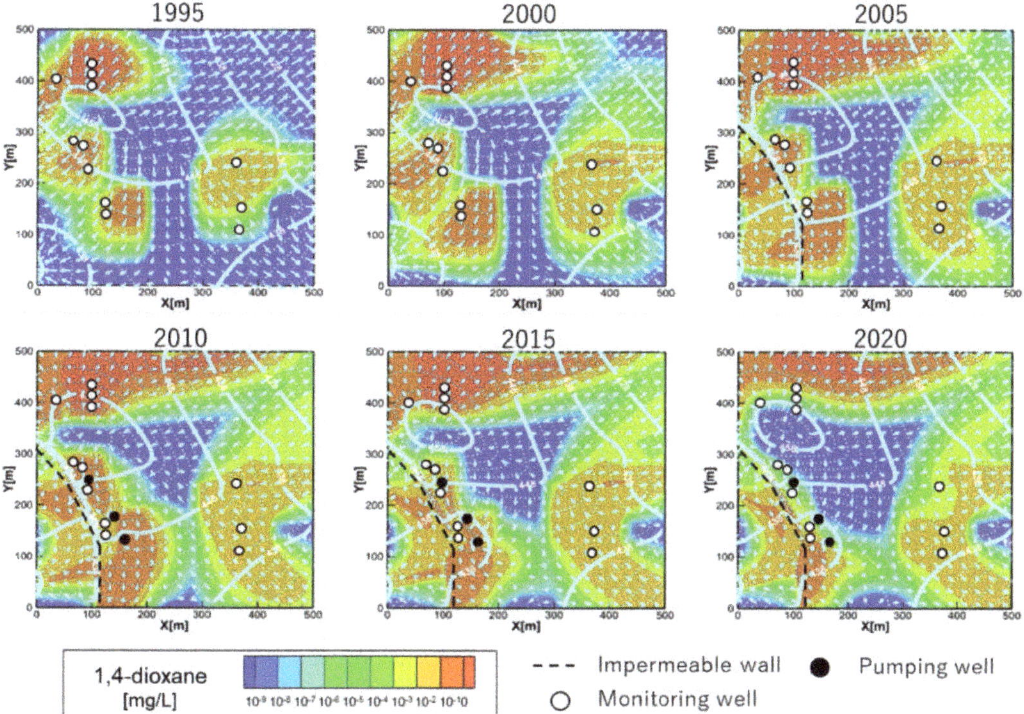

Figure 3. Distribution of groundwater pollution of 1,4-dioxane in the site over time (1995–2020). Contour lines indicate groundwater levels and vectors indicate flow directions. The color of the legend indicates the concentration of 1,4-dioxane.

From 2005, an impermeable wall was constructed along the border between the two prefectures to stop the movement of groundwater as well as the pollutants within it. As a result, the groundwater contours shown in Figure 4 overlapped along the impermeable wall, and the flow direction near this area changed from north to south. As expected, 1,4-dioxane was trapped on the east side of the impermeable wall to some extent.

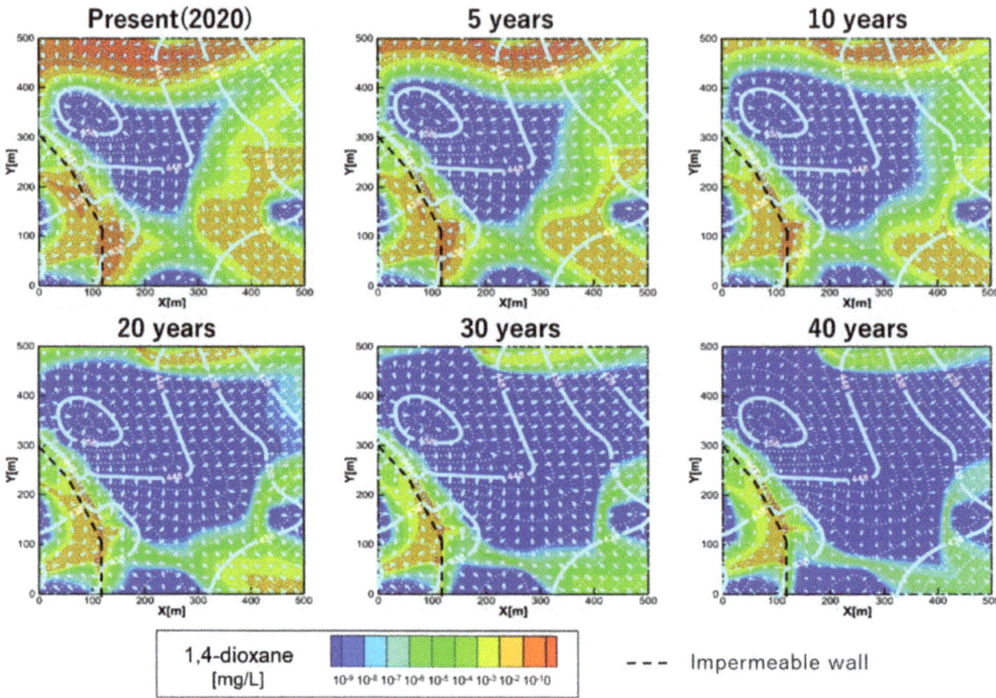

Figure 4. 1,4-Dioxane concentrations in groundwater by passive treatment over time. Contour lines indicate groundwater levels and vectors indicate flow directions. The color of the legend indicates the concentration of 1,4-dioxane.

In 2010, three pumping wells were installed near the impermeable wall at the waste dumping site for groundwater treatment. This action again increased the complexity of groundwater flow and groundwater levels. The groundwater near the pumping wells started to flow toward the wells, and the groundwater levels near the pumping wells declined. These changes imply accelerated groundwater flow and the transportation of pollutants in the groundwater. Interestingly, an obvious accumulation of 1,4-dioxane, along with the impermeable wall, was observed as a result of these artificial activities, that is, the impermeable wall and pumping well installments.

3.2. Evaluation of Different Methods for Groundwater Treatment

Based on the simulated current distribution of 1,4-dioxane in groundwater at the study site, two possible treatment methods, that is, passive and active treatments, were proposed and evaluated.

Passive treatment refers to the natural attenuation of 1,4-dioxane in groundwater. The simulation results of the distribution of 1,4-dioxane in the study site under natural attenuation over the next 40 years (2020–2060) are shown in Figure 4. The hydraulic environment of this site is relatively stable because no artificial activities have been conducted. Figure 4 suggests that as time passes, 1,4-dioxane moves with the groundwater flows, its concentration gradually decreasing. After 10 years of natural attenuation, most areas of the dumping site have 1,4-dioxane concentrations $\leq 10^{-3}$ mg/L, with very few areas in the range of 10^{-3}–10^{-1} mg/L. The concentration of 1,4-dioxane is relatively high near the impermeable wall, even after 40 years of natural attenuation, which may be attributed to the stagnation of groundwater flow near this area.

With the active treatment method, groundwater pumping, followed by treatment technologies, is applied. It is recommended to pump water through the monitoring wells, which are more widely installed than the current pumping wells—monitoring wells 7, 8, and 9 are close to the impermeable wall, which has an accumulation od 1,4-dioxane. A constant pumping rate of 250 m^3/day is suggested.

The distribution of 1,4-dioxane under active treatment is shown in Figure 5. The groundwater levels and groundwater directions near the monitoring wells and the impermeable wall were significantly influenced by groundwater pumping actions. These changes make the groundwater more dynamic and the groundwater environment becomes more complex, which also influences the movement of 1,4-dioxane. The northern areas with relatively high 1,4-dioxane concentrations (10^{-2}–10^{-1} mg/L) were quickly remediated within five years, with the 1,4-dioxane concentration decreasing to <10^{-3} mg/L. After 30 years of active treatment, the 1,4-dioxane concentration in the groundwater in most areas of the site was $\leq 10^{-3}$ mg/L, and did not accumulate near the impermeable wall.

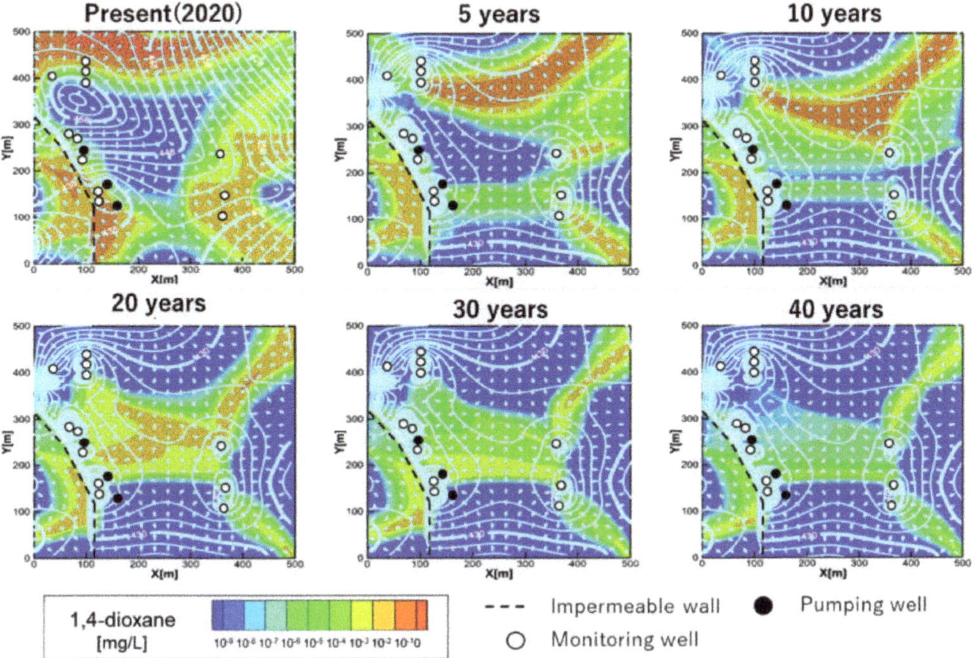

Figure 5. 1,4-Dioxane concentrations in groundwater by active treatment over time. Contour lines indicate groundwater levels and vectors indicate flow directions. The color of the legend indicates the concentration of 1,4-dioxane.

Comparing the simulation results of 1,4-dioxane distribution in the groundwater shown in Figures 4 and 5, a wider 1,4-dioxane distribution may occur in the case of the active treatment (e.g., after 30 years of treatment). It has been proposed that the pumping actions during active treatment may result in a more dynamic groundwater flow than during passive treatment, which promotes 1,4-dioxane transport to areas where it would otherwise be difficult to reach. This phenomenon was unanticipated and would influence the groundwater treatment efficiency; therefore, when designing pumping well locations, both the natural hydraulic environment and artificially induced groundwater flows should be taken into consideration. Finally, we suggest further research to study better prediction models and remediation strategies for groundwater pollution based on the findings of the present study.

4. Conclusions

The objective of this study was to clarify the possibility of numerically predicting pollutant transport in a dynamic and complex hydrologic environment and to investigate the characteristics of pollutant transport under both naturally and artificially induced groundwater flows. We attempted to reproduce the changes in 1,4-dioxane concentrations in groundwater observed in the monitoring wells using a quasi-3D flow and transport simulation that considered the multiple sources and spatiotemporal changes in hydraulic conditions.

The reproduced data and monitoring data (over a period of five years) for all monitoring wells generally exhibited good fitting results, suggesting that it is possible to predict the 1,4-dioxane concentration by considering groundwater flow. From the distribution of 1,4-dioxane, an obvious accumulation of 1,4-dioxane along with the impermeable wall was observed in these artificial activities, impermeable walls, and pumping well installments.

Passive treatment suggested that 1,4-dioxane moves with groundwater flows, and its concentration gradually decreases over time. The concentration of 1,4-dioxane was relatively high near the impermeable wall. With active treatment, the 1,4-dioxane concentration in the groundwater in most areas was $\leq 10^{-3}$ mg/L, and no longer accumulated near the impermeable wall. It is understood that the pumping actions during active treatment may result in a more dynamic groundwater flow than during passive treatment, promoting 1,4-dioxane transport to areas where it would otherwise be difficult to reach.

The study suggests the potential for the prediction and remediation of pollution at illegal waste dumping sites. However, further extensive studies on the complex geological structure of illegal waste dumping sites, dynamic seasonal rainfall, and groundwater changes are encouraged. In order to evaluate such complex environments, it is essential to improve prediction accuracy using more advanced models [39] and to analyze detailed field data.

Author Contributions: Conceptualization, T.P. and K.N.; Methodology, T.P. and K.N.; Software, T.P.; Validation, T.P.; Formal analysis, T.P.; Investigation, T.P.; Resources, T.P.; Data curation, T.P. and K.N.; Writing—original draft preparation, K.N. and T.P.; Writing—review and editing, N.W., J.W. and T.K.; Visualization, T.P.; Supervision, T.K.; Project administration, T.P. and K.N.; Funding acquisition, T.K. All authors have read and agreed to the published version of the manuscript.

Funding: This research was supported by DOWA Holdings funding and GP-RSS International join graduate program of Tohoku University, and the authors acknowledge the academic support of Prof. Toshikazu Shiratori at Tohoku University.

Institutional Review Board Statement: Not applicable.

Informed Consent Statement: Not applicable.

Data Availability Statement: The original contributions presented in the study are included in the article, and further inquiries can be directed to the corresponding author.

Acknowledgments: This research was supported by the International Join Graduate Program in Resilience and Safety Studies, Tohoku University, Japan.

Conflicts of Interest: The authors declare no conflict of interest.

References

1. Dieter, C.A.; Maupin, M.A.; Caldwell, R.R.; Harris, M.A.; Ivahnenko, T.I.; Lovelace, J.K.; Barber, N.L.; Linsey, K.S. Estimated use of water in the United States in 2015. *U.S. Geol. Surv. Circ.* **2018**, *1441*, 65.
2. World Water Assessment Programme. *The United Nations World Water Development Report 3: Water in a Changing World*; UNESCO: Paris, France; Earthscan: London, UK, 2009.
3. Zhao, Z. *A Global Assessment of Nitrate Contamination in Groundwater*; Report IGRAC: Westvest, The Netherlands, 2015.
4. USGS. Groundwater Quality. Available online: https://www.usgs.gov/special-topic/water-science-school/science/groundwater-quality?qt-science_center_objects=0#qt-science_center_objects (accessed on 2 August 2021).
5. Siebert, S.; Burke, J.; Faures, J.M.; Frenken, K.; Hoogeveen, J.; Doll, P.; Portmann, F.T. Groundwater use for irrigation-a global inventory. *Hydrol. Earth Syst. Sci. Discuss.* **2010**, *14*, 3977–4021.

6. Naveen, B.P.; Sumalatha, J.; Malik, R.K. A study on contamination of ground and surface water bodies by leachate leakage from a landfill in Bangalore, India. *Geo-Engineering* **2018**, *9*, 27. [CrossRef]
7. Almasri, M.N. Nitrate contamination of groundwater: A conceptual management framework. *Environ. Impact Assess. Rev.* **2007**, *27*, 220–242. [CrossRef]
8. Paz, D.H.F.; Lafayette, K.P.V.; de Holanda, M.J.O.; do Sobral, M.C.M.; de Costa, L.A.R.C. Assessment of environmental impact risks arising from the illegal dumping of construction waste in Brazil. *Environ. Dev. Sustain.* **2020**, *22*, 2289–2304. [CrossRef]
9. Liu, Y.; Kong, F.; Santibanez Gonzalez, E.D.R. Dumping, waste management and ecological security: Evidence from England. *J. Clean. Prod.* **2016**, *167*, 1425–1437. [CrossRef]
10. Mazza, A.; Piscitelli, P.; Neglia, C.; Rosa, G.D.; Iannuzzi, L. Illegal Dumping of Toxic Waste and Its Effect on Human Health in Campania, Italy. *Int. J. Environ. Res. Public Health* **2015**, *12*, 6818–6831. [CrossRef] [PubMed]
11. Santos, A.C.; Mendes, P.; Teixeiro, M.R. Social life cycle analysis as a tool for sustainable management of illegal waste dumping in municipal services. *J. Clean. Prod.* **2019**, *210*, 1141–1149. [CrossRef]
12. Yang, Y.; Luo, H.; Liu, R.; Li, G.; Yu, Y.; An, T. The exposure risk of typical VOCs to the human beings via inhalation based on the respiratory deposition rates by proton transfer reaction-time of flight-mass spectrometer. *Ecotoxicol. Environ. Saf.* **2020**, *197*, 110615. [CrossRef] [PubMed]
13. Du, L.; Xu, H.; Zuo, J. Status Quo of Illegal dumping research: Way forward. *J. Environ. Manag.* **2021**, *290*, 112601. [CrossRef]
14. Carriero, G.; Neri, L.; Famulari, D.; Lonardo, S.D.; Piscitelli, D.; Manco, A.; Esposito, A.; Chirico, A.; Facini, O.; Finardi, S.; et al. Composition and emission of VOCs from biogas produced by illegally managed waste landfill in Giugliano (Campania, Italy) and potential impact on the local population. *Sci. Total Environ.* **2018**, *640–641*, 377–386. [CrossRef]
15. Vitali, M.; Castellani, F.; Fragassi, G.; Mascitelli, A.; Martellucci, C.; Dilitti, G.; Scamosci, E.; Astolfi, M.L.; Fabiani, L.; Mastrantonio, R.; et al. Environment status of an Italian site highly polluted by illegal dumping of industrial waste: The situation 15 years after the judicial intervention. *Sci. Total Environ.* **2021**, *762*, 144100. [CrossRef]
16. Qi, S.; Luo, J.; O'Connor, D.; Cao, X.; Hou, D. Influence of groundwater table fluctuation on the non-equilibrium transport of volatile organic contaminants in the vadose zone. *J. Hydrol.* **2020**, *580*, 124353. [CrossRef]
17. Shi, X.; Ren, B. Predict three-dimensional soil manganese transport by HYDRUS-1D and spatial interpolation in Xiangtan manganese mine. *J. Clean. Prod.* **2021**, *292*, 125879. [CrossRef]
18. Oliveira, L.A.; Honorio, M.J.; Grecco, K.L.; Tornisielo, V.L.; Woodbury, B.L. Atrazine movement in corn cultivated soil using HYDRUS-2D: A comparison between real and simulated data. *J. Environ. Manag.* **2019**, *248*, 109311. [CrossRef]
19. Iwate Prefecture, Record of Lllegal Dumping on the Border between Lwate and Aomori Prefectures. Available online: http://www2.pref.iwate.jp/~{|}hp0315/haikibutu/kenkyo_archive/dioxane.html (accessed on 10 October 2020).
20. Baraias-Rodriguez, F.J.; Murdoch, L.C.; Falta, R.W.; Freedman, D.L. Simulation of site biodegradation of 1,4-dioxane under metabolic and co-metabolic conditions. *J. Contam. Hydrol.* **2019**, *223*, 103464. [CrossRef] [PubMed]
21. Milavec, J.; Tick, G.J.; Brusseau, M.L.; Carroll, K.C. 1,4-Dioxane co-solvency impacts on trichloroethene dissolution and sorption. *Environ. Pollut.* **2019**, *252*, 777–783. [CrossRef] [PubMed]
22. U.S. EPA. Integrated Risk Information System (IRIS) on 1,4-Dioxane. 2013. Available online: https://iris.epa.gov/static/pdfs/0326_summary.pdf (accessed on 1 October 2021).
23. IARCIARC. *Re-Evaluation of Some Organic Chemicals, Hydrazine, and Hydrogen Peroxide*; IARC: Lyon, France, 1999.
24. Song, K.; Ren, X.; Mohamed, A.K.; Liu, J.; Wang, F. Research on drinking-groundwater source safety management based on numerical simulation. *Sci. Rep.* **2020**, *10*, 15481. [CrossRef] [PubMed]
25. Ozel, H.U.; Gemici, B.T.; Gemici, E.; Ozel, H.B.; Cetin, M.; Sevik, H. Application of artificial neural networks to predict heavy metal contamination in the Bartin River. *Environ Sci. Pollut. Res.* **2020**, *27*, 42495–42512. [CrossRef]
26. Cheng, Y.; Zhu, J. Significance of mass –concentration relation on the contaminant source depletion in the nonaqueous phase liquid (NAPL) contaminated zone. *Transp. Porous Media* **2021**, *137*, 399–416. [CrossRef]
27. Sakamoto, Y.; Yasutaka, T.; Shirakawa, T.; Yamamura, M. Reproduction of latest contamination status through history matching for soil and groundwater contamination due to hexavalent chromium and risk assessment based on long-term prediction. *Proc. Civil Eng. Soc. G (Environ.)* **2017**, *73*, 81–100.
28. Bayer, P.; Finkel, M. Life cycle assessment of active and passive groundwater remediation technologies. *J. Contam. Hydrol.* **2006**, *83*, 171–199. [CrossRef]
29. Sakamoto, Y.; Nishiwaki, J.; Hara, J.; Kawabe, Y.; Sugai, Y.; Komai, T. Development of geo-environment risk assessment system on soil contamination due to mineral oil –numerical analysis for transport phenomena of oil in soil and groundwater and quantitative evaluation of risk level due to multi-component. *Proc. JSCE G (Environ.)* **2010**, *66*, 159–178. [CrossRef]
30. Sakamoto, Y.; Nishiwaki, J.; Hara, J.; Kawabe, Y.; Sugai, Y.; Komai, T. Development of multi-phase and multi-component flow model with reaction in porous media for risk assessment on soil contamination due to mineral oil. *Proc. JSCE G (Environ.)* **2011**, *67*, 78–92. [CrossRef]
31. US EPA. Technical Fact Sheet of 1,4-Dioxane. 2017. Available online: https://www.epa.gov/sites/default/files/2014-03/documents/ffrro_factsheet_contaminant_14-dioxane_january2014_final.pdf (accessed on 2 August 2021).
32. US EPA (Environmental Protection Agency). *Alternative Disinfectants and Oxidants Guidance Manual*; US EPA (Environmental Protection Agency) Office of Water (4607): Washington, DC, USA, 1999.

33. *International Agency for Research on Cancer (ICRC): Reevaluation of Some Organic Chemicals, Hydrazine and Hydrogen Peroxide;* World Health Organization: Lyon, France, 1999.
34. Nakamura, K.; Ito, H.; Kawabe, Y.; Komai, T. Study on partitioning and distribution in soil-water phases for 1,4-dioxane in geo-environment. *Proc. JSCE G (Environ.)* **2018**, *74*, 59–66. [CrossRef]
35. Iguro, C.; Furuichi, T.; Ishii, K.; Kim, S. Prediction of 1,4-dioxane dispersion based on change in groundwater flow after remedial measures by three-dimensional numerical simulation: Toward a permanent measure for Aomori-Iwate Illegal dumping site. *Proc. JSCE G (Environ.)* **2012**, *68*, II_265–II_272.
36. Trumm, D. Selection of active and passive treatment systems for AMD—flow charts for New Zealand conditions. *New Zealand J. Geol. Geophys.* **2010**, *53*, 195–210. [CrossRef]
37. Kewen, L.; Horne, R.N. Comparison of methods to calculate relative permeability from capillary pressure in consolidated water-wet porous media. *Water Resour. Res.* **2006**, *42*, W06405. [CrossRef]
38. Akamine, K.; Tanaka, S.; Arihara, N. A streamline-based 3-Phase equilibrium compositional model with adaptive implicit method. *J. Jpn. Assoc. Pet. Technol.* **2009**, *74*, 211–224. [CrossRef]
39. Iordache, A.; Iordache, M.; Sandru, C.; Voica, C.; Stegarus, D.; Zgavarogea, R.; Ionete, R.E.; Cotorcea(Ticu), S.; Miricioiu, M.G. A Fugacity Based Model for the Assessment of Pollutant Dynamic Evolution of VOCS and BTEX in the Olt River Basin (Romania). *Revista de Chimie (Rev. Chim.)* **2019**, *70*, 3456–3463. [CrossRef]

Article

A Comparative Test on the Sensitivity of Freshwater and Marine Microalgae to Benzo-Sulfonamides, -Thiazoles and -Triazoles

Luca Canova [1], Michela Sturini [1], Federica Maraschi [1], Stefano Sangiorgi [2] and Elida Nora Ferri [2,*]

[1] Department of Chemistry, University of Pavia, Via Taramelli 12, 27100 Pavia, Italy; luca.canova@unipv.it (L.C.); michela.sturini@unipv.it (M.S.); federica.maraschi@unipv.it (F.M.)
[2] Department of Pharmacy and Biotechnology, University of Bologna, Via S. Donato 15, 40127 Bologna, Italy; stefano.sangiorgi9@unibo.it
* Correspondence: elidanora.ferri@unibo.it

Featured Application: These preliminary data on the differences among aquatic microoorganisms in their response to benzo-fused nitrogen heterocyclic pollutants can help in selecting the most suitable biotest for environmental toxicity monitoring activities.

Abstract: The evaluation of the ecotoxicological effects of water pollutants is performed by using different aquatic organisms. The effects of seven compounds belonging to a class of widespread contaminants, the benzo-fused nitrogen heterocycles, on a group of simple organisms employed in reference ISO tests on water quality (unicellular algae and luminescent bacteria) have been assessed to ascertain their suitability in revealing different contamination levels in the water, wastewater, and sediments samples. Representative compounds of benzotriazoles, benzothiazoles, and benzenesulfonamides, were tested at a concentration ranging from 0.01 to 100 mg L^{-1}. In particular, our work was focused on the long-term effects, for which little information is up to now available. Species-specific sensitivity for any whole family of pollutants was not observed. On average, the strongest growth rate inhibition values were expressed by the freshwater *Raphidocelis subcapitata* and the marine *Phaeodactylum tricornutum* algae. *R. subcapitata* was the only organism for which growth was affected by most of the compounds at the lowest concentrations. The tests on the bioluminescent bacterium *Vibrio fisheri* gave completely different results, further underlining the need for an appropriate selection of the best biosensors to be employed in biotoxicological studies.

Keywords: microalgae; *Vibrio fisheri*; benzenesulfonamide; benzothiazole; benzotriazole; biotoxicity tests

Citation: Canova, L.; Sturini, M.; Maraschi, F.; Sangiorgi, S.; Ferri, E.N. A Comparative Test on the Sensitivity of Freshwater and Marine Microalgae to Benzo-Sulfonamides, -Thiazoles and -Triazoles. *Appl. Sci.* **2021**, *11*, 7800. https://doi.org/10.3390/app11177800

Academic Editor: George Aggelis

Received: 9 July 2021
Accepted: 19 August 2021
Published: 25 August 2021

Publisher's Note: MDPI stays neutral with regard to jurisdictional claims in published maps and institutional affiliations.

1. Introduction

The presence of any xenobiotic compounds in fresh, ground, or marine water represents a threat to organisms living in these ecosystems and, therefore, to human health. Among organic pollutants, a group of high production volume chemicals, benzotriazole (BTRs), benzothiazoles (BTs), and benzenesulfonamides (BSAs), benzo-fused nitrogen heterocyclic compounds containing a benzene ring, has been used for several decades in a large number of industrial activities and consuming products [1].

BTRs are corrosion inhibitors, antifreeze fluids, and aircraft de-icers, and they are employed in photography, plastic production, and dishwasher detergents [2–5]. BTs include vulcanization accelerators, corrosion inhibitors, biocides, UV light stabilizers in textiles and plastics, and precursors in the production of several kinds of pharmaceuticals [6]. BSAs are used for synthesizing dyes, photo-chemicals, disinfectants, and intermediates in the production of pharmaceuticals [7,8]. These polar compounds, some of them resistant to biodegradation, can constitute an environmental threat due to their widespread application and consequent ubiquitous occurrence in various environmental compartments [1,7].

Benzothiazoles, for example, are pervasive and spread into the atmosphere through tire chemical leaching [9]; they have been detected in human matrices, including exhaled breath, adipose tissue, urine, and amniotic samples [10–13]. They can be carcinogenic to humans [14], act as dermal sensitizers [15], and produce allergenic effects and respiratory tract irritation at sufficient exposure [16–18]. Several studies suggested that they show acute toxicity to fish, aquatic plants, and invertebrates at relatively high concentrations [19,20].

Benzotriazoles are spread through the progressive corrosion of metals, act as potent air pollutants, and may give rise to water pollution because of run-off from urban areas and roads [9,21]. Together with benzothiazoles, they were found in the watershed, surface freshwater, drinking water, and at high concentrations in primary and secondary wastewaters [22–26]. Their concentrations range from a few ng L^{-1} in water bodies to hundreds of µg L^{-1} in sewage and indoor dust [6,17,22,26]. Existing evidence suggests that benzotriazoles harm aquatic plants and animals and have estrogenic effects in fish and *Daphnia magna* [6,19,27–29].

Benzenesulfonamides were found in rivers and sewage plants in concentrations up to µg L^{-1} [7,30]. Concerning BSAs, studies exist only for *p*-toluene-sulfonamide (*p*-TSA), which was defined as moderately toxic [8,30], but the large amounts currently used recommend additional tests.

The toxicological and ecotoxicological information is estimated to be scarce, especially concerning chronic effects, although acute aquatic toxicity has been repeatedly reported [19,28,30–34]. Information about toxicity on wildlife is scarce, particularly concerning reptiles, birds, and marine mammals. Recently, benzotriazole UV filters have been found in the blood plasma of fishes, snapping turtles, double-crested cormorants, and bottlenose dolphins from various locations in North America and Europe, thus confirming the widespread diffusion of such compounds in the aquatic environments [35,36].

The increasing amount of potentially harmful pollutants in freshwater and marine environments requires fast and low-cost methods of analysis to carry out effective monitoring activities on the contaminants level. As recommended by the Organization for Economic Co-operation and Development (OECD), additional tests on the bioactivity of new pollutants are mandatory [37]. Advanced chemical technologies are usually the first choice because of their high sensitivity and selectivity, but they are time-consuming, require expensive instruments, and can reveal just the searched compounds [1,7,35]. On the contrary, biomonitoring methods are cheaper, faster, and extremely useful in evaluating the xenobiotics' overall biological effects on the ecosystems [34,38]. Nevertheless, the results obtained from the various employed organisms, even belonging to the same trophic level, must be carefully evaluated and interpreted always keeping in mind that each one will represent just a single aspect of the possible scenario, and then multiple organisms must be investigated.

This paper aimed to carry out a comparative evaluation of the response of four organisms, three microalgae and a marine bioluminescent bacterium, for the presence of seven compounds, separately tested. The compounds belong to the above-mentioned three classes of chemicals: benzothiazole (BT), 2-methylthiobenzothiazole (MeSBT), 2-hydroxybenzothiazole (HOBT), benzotriazole (BTR), 5-methylbenzotriazole (5TTR)), benzenesulfonamide (BSA), and *p*-toluene-sulfonamide (*p*-TSA) (Figure 1).

In detail, we tested the freshwater alga *Raphidocelis subcapitata* (previously *Pseudokirchneriella subcapitata*) and two marine organisms, the diatomea *Phaeodactylum tricornutum*, and the green alga *Dunaliella tertiolecta*. These algal species, are widespread in rivers, deltas, lagoons, and coastal habitats and showed a high degree of habitat tolerance. They are relatively easy to maintain in laboratory cultures and are used as standard toxicity test organisms for organic chemicals [39–42].

D. tertiolecta (Chlorophyceae, Chlamydomonadales) is a biflagellate green marine microalga, able to grow in severe environments. It lacks a cell wall that may be a potential barrier to the passage of pollutants into the cell [43,44]. It is easy to cultivate, has rapid growth, and is considered a good indicator to evaluate the toxicity of contaminants present in marine water.

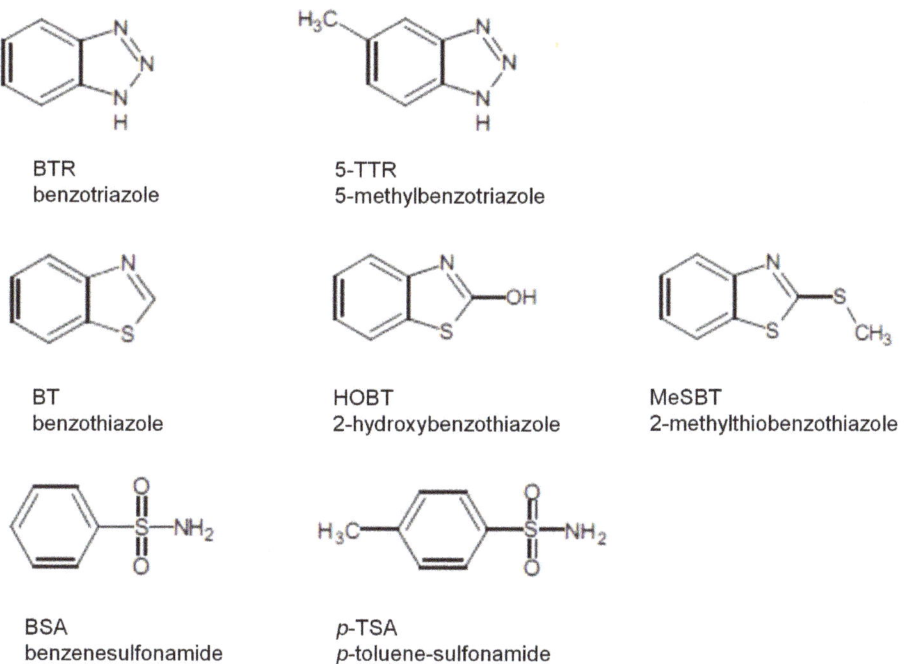

Figure 1. Chemical structures of the tested benzo-fused nitrogen heterocyclic compounds.

R. subcapitata (Chlorophyceae, Sphaeropleales) is a freshwater sessile unicellular green alga. It is easy to cultivate and shows a high growth rate and high sensitivity to several substances. It is one of the green algae representative of oligotrophic and eutrophic environments, a frequently used organism in toxicity studies [45].

P. tricornutum (Bacillariophyta, Bacillariophyceae) is a benthic pennate diatom with a rigid siliceous cell wall. It is the only standardized marine species for wastewater toxicity tests. It is characterized by easy cultivation and sensitivity to pollutants [46].

The well-established dependence of light emission intensity of the bioluminescent bacterium *Vibrio fisheri* from its wellness, i.e., from the presence of harmful or beneficial components, has been exploited to develop one of the most-used toxicity reference assays. *V. fisheri* is a bioluminescent marine bacterium for which light emission intensity is highly influenced by the conditions of its environment. Based on the ascertained inversely proportional reduction in light intensity with the increase in toxicants in solution, the test based on this bacterium has long represented the quicker and easier standardized assay for drinking water quality assessment [47,48].

2. Materials and Methods

2.1. Chemicals

BSA, *p*-TSA, BT, BTR, HOBT, MeSBT, and 5TTR pure compounds were supplied by Sigma-Aldrich (MI, Italy), as well as all chemicals requested to prepare the algae culture media and the bioluminescent bacteria nutrient broth. Thermo Scientific (Vantaa, Finland) supplied the 96-well "Black Cliniplate" microplates for luminescence detection, carried out on the "Victor light 1420" luminescence counter (Perkin-Elmer). Lyophilized aliquots of the luminescent bacteria *V. fischeri* were prepared from fresh cultures at our laboratory, starting from an original batch supplied by the Pasteur Institute (Paris, France). The Istituto

Zooprofilattico Sperimentale of Abruzzo and Molise "G. Caporale" (Teramo, Italy) supplied the three microalgae starting cultures.

2.2. Algal Growth Inhibition Assay

Culture media differed between the three species: for the *D. tertiolecta* the f/2 medium was prepared by adding to sterilized synthetic seawater (Instant Ocean®) a mixture of vitamin (cyanocobalamin + biotin + thiamine) and trace elements [49]. The *P. tricornutum* medium was prepared by adding sodium silicate (30 g L^{-1}) to the *D. tertiolecta* medium. The *R. subcapitata* was cultivated in Jaworski's culture medium [50].

The stock cultures were prepared by inoculating 0.5–1 mL of microalgae suspension on an Erlenmeyer flask containing 250 mL of the respective medium. The flasks were covered with a porous cotton plug and maintained under illumination (white lamp/red lamp Osram Daylight 2 × 36 W plus Osram Gro-Lux lamp 36 W) (8 h light/16 h dark) at 20 °C. The stock cultures were allowed to grow until the required cell density was reached to prepare a fresh stock culture. To perform the assay, a specimen of the stock culture in log phase growth was diluted, obtaining a cell suspension that contained no more than 1.10^3 cells mL^{-1}. Since physiological or metabolic defects can be better evaluated following a prolonged period of growth in spiked culture, a diluted inoculum was employed to let the algae grow for quite a long time before reaching the log phase again. The assays were performed not within the usual interval of 72–96 h, but the growth rate of algae was checked at different intervals after starting the test, about 20 ± 2 (interval A), 27 ± 2 (interval B), and 34 ± 2 days (interval C). The 10 mL test tubes were filled with the appropriate medium containing the desired pollutant's final concentration (100, 50, 25, 10, 5, 1, 0.1, and 0.01 mg L^{-1}), and then the algal culture inoculum (0.1 mL of the diluted suspension) was added. To prepare the controls, the algal inoculum was simply added into the appropriate medium. The tubes were covered with sterilized porous cotton and gauze and kept under the same conditions (light and temperature) as the stock cultures.

We evaluated the algal density, after gentle shaking to homogenize the suspension, by measuring the absorbance of the cultures at 684 nm, an indirect method for cell counting also mentioned in an ISO procedure [51]. The overall experiments were replicated on three independent samples.

The toxic effects were evaluated by calculating the percent of growth inhibition in the samples with respect to the controls according to the absorbance values (Equation (1)).

$$I\% = \frac{A_{control} - A_{sample}}{A_{control}} \times 100 \tag{1}$$

where A is the absorbance at 684 nm.

The EC_{50} values for each compound were calculated according to Finney, 1971 [52], by using the average value of the % samples inhibition, since the toxic effects are always relative to the respective controls in each experiment.

2.3. Bioluminescence Inhibition Assay

Lyophilized aliquots of *V. fischeri* containing 3% NaCl were reconstituted with 1 mL of distilled water and re-suspended in 10–30 mL of the nutrient broth (NaCl 15 g, peptone 2.5 g, yeast extract 1.5 g, glycerol 1.5 mL, HEPES 0.01 M in 500 mL, pH 7). An amount of 200 µL of the bacteria suspension and 100 µL of each sample in distilled water plus 3% NaCl or blank (i.e., 3% NaCl in distilled water) were added to the microplate wells. The emitted light was recorded at various intervals between 0–48 h after preparation of the microplate. Each sample was replicated on five independent wells. The concentrations tested on this organism were in the µg L^{-1} range, precisely 100, 50, and 0.1 µg L^{-1}, because of the high sensitivity usually displayed by this assay.

The bioluminescence inhibition percentage (*I%*) was used to express the toxicity of the tested samples and calculated according to the same equation employed for the algae, by using, in this case, the intensity of the emitted light (*L*):

$$I\% = \frac{L_{control} - L_{sample}}{L_{control}} \times 100$$

2.4. Statistical Analyses

Significant differences at 5% and 1% probability levels ($p \leq 0.05$ and 0.01, respectively) were determined via one-way/two-way non parametrical analysis of variance (Kruskal–Wallis test), with data reported as the mean ± standard error. The relationship between growth rate and light absorbance was tested by the Pearson's r correlation test; the predictive value of the growth-absorbance equation and explained variance were tested by the linear regression method. A non-parametrical U–Mann test for unmatched samples was adopted to evaluate pairs comparison. SPSS statistical software version 13.0 was used for all the tests [53].

3. Results

3.1. Preliminary Assessments

We adopted optical density measurements as the index of cells' density inside vessels, and the reliability of the relationship between the growth rate of target species and light absorbance was evaluated by regression analysis. In all cases, we found a highly significant correlation and a very high explained variance (R > 99% s), i.e., a highly predictive correlation between the two variables (Figure 2), as previously described [54]. Consequently, the differences in growth rate expressed as light absorbance between controls and samples have been considered a suitable parameter for describing the toxicant effect on target species. These effects were evaluated after a long time of contact, until one month, to highlight the possible chronic effects produced by these long-lasting pollutants. The environmental concentrations in water usually range from ng to µg L^{-1}, but we tested concentrations in the range 10 µg L^{-1} to 100 mg L^{-1} since, according to the literature, negligible toxic effects are generally reported at the environmental concentrations, so higher ones are employed to analyze the in vitro effects [19,31,34,55].

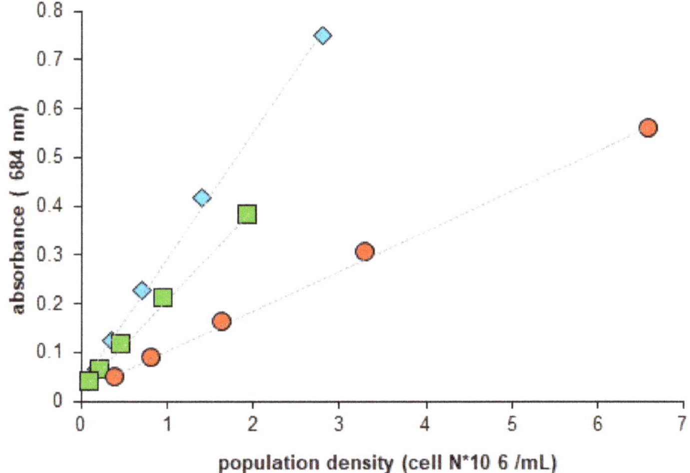

Figure 2. Relationship between target species cell density and light absorbance (◆ *P. tricornutum*, ■ *D. tertiolecta*, • *R. subcapitata*). All the regression functions are statistically significant at $p < 0.001$ with a high explained variance ($R^2_{Pt} = 0.998$; $R^2_{Dt} = 0.998$; $R^2_{Rs} = 0.998$).

3.2. Growth Response of Dunaliella tertiolecta

The response of the algae to the presence of the different compounds was expressed as the % growth inhibition with respect to the control. The values reported in Figures 3–5 were calculated by comparing the absorbance values recorded during the last measurement (34 ± 2 days).

- Benzenesulfonamides. The algae growth rate was affected by p-TSA at the tested concentrations down to 0.1 mg L^{-1} (Figure 3). This concentration reduced the growth rate by about 5%, a quite negligible value. BSA also showed a harmful effect on algae, but the growth rate was decreased with respect to the control only by the concentrations in the range 100–10 mg L^{-1} (K–Wallis χ^2_{BSA} = 5.21, df = 2, $p < 0.05$ χ^2_{pTSA} = 5.34, $p = 0.06$).
- Benzothiazoles. All the benzothiazoles affected *D. tertiolecta* growth rate at the higher concentrations (Mann–U$_{100}$ = 7, $p < 0.05$; Mann–U$_{50}$ = 8, $p < 0.05$), but the lowest effective concentration was different for the three compounds. Concerning BT, it corresponded to 50 mg L^{-1}, whereas 5 mg L^{-1} was the lowest concentration of HOBT, which reduced the population growth in a significant way. The lowest concentration of MeSBT producing a reduction in the growth rate was the 10 mg L^{-1} one (Figure 4).
- Benzotriazoles. The concentrations in the range 5–100 mg L^{-1} of BTR reduced the growth rate of the algal population, and 5TTR showed to be a little more toxic than BTR since all concentrations were able to produce a minimal but measurable reduction (K–Wallis test; χ^2_{BTR} = 5.99, df = 2, $p = 0.05$ and χ^2_{5TTR} = 7.65, $p < 0.05$) (Figure 5). The effect of some compounds, such as BT and BSA, became significant only at the B or C measuring time, confirming that the toxicity, as well as the possible eutrophic effects, is more frequently evident in these organisms after chronic exposure.

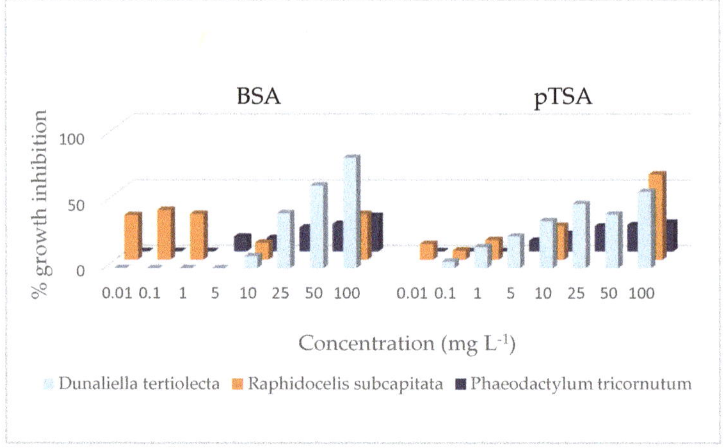

Figure 3. Values of the percentage of growth inhibition produced by the benzenesulfonamides on the three algal species.

3.3. Growth Response of Raphidocelis subcapitata

Benzenesulfonamides, benzothiazoles, and benzotriazoles. All the pollutants showed detrimental effects on the alga and inhibited, even if not always with a linear trend, the growth rate. In the whole tested range, 0.01–100 mg L^{-1}, a variable and apparent reduction in population viability was observed after exposure to benzenesulfonamides (K–Wallis test; χ^2_{BSA} = 5.1, $p = 0.072$; χ^2_{pTSA} = 5.31, $p = 0.06$; Figure 3) and benzothiazoles (K–Wallis test; χ^2_{BT} = 6.41, $p < 0.05$; χ^2_{HOBT} = 6.77, $p < 0.05$; χ^2_{MeSBT} = 5.99, $p = 0.05$; Figure 4). Differently from the other two organisms, this alga showed an unexpected higher sensitivity towards most parts of the compounds at lower concentrations. Only in the case of 5TTR, which significantly affected its growth (K–Wallis test; $\chi^2_{\text{BTR}} = 10.1$, $p < 0.01$; $\chi^2_{5\text{TTR}} = 8.05$, $p < 0.05$; Figure 5), the percent of inhibition values were in a relatively perfect linear relation with the respective concentrations.

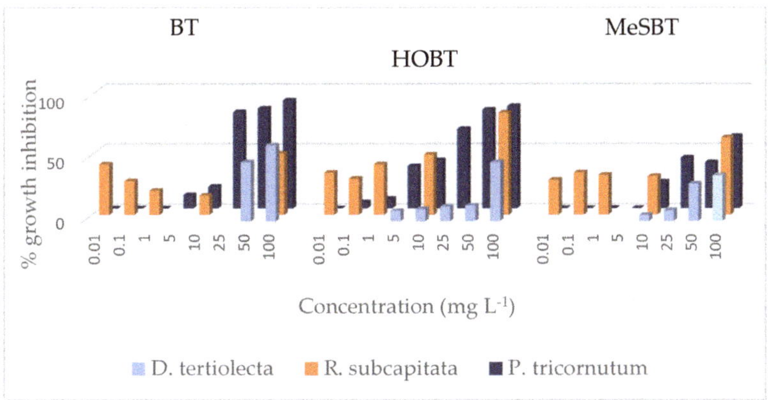

Figure 4. Values of the percentage of growth inhibition produced by the three benzothiazoles on the three algal species.

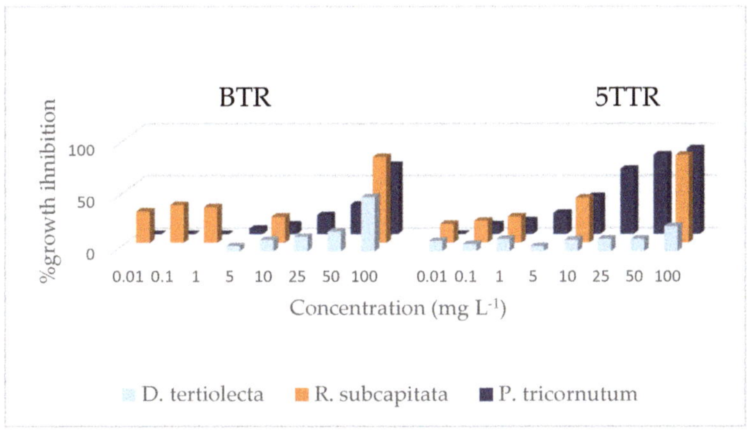

Figure 5. Values of the percentage of growth inhibition produced by the two benzotriazoles on the three algal species.

3.4. Growth Response of Phaeodactylum tricornutum

- Benzenesulfonamides. The toxicity of the two tested compounds on the marine diatom can be considered very similar in both the absolute values of growth inhibition and the lowest effective concentration, which was 5 mg L^{-1} in both cases (Figure 3).
- Benzothiazoles. These compounds showed quite different effects on algal growth. The BT and HOBT higher concentrations produced inhibition values above 80% (K–Wallis test; $\chi^2_{\text{BT}} = 6.20$, $p < 0.05$; $\chi^2_{\text{HOBT}} = 6.31$ $p < 0.05$; Figure 4). A measurable effect was produced until the 5 mg L^{-1} concentration, and marginal inhibition can be ascribed to the 1 mg L^{-1} solution in the case of HOBT. MeSBT resulted in being the least toxic, and the lowest effective concentration was the 10 mg L^{-1} one (K–Wallis test; $\chi^2_{\text{MeSBT}} = 5.84$ $p = $ ns; Figure 4).

- Benzotriazoles. To this alga, the 5TTR solutions resulted in being toxic at all concentrations between 0.1 and 100 mg L^{-1}, producing % inhibition values quite regularly, dependent on the benzothiazole concentration (K–Wallis test; χ^2_{5TTR} =7.15, $p < 0.05$; Figure 5).

BTR resulted in being slightly less toxic, and the 5 mg L^{-1} solution was the lowest one affecting the population growth (K–Wallis test; χ^2_{BTR} = 6.24, $p < 0.05$; Figure 5).

In Table 1, we reported the EC$_{50}$ values of the various compounds for the three algae. It is easier to understand that the less sensitive strain was the marine green algae *D. tertiolecta*. Better sensitivity was expressed by the freshwater alga and by the diatomea, the latter already known to be a sensitive microorganism among the marine ones.

Table 1. EC$_{50}$ values of the various compounds calculated for the three algae, in mg L^{-1}. The value "> 100" means that the maximum concentration tested (100 mg L^{-1}) produced an inhibition lower than 50%.

Compound	*Raphidocelis subcapitata*	*Dunaliella tertiolecta*	*Phaeodactylum tricornutum*
BSA	>100	57	>100
PTSA	92	>100	>100
BTR	67	>100	81
5TTR	38	>100	30
BT	>100	>100	41
MESBT	86	>100	75
HOBT	16	>100	32

3.5. Effects of the Compounds on Vibrio fisheri Light Emission Intensity

In Figure 6, the light emission of the various compounds at 0.1 µg L^{-1} was compared to that of the control at the typical acute (1 h) and chronic toxicity (24 h) intervals, respectively. Moreover, the last record of emission intensity was taken at 48 h after the bacteria–compound contact. It is possible to observe that the majority of the compounds at this concentration have a eutrophic effect on these organisms, or at least stimulate the light emission intensity. Indeed, their vitality was not affected by this concentration, even at the longest possible contact time for this assay. In detail, any significant differences in light emission between the control bacterial cultures and any of the cultures spiked with each compound (χ^2_{1h} = 7.53; df = 7; $p = 0.38$) were observed after 1 h of contact, while strong differences were observed after 24 h (χ^2_{24h} = 21.41; df = 7; $p = 0.003$) and 48 h (χ^2_{48h} = 21.32; df = 7; $p = 0.003$).

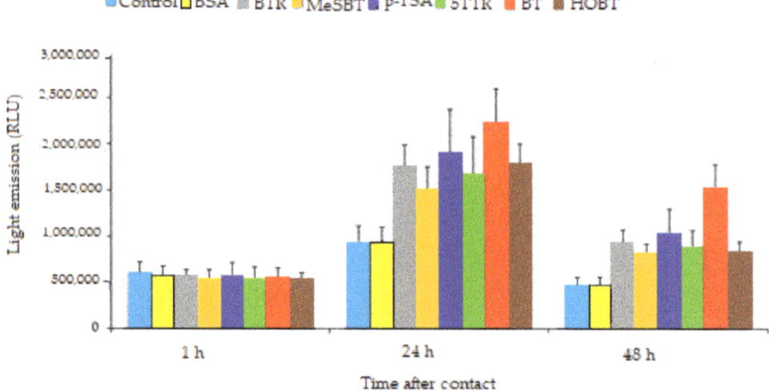

Figure 6. The light emission intensity of the *V. fisheri* suspensions in contact with the seven benzofused nitrogen heterocycles, all at the 0.1 µg L^{-1} concentration.

The observed results in the presence of the 50 and 100 µg L^{-1} solutions showed a more complicated behavior. At the usual acute toxicity interval, no clear inhibition effect was observed (Figure 7A); after 48 h of contact, the majority of the compounds produced a more or less pronounced reduction in light intensity (inhibition) in the range of 10–72%, but HOBT, at both concentrations, greatly stimulated it. BT produced a similar eutrophic effect only at the 50 µg L^{-1} concentration (Figure 7B).

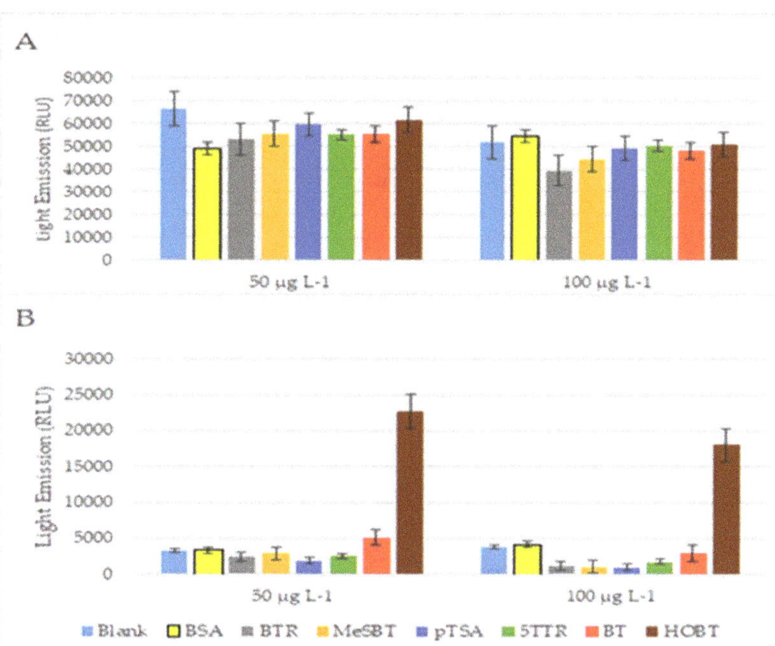

Figure 7. Bioluminescent bacteria light emission intensities recorded at short toxicity intervals (3 h (**A**) and 48 h (**B**)) after contact with the various compounds at the 50 and 100 µg L^{-1} concentration. With the exception of BSA, showing no effect, HOBT and BT at 50 µg L^{-1}, shown eutrophic effects, and other compounds resulted in a clearer toxic effect.

4. Discussion

As stated recently, toxicological data for microalgae are scanty when compared with other taxa [56]. Most of the information on the toxicity of several pollutants in the aquatic environments, the compounds under study included, comes from tests carried out on crustaceans and fish, although algae can be more sensitive than animals to aquatic toxicity [57–60]. Another problem is the prevalence in lab repositories of freshwater species with high tolerance to the pollution that have been frequently used as a model in ecotoxicological studies [56]. These species can be unsuitable surrogates for estuarine or marine microalgae [61].

A way to directly compare our data with those of the literature is not easy to find, since these three algae have never employed together, just two on three sporadically. Moreover, we focused our work on the long-term effects. Previous studies included different species of microalgae [19,34], or sometimes *Raphidocelis subcapitata* [62–64]. Very often, one of the tests in the panel employed to evaluate the biotoxicity is the Microtox$^{®}$ one [62,65,66] based on *Vibrio fisheri*, but it is an acute test (15 min), for which results are very far from our chronic toxicity data.

Nevertheless, we tested a wide range of concentrations of the investigated pollutants similarly to the most part of previous studies and, consistently with them, the first conclusion was that all the compounds had some effects at the higher concentrations, but at the levels detected in freshwaters, they did not pose a risk for aquatic ecosystems.

The tested algae should be applied, for example, to reveal the high concentrations found in wastewaters from sewage treatment plants. Sensitivity for specific families of pollutants was not observed, and each species showed different behaviors with respect to the three groups of compounds. It must be underlined that the differences in the response of microorganisms to the same or to similar pollutants may depend on various factors, ranging from lipid peroxidation to metabolic energy deficiency, decreased photosynthetic energy, or endocrine disrupting effects, and are therefore species-specific and related to the ecology and physiology of the organism [67]. Except for benzenesulphonamides, the freshwater alga, in particular, and the diatomea seemed to be the most sensitive to these compounds, not only concerning the degree of growth inhibition produced by the high concentration, but mainly with respect to the response to the lowest ones. In this family of pollutants, benzothiazoles and benzotriazoles showed the most marked effect on our algal species. According to our expectations, the freshwater *Raphidocelis subcapitata* was more sensitive to benzo-fused nitrogen heterocyclic pollutants than the marine ones, which are well adapted to fast changes in the chemical balance of their habitat.

On the contrary, an unexpected result was the absence of acute adverse effects on the *Vibrio fisheri* light emission, at least at the 50 and 100 μg L^{-1} concentrations. A detrimental effect was observed only at the chronic toxicity intervals (24 and 48 h), and was not produced by all compounds. In fact, two compounds showed at that time a stimulation of light emission, and this phenomenon is quite frequent when the tested compounds do not produce an acute toxicity effect toward the *V. fisheri*. Over time, the light induced degradation or the bacterial metabolism transform the organic pollutants into carbon sources for bacteria. This could be the effect produced by all compounds but BSA at the 0.1 μg L^{-1} concentration, or, in this case, it possibly involves the hormesis effect [68]. The *V. fisheri* test is considered highly sensitive, but, in this case, the response of acute toxicity tests was surprising: there was no toxicity, and no detected risks. Actually, benzotriazoles at the concentrations present in some WWTP effluents were estimated to pose a risk to *V. fisheri* [69], but it was not so evident by using solutions of the pure compounds. This result is a more evident example of how different the effects of any chemical on organisms belonging to the various, and even the same, trophic levels can be.

On the other hand, we estimate that it is important to underline that the most significant results from the compounds under study were obtained from all organisms after chronic exposure intervals, confirming the rationale of our experimental design and the need for long-term assays in evaluating the environmental impact. The selection of the most suitable organism and bioassay(s) design must be accurate and, in any case, based on multiple tests.

The results of this investigation should be helpful in determining the suitability of the various available organisms as environmental biomarkers of these and similar chemicals. Based on these preliminary data, we can state that the three microalgae species are unsuitable for rapidly evaluating the effect of the pure benzo-fused nitrogen heterocyclic compounds at the low contamination levels of surface and drinking waters.

However, the continuous and large use of these compounds can lead to their accumulation in particular compartments, such as sludge and sediments, and their hydrophilicity helps in the transfer to the water. Our plan is to investigate the possible onset of physiological and/or somatic effects of these benzo-fused nitrogen heterocycles in algae grown for various cycles in media containing the low concentration detected in freshwater or coastal environments. These effects, when permanent, would be employed as witnesses of long-term pollution. Moreover, the same experiments must be repeated at similar concentrations in real samples, to exactly evaluate the influence of the co-pollutant traces.

Author Contributions: Conceptualization, M.S., E.N.F., L.C.; investigation, E.N.F., S.S.; formal analysis, L.C., M.S., F.M.; data curation S.S., F.M.; writing—original draft preparation, E.N.F., L.C., M.S.; writing—review and editing, E.N.F., S.S. All authors have read and agreed to the published version of the manuscript.

Funding: This research received no external funding.

Institutional Review Board Statement: Not applicable.

Informed Consent Statement: Not applicable.

Data Availability Statement: The data presented in this study are available on request from the corresponding author.

Conflicts of Interest: The authors declare no conflict of interest.

References

1. Hidalgo-Serrano, M.; Borrull, F.; Marcé, R.M.; Pocurull, E. Presence of benzotriazoles, benzothiazoles, and benzensulfonamides in surface water samples by liquid chromatography coupled to high-resolution mass spectrometry. *Sep. Sci. Plus* **2019**, *2*, 72–80. [CrossRef]
2. Richardson, S.D.; Kimura, S.Y. Water analysis: Emerging contaminants and current issues. *Anal. Chem.* **2016**, *88*, 546–582. [CrossRef]
3. Bruzzoniti, M.C.; Sarzanini, C.; Rivoira, L.; Tumiatti, V.; Maina, R. Simultaneous determination of five common additives in insulating mineral oils by high-performance liquid chromatography with ultraviolet and coulometric detection. *J. Sep. Sci.* **2016**, *39*, 2955–2962. [CrossRef]
4. Janna, H.; Scrimshaw, M.D.; Williams, R.J.; Churchley, J.; Sumpter, J.P. From dishwasher to tap? Xenobiotic substances benzotriazole and tolyltriazole in the environment. *Environ. Sci. Technol.* **2011**, *45*, 3858–3864. [CrossRef]
5. Careghini, A.; Mastorgio, A.F.; Saponaro, S.; Sezenna, E. Bisphenol A, nonylphenols, benzophenones, and benzotriazoles in soils, groundwater, surface water, sediments, and food: A review. *Environ. Sci. Pollut. Res. Int.* **2015**, *62*, 5711–5741. [CrossRef]
6. Liao, C.; Kim, U.-J.; Kannan, K. A review of environmental occurrence, fate, exposure, and toxicity of benzothiazoles. *Environ. Sci. Technol.* **2018**, *52*, 5007–5026. [CrossRef] [PubMed]
7. Herrero, P.; Borrull, F.; Pocurrull, E.; Marcé, R.M. An overview of analytical methods and occurrence of benzotriazoles, benzothiazoles and benzosulfonamides in the environment. *Trends Anal. Chem.* **2014**, *62*, 46–55. [CrossRef]
8. Richter, D.; Massmann, G.; Taute, T.; Duennbier, U. Investigation of the fate of sulfonamides downgradient of a decommissioned sewage farm near Berlin, Germany. *J. Contam. Hydrol.* **2009**, *106*, 183–194. [CrossRef] [PubMed]
9. Asheim, J.; Vike-Jonas, K.; Gonzalez, S.V.; Lierhagen, S.; Venkatraman, V.; Veivåg, I.L.S.; Snilsberg, B.; Flaten, T.P.; Asimakopoulos, A.G. Benzotriazoles, benzothiazoles and trace elements in an urban road setting in Trondheim, Norway: Re-visiting the chemical markers of traffic pollution. *Sci. Total Environ.* **2019**, *649*, 703–711. [CrossRef] [PubMed]
10. Ferrario, J.B.; DeLeon, I.R.; Tracy, R.E. Evidence for toxic anthropogenic chemicals in human thrombogenic coronary plaques. *Arch. Environ. Contam. Toxicol.* **1985**, *14*, 529–534. [CrossRef]
11. Asimakopoulos, A.G.; Wang, L.; Thomaidis, N.S.; Kannan, K. Benzotriazoles and benzothiazoles in human urine from several countries: A perspective on occurrence, biotransformation, and human exposure. *Environ. Int.* **2013**, *59*, 274–281. [CrossRef]
12. Wang, L.; Asimakopoulos, A.G.; Kannan, K. Accumulation of 19 environmental phenolic and xenobiotic heterocyclic aromatic compounds in human adipose tissue. *Environ. Int.* **2015**, *78*, 45–50. [CrossRef]
13. Li, X.; Wang, L.; Asimakopoulos, A.G.; Sun, H.; Zhao, Z.; Zhang, Z.; Zhang, L.; Wang, Q. Benzotriazoles and benzothiazoles in paired maternal urine and amniotic fluid samples from Tianjin, China. *Chemosphere* **2018**, *199*, 524–530. [CrossRef]
14. Sorahan, T. Cancer risks in chemical production workers exposed to 2-mercaptobenzothiazole. *Occup. Environ. Med.* **2009**, *66*, 269–273. [CrossRef]
15. Wang, X.; Suskind, R.R. Comparative studies of the sensitization potential of morpholine, 2-mercaptobenzothiazole and 2 of their derivatives in guinea pigs. *Contact Dermat.* **1988**, *19*, 11–15. [CrossRef] [PubMed]
16. U.S. Consumer Product Safety Commission (CPSC). Sensory and Pulmonary Irritation Studies of Carpet System Materials and Their Constituent Chemicals. 2006. Available online: http://www.cpsc.gov/LIBRARY/FOIA/FOIA98/os/3519926D.pdf (accessed on 7 July 2021).
17. Wan, Y.; Xue, J.; Kannan, K. Benzothiazoles in indoor air from Albany, New York, USA, and its implications for inhalation exposure. *J. Hazard. Mater.* **2016**, *311*, 37–42. [CrossRef] [PubMed]
18. Wang, L.; Asimakopoulos, A.G.; Moon, H.B.; Nakata, H.; Kannan, K. Benzotriazole, benzothiazole, and benzophenone compounds in indoor dust from the United States and East Asian countries. *Environ. Sci. Technol.* **2013**, *47*, 4752–4759. [CrossRef] [PubMed]
19. Seeland, A.; Oetken, M.; Kiss, A.; Fries, E.; Oehlmann, J. Acute and chronic toxicity of benzotriazoles to aquatic organisms. *Environ. Sci. Pollut. Res.* **2012**, *19*, 1781–1790. [CrossRef]
20. Durjava, M.K.; Kolar, B.; Arnus, L.; Papa, E.; Kovarich, S.; Sahlin, U.; Peijnenburg, W. Experimental assessment of the environmental fate and effects of triazoles and benzotriazole. *Altern. Lab. Anim.* **2013**, *41*, 65–75. [CrossRef]

21. Han, X.; Xie, Z.Y.; Tian, Y.H.; Yan, W.; Miao, L.; Zhang, L.L.; Zhu, X.W.; Xu, W. Spatial and seasonal variations of organic corrosion inhibitors in the Pearl River, South China: Contributions of sewage discharge and urban rainfall runoff. *Environ. Pollut.* **2020**, *262*, 114321. [CrossRef] [PubMed]

22. Matamoros, V.; Jover, E.; Bayona, J.M. Occurrence and fate of benzothiazoles and benzotriazoles in constructed wetlands. *Water Sci. Technol.* **2010**, *61*, 191–198. [CrossRef]

23. Wang, L.; Zhang, J.; Sun, H.; Zhou, Q. Widespread occurrence of benzotriazoles and benzothiazoles in tap water: Influencing factors and contribution to human exposure. *Environ. Sci. Technol.* **2016**, *50*, 2709–2717. [CrossRef]

24. Zhao, X.; Zhang, Z.F.; Xu, L.; Liu, L.Y.; Song, W.W.; Zhu, F.J.; Li, Y.F.; Ma, W.L. Occurrence and fate of benzotriazoles UV filters in a typical residential wastewater treatment plant in Harbin, China. *Environ. Pollut* **2017**, *227*, 215–222. [CrossRef]

25. Felis, E.; Sochacki, A.; Magiera, S. Degradation of benzotriazole and benzothiazole in treatment wetlands and by artificial sunlight. *Water Res.* **2016**, *104*, 441–448. [CrossRef]

26. Karthikraj, R.; Kannan, K. Mass loading and removal of benzotriazoles, benzothiazoles, benzophenones, and bisphenols in Indian sewage treatment plants. *Chemosphere* **2017**, *181*, 216–223. [CrossRef]

27. Gatidou, G.; Oursousdou, M.; Stefanatou, A.; Stasinakis, A.S. Removal mechanisms of benzotriazoles in duckweed Lemna minor wastewater treatment systems. *Sci. Total Environ.* **2017**, *596–597*, 12–17. [CrossRef]

28. Giraudo, M.; Douville, M.; Cottin, G.; Houde, M. Transcriptomics, cellular and life-history responses of *Daphnia magna* chronically exposed to benzotriazoles: Endocrine distrupting potential and molting effects. *PLoS ONE* **2017**, *12*, e0171763. [CrossRef]

29. Duan, Z.; Xing, Y.; Feng, Z.; Zhang, H.; Li, C.; Gong, Z.; Wang, L.; Sun, H. Hepatotoxicity of benzotriazole and its effect on the cadmium induced toxicity in zebrafish *Danio rerio*. *Environ. Pollut.* **2017**, *224*, 706–713. [CrossRef] [PubMed]

30. Meffe, R.; Kohfahl, C.; Hamann, E.; Greskowiak, J.; Massmann, G.; Dünnbier, U.; Pekdeger, A. Fate of para-toluenesulfonamide (*p*-TSA) in groundwater under anoxic conditions: Modelling results from a field site in Berlin (Germany). *Environ. Sci. Pollut. Res.* **2014**, *21*, 568–583. [CrossRef]

31. Zeng, F.; Sherry, J.P.; Bols, N.C. Evaluating the toxic potential of benzothiazoles with the rainbow trout cell lines, RTgill-W1 and RTL-W1. *Chemosphere* **2016**, *155*, 308–318. [CrossRef]

32. Molins-Delgado, D.; Díaz-Cruz, M.S.; Barceló, D. Removal of polar UV stabilizers in biological wastewater treatments and ecotoxicological implications. *Chemosphere* **2015**, *119S*, S51–S57. [CrossRef]

33. Yang, T.; Mai, J.; Wu, S.; Liu, C.; Tang, L.; Mo, Z.; Zhang, M.; Guo, L.; Liu, M.; Ma, J. UV/chlorine process for degradation of benzothiazole and benzotriazole in water: Efficiency, mechanism, and toxicity evaluation. *Sci. Total Environ.* **2021**, *760*, 144304. [CrossRef] [PubMed]

34. Gatidou, G.; Anastopoulou, P.; Aloupi, M.; Stasinakis, A.S. Growth inhibition and fate of benzotriazoles in *Chlorella sorokiniana* cultures. *Sci. Total Environ.* **2019**, *663*, 580–586. [CrossRef] [PubMed]

35. Loos, R.; Tavazzi, S.; Mariani, G.; Suurkuusk, G.; Paracchini, B.; Umlauf, G. Analysis of emerging organic contaminants in water, fish, and suspended particulate matter (SPM) in the Joint Danube Survey using solid-phase extraction followed by UHPLC-MS-MS and GC-MS analysis. *Sci. Total Environ.* **2017**, *607–608*, 1201–1212. [CrossRef]

36. Lu, Z.; De Silva, A.O.; Zhou, W.; Tetreault, G.R.; de Solla, S.R.; Fair, P.A.; Houde, M.; Bossart, G.; Derek, C.; Muir, G. Substituted diphenylamine antioxidants and benzotriazole UV stabilizers in blood plasma of fish, turtles, birds and dolphins from North America. *Sci. Total Environ.* **2019**, *647*, 182–190. [CrossRef]

37. OEDC. Advisory Document of the Working Group on Good Laboratory Practice on the Management, Characterisation and Use of Test Items. Environment Directorate Joint Meeting of the Chemicals Committee and the Working Party on Chemicals, Pesticides and Biotechnology. 2018. Available online: https://www.oecd.org/officialdocuments/publicdisplaydocumentpdf/?cote=env/jm/mono(2018)6&doclanguage=en (accessed on 7 July 2021).

38. Camuel, A.; Guieysse, B.; Alcántara, C.; Béchet, Q. Fast algal eco-toxicity assessment: Influence of light intensity and exposure time on *Chlorella vulgaris* inhibition by atrazine and DCMU. *Ecotoxicol. Environ. Saf.* **2017**, *140*, 141–147. [CrossRef]

39. Chhetri, R.K.; Baun, A.; Andersen, H.R. Algal toxicity of the alternative disinfectants performic acid (PFA), peracetic acid (PPA) chlorine dioxide (ClO_2) and their by-products hydrogen peroxide (H_2O_2) and chlorite ($ClO_2{}^-$). *Int. J. Hyg. Environ. Health* **2017**, *220*, 570–574. [CrossRef]

40. DeLorenzo, M.E.; Keller, J.M.; Arthur, C.D.; Finnegan, M.C.; Harper, H.E.; Winder, V.L.; Zdankiewicz, D.L. Toxicity of the antimicrobial compound triclosan and formation of the metabolite methyl-triclosan in estuarine systems. *Environ. Toxicol.* **2008**, *23*, 224–232. [CrossRef]

41. ISO 14442:2006. Water Quality—Guidelines for Algal Growth Inhibition Tests with Poorly Soluble Materials, Volatile Compounds, Metals, and Wastewater. Available online: https://www.iso.org/standard/34814.html (accessed on 7 July 2021).

42. OEDC. Freshwater alga and cyanobacteria, growth Inhibition test (n.201). Guidelines for Testing of Chemicals, Section 2; OECD: Paris, France. Available online: https://www.oecd-ilibrary.org/environment/oecd-guidelines-for-the-testing-of-chemicals-section-2-effects-on-biotic-systems_20745761 (accessed on 18 August 2021).

43. Borowitzka, M.A.; Siva, C.J. The taxonomy of the genus *Dunaliella* (Chlorophyta, Dunaliellales) with emphasis on the marine and halophilic species. *J. Appl. Phycol.* **2007**, *19*, 567–590. [CrossRef]

44. Manzo, S.; Miglietta, M.L.; Rametta, G.; Buono, S.; Di Francia, G. Toxic effects of ZnO nanoparticles towards marine algae *Dunaliella tertiolecta*. *Sci. Total Environ.* **2013**, *445–446*, 371–376. [CrossRef] [PubMed]

45. Leliaert, F.; Smith, D.R.; Moreau, H.; Herron, M.D.; Verbruggen, H.; Delwiche, F.; De Clerck, O. Phylogeny and molecular evolution of the green algae. *Crit. Rev. Plant. Sci.* **2012**, *31*, 1–46. [CrossRef]
46. Singh, J.; Saxena, R.C. An Introduction to microalgae: Diversity and significance. In *Handbook of Marine Microalgae*, 1st ed.; Kim, S.K., Ed.; Academic Press: New York, NY, USA, 2015; pp. 11–24. [CrossRef]
47. ISO 11348-3:2007. Water Quality—Determination of the Inhibitory Effect of Water Samples on the Light Emission of Vibrio Fischeri (Luminescent Bacteria Test). Available online: https://www.iso.org/standard/40518.html (accessed on 7 July 2021).
48. Johnson, B.T. Microtox® Acute Toxicity Test. In *Small-Scale Freshwater Toxicity Investigations*; Blaise, C., Férard, J.F., Eds.; Springer: Dordrecht, Germany, 2005. [CrossRef]
49. Algae Research and Supply. F/2 Medium. Available online: https://algaeresearchsupply.com/pages/f-2-media (accessed on 7 July 2021).
50. CCLA. Culture Collection of Autotrophic Organisms. The Jaworskis' Medium. Available online: https://ccala.butbn.cas.cz/en/jaworskis-medium (accessed on 7 July 2021).
51. International Standards Organization. ISO N. 8692/2004 amended by ISO 8692/2012. Water Quality—Freshwater Algal Growth Inhibition Test with Unicellular Green Algae. Available online: https://www.iso.org/standard/54150.html (accessed on 7 July 2021).
52. Finney, D.J. *Probit Analysis*, 3rd ed.; Cambridge University Press: New York, NY, USA, 1971.
53. Steel, R.G.; Torrie, J.H. *Principles and Procedures of Statistics*, 3rd ed.; Mc-Graw Hill: New York, NY, USA, 2000.
54. Riberio Rodrigues, L.H.; Arenzon, A.; Raya-Rodriguez, M.T.; Ferreira Fontoura, N. Algal density assessed by spectrophotometry: A calibration curve for the unicellular algae *Pseudokirchneriella subcapitata*. *J. Environ. Chem. Ecotoxicol.* **2011**, *3*, 225–228. [CrossRef]
55. Montesdeoca-Esponda, S.; Álvarez-Raya, C.; Torres-Padrón, M.E.; Sosa-Ferrera, Z.; Santana-Rodríguez, J.J. Monitoring and environmental risk assessment of benzotriazole UV stabilizers in the sewage and coastal environment of Gran Canaria (Canary Islands, Spain). *J. Environ. Manag.* **2019**, *233*, 567–575. [CrossRef]
56. Flood, S.; Burkholder, J.; Cope, G. Assessment of atrazine toxicity to the estuarine phytoplankter *Dunaliella tertiolecta* (Chlorophyta), under varying nutrient conditions. *Environ. Sci. Pollut. Res.* **2018**, *25*, 11409–11423. [CrossRef] [PubMed]
57. Lewis, M.A. Chronic Toxicities of Surfactants and Detergents Builders to algae—A Review and Risk Assessment. *Ecotoxicol. Environ. Saf.* **1990**, *20*, 123–140. [CrossRef]
58. Lewis, M.A. Use of Freshwater Plants for Phytotoxicity Testing—A Review. *Environ. Pollut.* **1995**, *87*, 319–336. [CrossRef]
59. Blanck, H.; Wangberg, S.A. Pattern of Cotolerance in Marine Periphyton Communities established under Arsenate Stress. *Aquatic Toxicol.* **1991**, *21*, 1–14. [CrossRef]
60. Walsh, G.E. Principles of Toxicity Testing with Marine Unicellular Algae. *Environ. Toxicol. Chem.* **1988**, *7*, 979–987. [CrossRef]
61. Graymore, M.; Stagnitti, F.; Allinson, G. Impacts of atrazine in aquatic ecosystems. *Environ. Int.* **2001**, *26*, 483–495. [CrossRef]
62. Corsi, S.R.; Geis, S.W.; Loyo-Rosales, J.E.; Rice, C.P. Aquatic toxicity of nine aircraft deicer and anti-icer formulations and relative toxicity of additive package ingredients alkylphenol ethoxylates and 4,5-methyl-1H-benzotriazoles. *Environ. Sci Technol.* **2006**, *40*, 7409–7415. [CrossRef]
63. Azizian, M.F.; Nelson, P.O.; Thayumanavan, P.; Williamson, K.J. Environmental impact of highway construction and repair materials on surface and ground waters. Case study: Crumb rubber asphalt concrete. *Waste Manag.* **2003**, *23*, 719–728. [CrossRef]
64. European Chemicals Agency (ECHA). N-cyclohexylbenzothiazol-2-sulphenamide, CAS No: 95-33-0, EINECS No.202-411-2. *Summary Risk Assessment Report, Final Report, May.* 2008. Available online: https://echa.europa.eu/documents/10162/48e99dc2-8d09-47e5-b167-61e8d2b13fd1 (accessed on 7 July 2021).
65. Reemtsma, T.; Fiehn, O.; Kalnowski, G.; Jekel, M. Microbial transformations and biological effects of fungicide-derived benzothiazoles determined in industrial wastewater. *Environ. Sci. Technol.* **1995**, *29*, 478–485. [CrossRef]
66. Pillard, D.A.; Cornell, J.S.; Dufresne, D.L.; Hernandez, M.T. Toxicity of benzotriazole and benzotriazole derivatives to three aquatic species. *Wat Res.* **2001**, *35*, 557–560. [CrossRef]
67. Pérez, P.; Fernández, E.; Beiras, R. Use of Fast Repetition Rate Fluorometry on Detection and Assessment of PAH Toxicity on Microalgae. *Water Air Soil Pollut.* **2010**, *209*, 345–356. [CrossRef]
68. Ray, S.D.; Farris, F.F.; Hartmann, A.C. Hormesis. In *Encyclopedia of Toxicology*, 3rd ed.; Wexler, P., Ed.; Elsevier: Amsterdam, The Netherlands, 2014; pp. 944–948. [CrossRef]
69. Molins-Delgado, D.; Távora, J.; Diaz-Cruz, M.S.; Barceló, D. UV filters and benzotriazoles in urban aquatic ecosystems: The footprint of daily use products. *Sci. Total Environ.* **2017**, *601–602*, 975–986. [CrossRef]

applied
sciences

MDPI

Article

Metal Bioaccumulation by Carp and Catfish Cultured in Lake Chapala, and Weekly Intake Assessment

Claudia Alvarado [1,*], **Diego M. Cortez-Valladolid** [1], **Enrique J. Herrera-López** [2], **Ximena Godínez** [1] **and José Martín Ramírez** [3]

[1] Centre for Research and Assistance in Technology and Design of Jalisco State A.C., Food Technology Unit, Camino Arenero 1227, El Bajío del Arenal, Zapopan 45019, Jalisco, Mexico; d.cortez.v@outlook.com (D.M.C.-V.); x1m3n.n@gmail.com (X.G.)

[2] Centre for Research and Assistance in Technology and Design of Jalisco State A.C., Industrial Biotechnology, Bioelectronics LINBIA Laboratory, Camino Arenero 1227, El Bajío del Arenal, Zapopan 45019, Jalisco, Mexico; eherrera@ciatej.mx

[3] Centre for Research and Assistance in Technology and Design of Jalisco State A.C., Analytic and Metrologic Services Unit, Av. Normalistas 800, Colinas de la Normal, Guadalajara 44270, Jalisco, Mexico; mramirez@ciatej.mx

* Correspondence: calvarado@ciatej.mx; Tel.: +52-(33)-3345-5200 (ext. 2130)

Abstract: Aquaculture offers great potential for fish production in Lake Chapala, but reports of heavy metal contamination in fish have identified a main concern for this activity. In the present study, cultures of the species *Cyprinus carpio* and *Ictalurus punctatus* were grown in a net cage in Lake Chapala. The patterns of heavy metal accumulation (Cu, Zn, Cd, Hg, Pb, As) in muscle and liver were monitored in order to evaluate the level of metal incorporation in the fish. Estimates of weekly metal intake (EWI) were made based on the results of the concentrations in edible parts of fish of commercial size. The patterns of metal bioaccumulation between tissues and species showed that liver had a higher concentrating capacity for Zn, Cu, Cd, and Pb. In contrast, similar concentrations of Hg and As were found in the liver and muscle tissue. According to the EWI estimates, the heavy metals in these cultured fish do not represent a risk for human consumption.

Keywords: heavy metals; bioaccumulation; Lake Chapala; carp; catfish; PTWI

Citation: Alvarado, C.; Cortez-Valladolid, D.M.; Herrera-López, E.J.; Godínez, X.; Ramírez, J.M. Metal Bioaccumulation by Carp and Catfish Cultured in Lake Chapala, and Weekly Intake Assessment. *Appl. Sci.* **2021**, *11*, 6087. https://doi.org/10.3390/app11136087

Academic Editor: Elida Nora Ferri

Received: 28 April 2021
Accepted: 23 June 2021
Published: 30 June 2021

Publisher's Note: MDPI stays neutral with regard to jurisdictional claims in published maps and institutional affiliations.

1. Introduction

Lake Chapala is the largest freshwater lake in Mexico, with a total surface area of 114,659 ha [1]. A shallow, alkaline body of water, it has a mean depth of less than 7 m. Lake Chapala belongs to the Lerma-Chapala Basin. Its main tributary input comes from the Lerma River, which supplies about 80% of its waters (Figure 1). This river is also the main source of anthropogenic pollution, including heavy metals, because ~3500 industries pour their treated wastewaters into it [2,3]. Sediments from Lake Chapala are rich in heavy metals, and previous studies have shown that lead (Pb) and cadmium (Cd) are present in the exchangeable fraction of water, and possibly available for absorption by living organisms, including fish [4]. The sediments, however, have a metal adsorption–desorption dynamic that is strongly influenced by pH and marked seasonal fluctuations in water levels, so low concentrations of metals in the water have also been reported [5,6].

For humans, consuming fish is an excellent source of proteins and n-3 polyunsaturated fatty acids, but concern has arisen over the potential danger of heavy metals [7]. Human diets are a significant source of exposure to heavy metals that carry a toxicological risk because they cannot be degraded by biological processes. Due to this potential for harm, the Joint FAO/WHO Expert Committee on Food Additives (JECFA) has established a Provisional Tolerable Weekly Intake (PTWI) for metals that estimates the amount of a given contaminant that can be assimilated weekly per unit of body weight (bw) over a lifetime, without constituting an appreciable health risk [8].

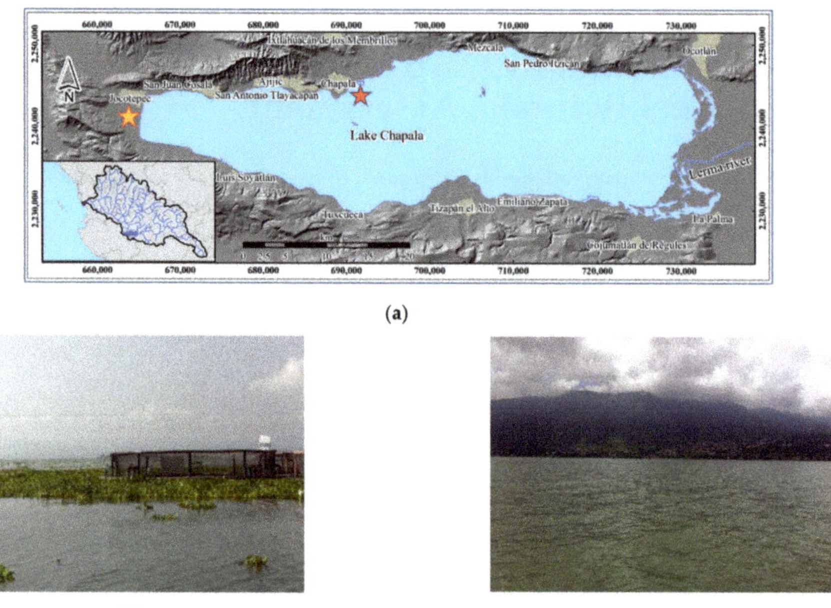

(a)

(b)   **(c)**

Figure 1. Location of the experimental cage in Lake Chapala. (**a**) Map of Lake Chapala showing with the red star the experimental cage and the yellow star the control group. (**b**) Experimental cage. (**c**) General view of Lake Chapala environment.

Contaminated fish are the principal source of heavy metals such as mercury, lead, and arsenic for humans. Previous reports in fish from Mexican water reservoirs have shown borderline-to-high quantities of heavy metals, making it essential to maintain constant monitoring of their intake through fish consumption in the diet [9–12].

The aim of the present study was to evaluate the bioaccumulation of metals (Zn, Cu, Cd, Hg, As, Pb) along time in fish cultured in a natural environment to evaluate the risk of dietary intake of metals in humans.

2. Materials and Methods

2.1. Study Design and Sampling

About one thousand catfish fingerlings (*Ictalurus punctatus*) aged 15 weeks (average weight 4.80 ± 2.96 g and length 8.51 ± 2.04 cm), and two hundred adult carps (*Cyprinus carpio*) aged 30 weeks (average weight 30.04 ± 15.55 g, and length 11.08 ± 4.69 cm) were cultured in the same floating net cage in Lake Chapala, Jalisco, Mexico (coordinates 20°17′2.9″ N, −103°10′27.2″ W) (Figure 1). These fish were called the experimental group. The cage measured 6 m in diameter and 2 m deep, with a surface area of 100.13 m^2 and volume of 56 m^3. The mesh size was 1 inch. It was placed 3 m above the bottom of the lake. During the fingerling age period, the catfish were kept in a nursery cage within the larger cage, and after they remained in it for 3 months, fish were released. In order to detect any factor that might affect metal bioaccumulation, a control group of fish with the same characteristics was cultured in an earth pond in Jocotepec, Jalisco, a lakeside community (Figure 1).

The fish were fed Winfish-Zeigler 3506 (3.5 mm) until they reached a weight of 50 g (6 months), when this was replaced with Winfish-Ziegler 2505 (5.5 mm). Culture performance was monitored monthly in terms of biometrics, morbidity, and mortality. The properties of the lake, monitored and recorded weekly, were pH, temperature, and dissolved oxygen. Table 1 shows the metals concentration of the feedstuffs.

Table 1. Metal concentration of fish feedstuffs and the maximum content allowed.

Metal	Winfish-Ziegler 3506 (3.5 mm)	Winfish-Ziegler 2505 (5.5 mm)	Maximum Content [13]
Cu (mg/kg)	21.77 ± 2.45	52.83 ± 1.58	-
Zn (mg/kg)	216.50 ± 10.89	94.86 ± 2.53	-
Cd (µg/kg)	0.36 ± 0.34	0.08 ± 0.01	1000
As (µg/kg)	0.90 ± 0.28	0.43 ± 0.02	6000
Pb (µg/kg)	1.24 ± 0.05	0.81 ± 0.08	5000
Hg (µg/kg)	17.89 ± 3.16	13.17 ± 0.52	500

The collection of fish samples involved taking 3–5 individuals from each species and for each experimental group at intervals of 6–7 weeks of exposure, concretely on days 0, 38, 81, 123, 179, 241, 298, 369, 424, and 473. For sampling, the fish were transferred to containers with 0.2 g/L of MS 222 anesthetic (Sigma Aldrich). When dead, the fish were placed into new plastic bags and transported to the laboratory in a container to maintain the chain of command. The length and weight of each fish and its tissues were recorded before dissection. Liver and muscle tissue were removed, weighed, and homogenized individually, except for catfish, at the three first sampling stages where the fish were very small, and the same tissues from different fish were placed in the same bag. Samples were frozen at −20 °C for further analysis.

2.2. Analysis of Metals

The concentrations of copper (Cu), zinc (Zn), cadmium (Cd), lead (Pb), mercury (Hg), and arsenic (As) were analyzed in the liver and muscles, as well as in the fish food. The samples of fish tissues and fish food were treated by microwave digestion with nitric acid, according to the 5BI-8 sample preparation note. Briefly, 1.00 ± 0.05 g of each homogenized sample was weighed in microwave vessels and HNO_3 was added. The vessels were placed in a CEM Marx microwave and heated at 1200 W, 200 psi, and 210 °C for 10 min, then allowed to cool at room temperature until the digestion time was completed. The samples were then gauged to 50 mL with deionized water.

All samples were analyzed in duplicate. Cu and Zn concentrations were measured by inductively coupled plasma optical emission spectroscopy (ICP-OES) in a Perkin Elmer Optima 8300DV (Shelton, CT) using internal method INS-SM/US-71, based on EPA method 6010B. For Cd, Pb, Hg, and As, internal method INS-SM/US-220 was followed using inductively coupled plasma mass spectrometry (ICP-MS) in a Perkin Elmer ELAN 9000 (Shelton, CT, USA).

Standard Perkin Elmer solutions (N9300174) were used to prepare the calibration curves, which presented R2 > 0.998. The recovery percentages performed by spiked samples of the fish ranged from 90% to 110%. Precision was measured by evaluating a Perkin Elmer N9300211 solution. The coefficients of variation did not exceed 9%. DORM-4 (fish protein) reference material from Canada's National Research Council (CNRC, Ottawa, Canada) and FAPAS T07213QC (crab meat) reference material from The Food and Environmental Research Agency (FAPAS, Sand Hutton, UK) were digested and analyzed in triplicate for quality control. Table 2 shows the recoveries of heavy metals and the corresponding certified values for the reference materials, according to the method used.

Table 2. Certified concentration and measured values for reference material DORM-4 and FAPAS T07213QC by method.

Element	Reference Material	Certified Conc.	Units	Measured Concentration	Detection Limit	Method
Hg	DORM-4	410	µg/kg	385	1.9	ICP-MS
Hg	FAPAS T07213QC	93.5	µg/kg	76.7	1.9	ICP-MS
Cu	DORM-4	15.9	mg/kg	14.8	0.32	ICP-OES
Zn	DORM-4	52.2	mg/kg	57.37	1.16	ICP-OES
Pb	DORM-4	416	µg/kg	423	0.287	ICP-MS
Pb	FAPAS T07213QC	50.1	µg/kg	47.0	0.287	ICP-MS
Cd	DORM-4	306	µg/kg	301	1.3	ICP-MS
Cd	FAPAS T07213QC	5.53	µg/kg	5.78	1.3	ICP-MS
As	DORM-4	6800	µg/kg	7158	1.8	ICP-MS
As	FAPAS T07213QC	13.9	µg/kg	14.0	1.8	ICP-MS

2.3. Estimated Weekly Intake

The estimated weekly intake (EWI) of essential (Cu, Zn) and non-essential metals (Cd, Pb, As, Hg) through consumption of the cultured catfish and carp were calculated, assuming a weekly consumption of 200 g for Latin American populations [14] and an average body weight of 70 kg for Mexican people [15]. Calculations were performed by applying the mean and maximum concentrations for each metal analyzed in the study in the following equation:

$$\text{EWI} = \frac{\textbf{Metal Concentration} \times \textbf{Fish intake } (0.2 \textbf{ kg})}{\textbf{Body weight } (70 \textbf{ kg})}$$

Results were compared as the percentage of contribution to their respective Provisional Tolerable Weekly Intake (PTWI) value established by JECFA [8]:

$$\% \textbf{ PTWI} = \frac{\textbf{EWI}}{\textbf{PTWI}} \times 100$$

2.4. Data Processing

The data from the catfish and carp were stored in Excel® software. Data were ordered according to the metals Hg, Cu, Zn, Pb, Cd, and As evaluated in the liver and muscle of the fish species.

The liver concentration factor (LCF) was obtained by dividing the mean concentration of every metal in the liver over the mean concentration of the metal in muscle, for every time in the experimental group. Diverse trendlines—linear, exponential, and logarithmic, among others—were evaluated to properly fit the catfish and carp data. The R2 coefficient and visual inspection were used as criteria to select the most appropriate function for representing the shape of the fitted data. In a second approach, the data were analyzed, and the mathematical expressions selected for the fitted data were imported into Matlab® 2015 software, which generated numerical vectors for each data category as a function of sampling time to represent the concentrations of metals that had accumulated in the organs examined. Our experimental data were then plotted versus the interpolated curves generated by the mathematical equations obtained during the fitting procedure.

2.5. Statistical Analyses

Means and standard deviations of every point in time by fish were then determined for every data vector, experimental versus control. F-tests run in STATGRAPHICS® 2018 software were used to determine significant differences between the standard deviations of the two samples at a confidence level of 95.0%.

3. Results

3.1. Pattern of Metal Bioaccumulation

3.1.1. Essential Metals: Cu and Zn

Table 3 shows the range of metal concentrations in catfish and carp obtained for every metal and the ranges of LCF for fish cultured in Lake Chapala, named the experimental group. Figures 2 and 3 show the pattern of nutritive metal bioaccumulation in muscle and liver of the cultured carp and catfish in experimental and control groups. Table 4 shows the differences between experimental and control groups, for every metal, species, and tissue. Cu and Zn did not show statistical differences between groups.

Table 3. Range of metal concentrations in muscle and liver of cultured fish, and range of liver concentration factor (LCF) by fish species in the experimental group along time.

Metal	Carp (*Cyprinus carpio*)			Catfish (*Ictalurus punctatus*)		
	Muscle	Liver	LCF	Muscle	Liver	LCF
Cu (mg/kg)	0.36–2.39	6.44–47.81	9–37	0.56–3.83	1.48–5.84	1–9
Zn (mg/kg)	8.84–16.05	38.44–265.00	2–28	5.66–12.64	17.39–28.79	2–4
Cd (µg/kg)	<1.3–3.67	13.25–472.40	8–220	<1.3–2.14	5.39–51.56	1–42
Pb (µg/kg)	3.25–17.44	9.70–76.38	0.6–23	3.51–15.23	2.87–23.90	0.5–3
As (µg/kg)	43.73–192.78	38.35–126.08	0.3–3	10.57–126.40	11.47–87.34	0.3–3
Hg (µg/kg)	<1.9–5.88	<1.9–18.59	0.3–5	<1.9–34.56	<1.9–24.60	0.5–1

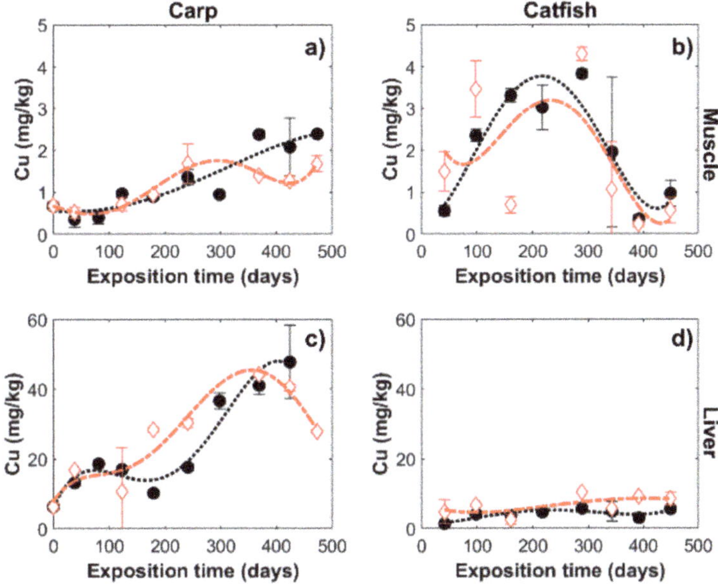

Figure 2. Cu accumulation in the experimental (•) and control groups (◇), in front of (**a**) muscle in carp, (**b**) muscle in catfish, (**c**) liver in carp, and (**d**) liver in catfish.

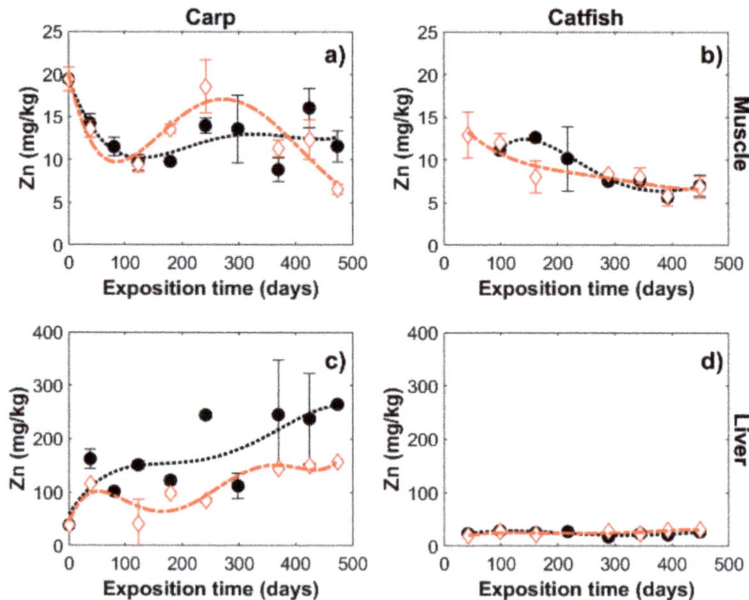

Figure 3. Zn accumulation in the experimental (●) and control groups (◊), in front of (**a**) muscle in carp, (**b**) muscle in catfish, (**c**) liver in carp, and (**d**) liver in catfish.

Table 4. F-test to compare the variances of the experimental and control groups for carp and catfish, in muscle and liver.

Metal	Fish	Muscle		Liver	
		F	*p*-Value	F	*p*-Value
Cu	Carp	3.017	0.168	1.213	0.820
	Catfish	0.747	0.732	0.317	0.188
Zn	Carp	0.722	0.679	2.797	0.198
	Catfish	1.661	0.591	0.711	0.689
Cd	Carp	0.361	0.240	156.912	0.000 *
	Catfish	1.038	0.965	0.301	0.213
Pb	Carp	0.582	0.528	0.368	0.249
	Catfish	0.008	0.000 *	0.605	0.557
As	Carp	0.793	0.786	0.253	0.158
	Catfish	2.130	0.380	1.193	0.851
Hg	Carp	1.247	0.796	0.429	0.433
	Catfish	1.112	0.921	2.240	0.454

* Significant differences between the control and experimental group.

- Copper

Figure 2a,b show the Cu accumulation patterns in the muscle of both species and in both groups. A similar concentration range was observed in the muscle of both species (Table 3), but the pattern of accumulation showed large differences, as the concentration in the carp showed a steady increase over time, while in the catfish, a curve with a maximum on day 288 was observed. This behavior in carp muscle was observed in both groups, experimental and control.

Figure 2c,d show the accumulation patterns of Cu in the fish livers. Both species showed higher concentrations in the liver, but the carp had levels 22 times higher than in the muscle tissue (Table 3), 6.44–47.81 mg/kg in liver versus 0.36–2.39 mg/kg in muscle. The catfish liver, in contrast, showed an accumulation only three times as large as that of the muscle tissue (1.48–5.84 mg/kg in liver versus 0.56–3.83 mg/kg in muscle). The carp liver (6.44–47.81 mg/kg) also had a concentration 10 times higher than the catfish liver of the experimental group (1.48–5.84 mg/kg).

- Zinc

Figure 3a,b show the Zn accumulation pattern in the muscles of both species. The range of concentrations for experimental groups remained in a range between 5.66 and 16.05 mg/kg for both species, and the general pattern of concentration described a descending trend. The experimental and control groups showed similar patterns of bioaccumulation.

Zn accumulation in the liver of both species was higher than in the muscle (Figure 3c,d). Zn concentrations in the liver began at similar levels in the two species, but a constant increase was seen in the carp. At the end of the experiment, the carp liver had 10 times more Zn than the catfish liver (265.00 versus 28.79 mg/kg) in experimental groups.

3.1.2. Non-Essential Metals: Cd, As, Pb, and Hg

- Cadmium

Figure 4a,b show the Cd accumulation pattern in the muscle of the catfish and carp. Muscle concentrations in both species were in the same range (1.3–3.67 µg/kg). The accumulation pattern in the carp showed a slight rise over time, but in the catfish, it was rather flat. Only carp showed significant differences in liver between experimental and control groups (Table 4).

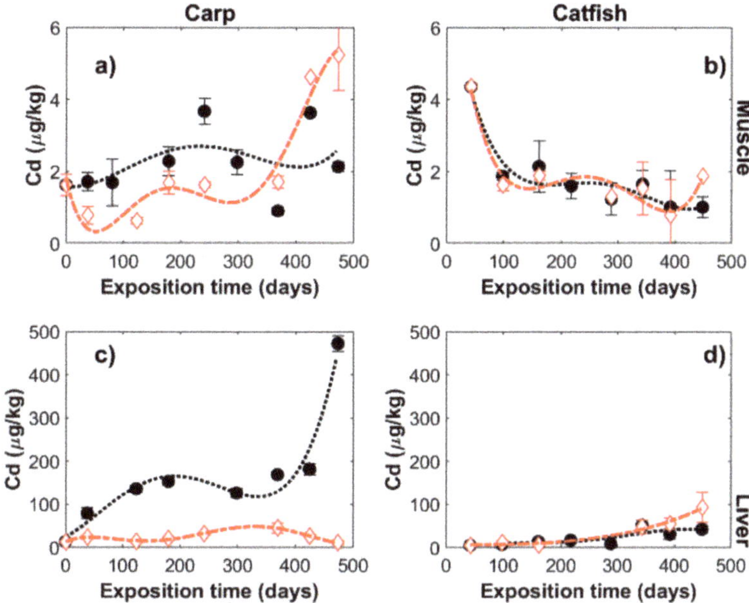

Figure 4. Cd accumulation in the experimental (●) and control groups (◇), in front of (**a**) muscle in carp, (**b**) muscle in catfish, (**c**) liver in carp, and (**d**) liver in catfish.

On average, Cd concentrations in carp were 78 times higher in the liver than muscle and 17 times higher in the case of catfish in the experimental group, revealing the concentrating function of liver in both species (Figure 4c,d). The Cd concentration in the carp liver was twice as high as that of the catfish at the beginning of the experiment, but by the end, it was almost 10 times higher (472.40 versus 51.56 µg/kg) for the carps cultured in Lake Chapala only. The general pattern of metal accumulation had an upward trend in the liver of the experimental fish that was more prominent in the carp. The carp cultured in Lake Chapala had a significantly high concentration in liver with respect to the carp in the earth pond.

- Lead

Figure 5a,b show the Pb accumulation pattern in the muscle of the fish. The concentration ranges were very similar between both species and organs over time (Table 3). Only in the catfish muscle did the control group show a significantly higher concentration at the beginning of the experiment (Table 4).

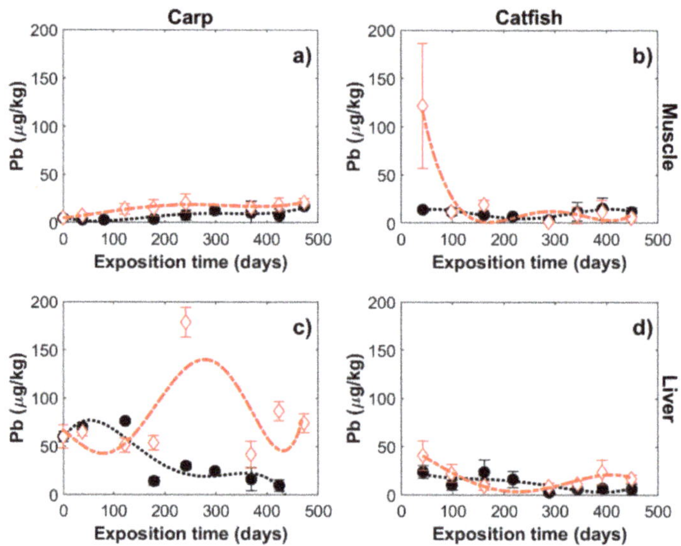

Figure 5. Pb accumulation in the experimental (•) and control groups (◇), in front of (**a**) muscle in carp, (**b**) muscle in catfish, (**c**) liver in carp, and (**d**) liver in catfish.

In the catfish liver, the Pb concentrations were on average 1.2 times higher than in the muscle. LCF between 11 to 23 was observed in carp liver, which gradually decreased after day 123 (Figure 5c,d).

- Arsenic

Figure 6a,b show the As accumulation pattern in the muscles of the catfish and carp. In this case, a downward trend of bioaccumulation was observed in the muscle of both species. Concentrations were slightly higher in the carp than the catfish, but within the same range (Table 3). As did not show statistical differences between groups (Table 4). The carps of the experimental group had on average 15 µg/kg of As higher than the control group, but after 298 days of exposition, they reversed the trend. No liver concentration effect was observed for As in either species, as the concentrations were slightly lower than in the muscle throughout the study period (Table 2; Figure 6c,d). The catfish and carp livers had ranges of 11.47–87.34 and 38.35–126.08 µg/kg, respectively. The general trend in the carp liver showed a soft upward shape, but the trend in the catfish was flat both in the experimental and control groups (Figure 6d).

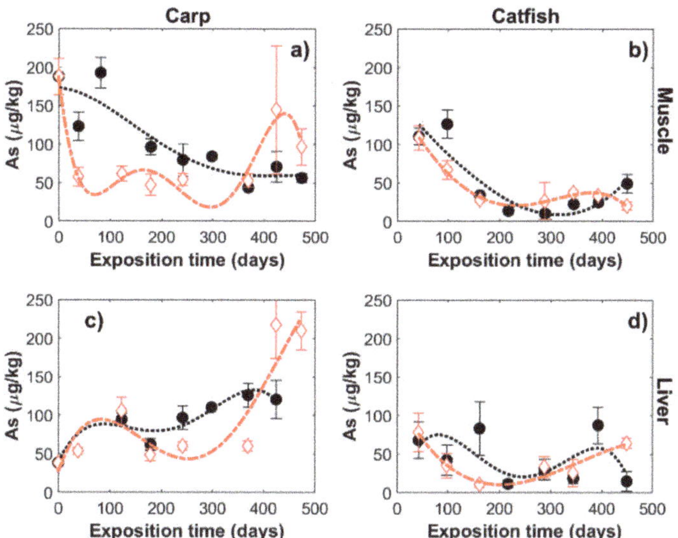

Figure 6. As accumulation in the experimental (●) and control groups (◇), in front of (**a**) muscle in carp, (**b**) muscle in catfish, (**c**) liver in carp, and (**d**) liver in catfish.

- Mercury

Figure 7 shows the Hg bioaccumulation in the muscle and liver of the two species. Hg did not show statistical differences between groups (Table 4). No liver accumulation effect was observed in catfish, and soft LCF was observed in carp (0.3 to 5). In fact, in many samples, concentrations were below the method's detection limit (1.9 µg/kg). Concentration ranges were similar between the tissues (Table 2), and lower in the carp (<1.9–18.59 µg/kg) than the catfish (<1.9–34.56 µg/kg). A slightly upward trend of bioaccumulation was observed in both species in muscle.

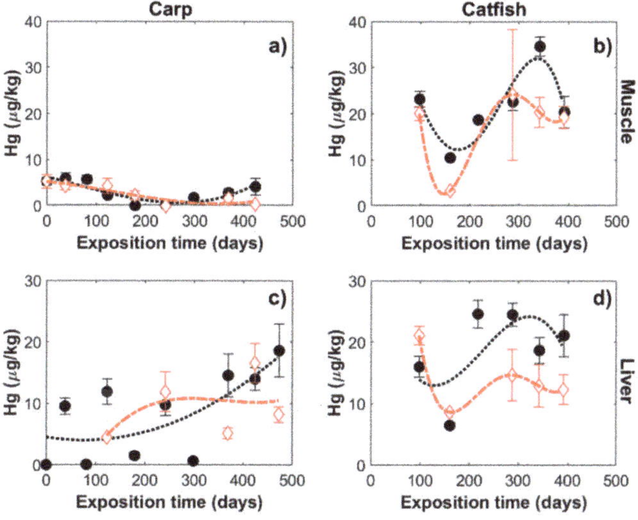

Figure 7. Hg accumulation in the experimental (●) and control groups (◇), in front of (**a**) muscle in carp, (**b**) muscle in catfish, (**c**) liver in carp, and (**d**) liver in catfish.

3.2. Estimated Weekly Intake

The means of the metal concentrations obtained from the muscle tissue of commercial-sized fish in experimental and control groups are summarized in Table 5. This information was used to calculate the EWI and the percentages of contribution to the PTWI of each metal (Table 5).

Table 5. Mean concentration (ww) of metals in fish muscle at consumption size in experimental and control groups. Estimated weekly intake (EWI) and contribution to Provisional Tolerable Weekly Intake (%PTWI) through consumption of cultured fish.

Metal	Fish	Control Group			Experimental Group			Reference PTWI [†]
		Conc.	EWI *	% PTWI	Conc.	EWI *	% PTWI	
Cu (mg/kg)	Carp	1.45 ± 0.21	0.004	0.12	2.28 ± 0.18	0.007	0.19	0.5
	Catfish	0.63 ± 0.42	0.002	0.05	1.09 ± 0.80	0.003	0.09	
Zn (mg/kg)	Carp	10.08 ± 3.07	0.029	0.41	12.16 ± 3.64	0.035	0.50	1.0
	Catfish	6.93 ± 1.11	0.020	0.28	6.78 ± 1.02	0.019	0.28	
Cd (µg/kg)	Carp	3.86 ± 1.88	0.011	0.18	2.23 ± 1.36	0.006	0.11	6.0
	Catfish	1.40 ± 0.56	0.004	0.07	1.23 ± 0.37	0.004	0.06	
Pb (µg/kg)	Carp	18.85 ± 2.85	0.054	0.22	11.97 ± 4.98	0.034	0.14	25
	Catfish	8.97 ± 3.02	0.026	0.10	13.03 ± 1.93	0.037	0.15	
As (µg/kg)	Carp	98.05 ± 45.76	0.280	1.87	57.01 ± 13.55	0.163	1.09	15
	Catfish	30.54 ± 8.74	0.087	0.58	32.39 ± 14.74	0.093	0.62	
Hg (µg/kg)	Carp	0.98 ± 0.91	0.003	0.18	3.44 ± 0.93	0.010	0.61	1.6
	Catfish	19.78 ± 0.82	0.057	3.53	27.46 ± 10.04	0.078	4.90	

* Estimated weekly intake for a 70 kg person (µg/kg bw per week). [†] Provisional Tolerable Weekly Intake (µg/kg bw per week) [8].

The cultured catfish and carp showed low concentrations of Cu and Zn, and very low EWI values, with maximums of 0.007 and 0.068 µg/kg of bw respectively, as shown in Table 5.

The estimates of weekly intakes of non-essential metals in the experimental group were below the respective PTWI for both species. The contributions of Cd, Pb, and As represented less than 1.5% of the total PTWI, while the values for Hg ranged from 0.61% to 4.90% (Table 5).

4. Discussion

The between-tissue and between-species analyses of the metal bioaccumulation patterns showed a higher concentrating capacity of the liver for Zn, Cu, Cd, and Pb compared to the muscle tissue. In contrast, the liver showed similar or lower concentrations of Hg and As than the muscle. These findings of the liver's higher capacity for concentrating Zn, Cu, and Cd agree with previous reports on the sequestrating–detoxifying function of the liver, which involves the action of the protein (Cd, Zn)-metallothionein [16–18]. The liver's high capacity for metal accumulation has been documented extensively in cyprinids for these three metals [17,19] and in several families of catfish [20–23].

The comparison of the liver's accumulation capacity for Cu and Zn between the two species analyzed showed that for the carp, it was 10 times greater than that of the catfish. The findings were observed in both the experimental and the control groups, and the results disagree with previous reports, since a higher capacity in catfish liver compared to *Cyprinidae* species has been reported in wild fish in natural environments [24,25]. The explanation of the higher concentration obtained in the carp in our study could be attributable to two factors. Studies of fish in the same aquatic environment have reported feeding behavior and trophic position as the main sources of metals [26,27]. In our work, however, both species cultured in Lake Chapala were fed the same commercial food in the same cage, so they had only limited access to food from the environment. As is observed in the results section, metal concentrations in the commercial food were low, so under conditions in which the diet of the fish is controlled, their natural feeding habits and their effect on metal accumulation will not be reflected.

On the other hand, diverse studies have reported that the age of the fish correlates positively with metal accumulation [24,28]. In the present work, the age of the catfish (98–545 days) was lower than that of the carp (240–713 days). We have one moment at the same age of fish that could be comparable, at 538 days of carp (288 days of exposure) and the catfish aged 545 days (449 days of exposure), the Zn and Cd concentration factors (LCF) were closer. Only Cu maintained a large between-species difference, which was observed from the early stages of the accumulation pattern. We conclude, therefore, that at the same age, and under similar dietary and environmental conditions, the livers of the catfish and carp maintained similar concentrating capacities for Zn and Cd, but not for Cu. The catfish liver showed a different bioaccumulation pattern for Cu than the carp, with the latter generating an accumulation factor lower than the former.

The Cd pattern of accumulation in liver showed a clear difference between experimental and control groups only in the carp case, as is observed in Figure 4c. The Cd concentration at the end of the exposition time was 472.40 µg/kg in the experimental group versus only 23.42 µg/kg for the control. The effect is attributable to the lake, although the Cd concentration in water was acceptable (0.01 ± 0.00 µg/L) [6]. The Cd concentration in liver was far from toxic limits according to previous studies in sub-lethal exposition, where the liver Cd concentrations were between 39,300.00 and 46,100.00 µg/L [16,17]. However, it is advisable to continue monitoring the fish in the lake to determine the maximum level reached by Cd in the liver. The concentrations of As, Hg, and Pb found in the tissues analyzed in this study tended to be lower than those reported previously for the same fish species in the wild [9,12,24,25,29]. With respect to aquaculture fish, our literature review only identified reports on carp species. The concentrations of As and Pb found in the muscle in those studies [30,31] were similar to our results. In another finding, the Hg concentrations in the muscle of carp in the present study were 400-fold less than those reported in [31]. The difference could be attributed to the fact that that work was conducted in wild carp, with free contact with sediments and contaminated sources of food. Therefore, maintaining control of the feed administered reduced the possibility of the fish accumulating heavy metals.

The liver from both species showed a 1.5–4-fold capacity for accumulating Pb compared to the muscle, capacities much lower than those of Cu, Zn, and Cd (Table 2). The concentrations of As and Hg, however, were either similar or lower in the liver than in the muscle in both species. Here, our results are consistent with the patterns of Pb, As, and Hg accumulation cited in previous studies of catfish [22–24], and carp tissues [24,25,30].

The main uptake pathway for Pb is through the water [25]. Thus, the concentration of Pb in the water is decisive for bioaccumulation in fish. The Pb concentrations in the water samples taken from Lake Chapala during the experiment (0.33–0.37 µg/L) [6] were below Mexican standards (10.00 µg/L) [32]. This could account for the low Pb concentrations found in the tissues of both species.

Studies of the mechanism of As and Hg accumulation in fish have been conducted at different trophic levels. The form of the chemical is another key factor in fish metal uptake. V (arsenate) and methylmercury (MeHg), respectively, are the bioavailable forms of As and Hg [33,34]. Food is the main pathway of metal intake by fish, but while the biomagnification phenomenon has been observed for Hg through the tropic chain with higher concentrations in predator species, As has not been associated with trophic position [33,35,36]. The heavy metals in the fish commercial food were always lower than international standards [13]. This explains the low As and Hg concentrations found in the fish in our study, since food was provided as in standard aquaculture systems with limited access to natural food sources.

Another factor that plays a role in the uptake of metals in aquatic organisms is pH. Lake Chapala maintained a pH of 8.6–9.5 during our experiment [6]. This alkaline pH promotes the aggregation of metals into particles that settle, thus reducing their dissolution in the liquid phase [5]. This explains the low metal concentration found in the water of Lake Chapala despite the high concentration of its sediments [6].

The data in Table 5 indicate the EWI levels for the fish from Lake Chapala in the experimental and control groups, and all of them were below 1% of the PTWI. Regarding the intake of Cu and Zn by fish consumption as essential metals, the values obtained are part of normal human dietary requirements.

The risk of consuming fish revolves around their content of non-essential heavy metals. Table 6 presents the comparison of the EWI and % PTWI estimated during the present study from the fish cultivated in Lake Chapala, with other reports of cultured and wild freshwater fish.

Table 6 shows that the EWI and % PTWI values for Cd in the present work were similar (0.11%), although lower than the unique report of cultured carp (0.35%) [30], as well as others reports of wild fish from lakes [30,37]. On the other hand, some previous works showed EWI up to two orders of magnitude higher than our work (0.395 and 0.743), and PTWI higher than 5%, which is indicative of caution [12,38].

The EWI values for lead in the present work were one or two orders of magnitude lower than previous reports, as shown in Table 5. Nevarez et al. [12] showed the highest EWI values for lead in native catfish (7.307 µg/kg of bw) living in a dam environment. According to the authors, the causes were a natural, local source of Pb, runoff from rain, and residue from extractive mining. It is interesting to notice that Nevarez et al. [12] and Alipour and Banagar [38] reported the highest % PTWI in two metals, Cd and Pb.

The EWI for As obtained in the present work for carp (0.163 µg/kg of bw) was lower, although it resulted in the same range as those reported by Alam et al. [30] for native and cultured carp (0.271 and 0.511 µg/kg of bw, respectively). The water concentrations during experiments were slightly higher in Lake Chapala (9.22–11.0 µg/L) [6] compared to the values reported by Alam et al. [39] (0.72 to 3.1 µg/L). Regarding EWI for catfish, the obtained 0.093 µg/kg of bw agrees with Nevarez et al. [9], who reported 0.189 µg/kg of bw in the same type of fish. The As water concentration reported by Nevarez et al. [9] (1.34–5.65 µg/L) was in a similar range as that in the present study. The % PTWI was less than 5% in the previous reports and agrees with their water and fish As concentration.

Mercury is the principal toxic metal of concern in terms of consuming fish and seafood, which constitute the primary sources of this metal [40]. In the present study, Hg had 0.61% and 4.9% of the PTWI in carp and catfish, respectively. Based on the mean mercury concentrations reported by Trasande et al. [41], Stong et al. [10], and Torres et al. [11] for native carp from Lake Chapala, a decrease in the concentration is noted over time that results in a lower weekly intake from native fish that represented a contribution to the PTWI of 155.36% in 2010 and 40.89% in 2014. Clearly, these figures are much higher than the results of our study, emphasizing the decrease in metal accumulation that is evident in fish cultured in floating cages. In another study, Nevarez et al. [12] reported an EWI of a dam catfish (0.124 µg/kg of bw), while Łuczyńska and Paszczyk [42] reported a figure of 0.467 µg/kg of bw for native lake roach from the family *Cyprinidae*, representing contributions to the PTWI of 9.49% and 26%, respectively.

Based on the estimated weekly intake values obtained in the present study, the consumption of cultured carp or catfish do not represent health risks for the entire population, even with a weekly intake above 200 g (or up to 1 kg), due to the low bioaccumulation of metals in fish that are cultured in cages and fed commercial feed. Under these conditions, the fish present a contribution to PTWI < 1.5% for Cd, As, and Pb. Since Hg represents a greater health risk, and the main source of intake of this harmful metal is fish and shellfish consumption, the EWI values calculated also reflect a low risk (% PTWI < 10.0), not only for the general population, but even more so for children and pregnant woman. Consumption of native fish, in contrast, represents a moderate-to-high risk (40.89–155.36%). As a precaution, in populations with higher weekly intakes, consumption of free fish by children and pregnant women should be monitored and limited.

According to the results obtained, consuming fish raised under conditions of aquaculture in Lake Chapala does not represent a health risk for heavy metal consumption. The key to keeping fish under acceptable levels of metal concentrations was to provide metal-free supplementary feed. It is, however, advisable to monitor metal concentrations in wild fish as a means of control.

Table 6. Mean concentration (μg/kg ww), estimated weekly intake (μg/kg bw/week), and contribution to PTWI from native or cultured freshwater fish reported in previous works, compared with the present study.

Metal	Fish	Origin	Country	Mean	EWI *	% PTWI	Reference
	Carp	Lake Chapala	Mexico	2.23	0.006	0.11	Present Study
	Catfish			1.23	0.004	0.06	
	Catfish	El Rejon Dam		148.00	0.395	6.58	[12]
Cd	Cultured Carp	Lake Kasumigaura	Japan	7.40	0.021	0.35	[30]
	Wild Carp			9.00	0.026	0.43	
	Crucian carp	Honghu Lake	China	8.70	0.028	0.47	[37]
	Yellow Catfish			5.60	0.018	0.30	
	Silver Carp	Chah Nime Lake	Iran	31.20[†]	0.096	1.60	[43]
	Carp	Gorgan Bay		260.00	0.743	12.38	[38]
	Carp	Lake Chapala	Mexico	11.97	0.034	0.14	Present Study
	Catfish			13.03	0.037	0.15	
	Catfish	El Rejon Dam		2740.00	7.307	29.23	[12]
Pb	Crucian carp	Honghu Lake	China	93.80	0.305	1.22	[37]
	Yellow catfish			124.20	0.403	1.61	
	Silver carp	Chah Nime Lake	Iran	47.84[†]	0.47	0.59	[43]
	Carp	Gorgan Bay		430.00	1.229	4.91	[38]
	Carp	Lake Chapala	Mexico	57.01	0.163	1.09	Present Study
	Catfish			32.39	0.093	0.62	
As	Catfish	El Rejon Dam		66.00[†]	0.189	1.26	[9]
	Cultured Carp	Lake Kasumigaura	Japan	178.90	0.511	3.41	[30]
	Wild carp			95.00	0.271	1.81	
	Yellow catfish	Honghu Lake	China	4.00	0.013	0.09	[37]
	Carp			3.44	0.010	0.61	Present Study
	Catfish			27.46	0.078	4.90	
	Carp	Lake Chapala		870.00	2.486	155.36	[41]
	Carp		Mexico	390.00	1.114	69.64	[10]
Hg	Carp			229.00	0.654	40.89	[11]
	Carp	San Antonio Dam		72.50	0.207	12.95	
	Catfish	El Rejon Dam		46.50	0.124	7.75	[12]
	Roach	Olsztyn Lake	Poland	140.00	0.467	29.17	[42]

* Estimated weekly intake for an adult of 70 kg or according to author, and fish consumption of 200 g. [†] Mean concentration adjusted to ww, 70% humidity assumed.

5. Conclusions

Observations from this study indicate that metal accumulation in fish varies markedly among different tissues and metals, but not between species, with the sole exception of Cu. Fish aquaculture in Lake Chapala is thus a viable option for producing fish with a low risk of heavy metal consumption by humans. Nonetheless, maintaining strict monitoring of metals in fish muscle is crucial because this lake is characterized by highly dynamic activity that causes frequent variations in volume and pH that affect the concentrations of metals in its waters. Finally, Hg in catfish showed the highest risk in terms of their PTWI, though even in this case, the % PTWI was below 5 %.

Author Contributions: Conceptualization, C.A.; methodology, C.A. and E.J.H.-L.; software, E.J.H.-L.; formal analysis, X.G., D.M.C.-V., and J.M.R.; resources, C.A.; data curation, writing—original draft preparation, C.A. and D.M.C.-V. All authors have read and agreed to the published version of the manuscript.

Funding: This research was funded by Mexico's National Science and Technology Council (CONA-CYT), grant number 216110.

Institutional Review Board Statement: The study was conducted according to the guidelines of Mexican NOM-033-ZAG/ZOO-1994 and World Organization Aquatic Animals Health Code for slaughter of domestic animals.

Data Availability Statement: The data presented in this study are available on request from the corresponding author.

Acknowledgments: The authors acknowledge Domingo Ausin, Jesús Arriaga, and Sergio Escobar for their support in the fish culture and sampling work.

Conflicts of Interest: The authors declare no conflict of interest.

References

1. Comisión Estatal de Agua Jalisco. Available online: https://www.ceajalisco.gob.mx/contenido/chapala/ (accessed on 6 April 2021).
2. Hansen, A.M.; Zavala, A.L.; Inclan, L.B. Fuentes de contaminación y enriquecimiento de metales en sedimentos de la cuenca Lerma-Chapala. *Ing. Hidraul. Mex.* **1995**, *10*, 55–69.
3. De Anda, J.; Shear, H.; Maniak, U.; Zárate-del Valle, P.F. Solids distribution in Lake Chapala, Mexico. *J. Am. Water Resour. Assoc.* **2004**, *40*, 97–109. [CrossRef]
4. Trujillo-Cárdenas, J.L.; Saucedo-Torres, N.P.; Del Valle, P.F.Z.; Ríos-Donato, N.; Mendizábal, E.; Gómez-Salazar, S. Speciation and sources of toxic metals in sediments of Lake Chapala, Mexico. *J. Mex. Chem.* **2010**, *54*, 79–87. [CrossRef]
5. Hansen, A.M. Adsorption-desorption behaviors of Pb and Cd in Lake Chapala, Mexico. *Environ. Int.* **1997**, *23*, 553–564. [CrossRef]
6. Alvarado, C.; Ramírez, J.M.; Herrera-López, E.J.; Cortez-Valladolid, D.; Ramírez, G. Bioaccumulation of Metals in Cultured Carp (*Cyprinus carpio*) from Lake Chapala, Mexico. *Biol. Trace Elem. Res.* **2020**, *195*, 226–238. [CrossRef]
7. Mozaffarian, D.; Rimm, E.B. Fish Intake, Contaminants, and Human Health: Evaluating the Risks and the Benefits. *JAMA* **2006**, *296*, 1885–1899. [CrossRef]
8. World Health Organization. Evaluations of the Joint FAO/WHO Expert Committee on Food Additives (JEFCA). Available online: https://apps.who.int/food-additives-contaminants-jecfa-database/search.aspx (accessed on 28 January 2021).
9. Nevárez, M.; Moreno, M.V.; Sosa, M.; Bundschuh, J. Arsenic in freshwater fish in the Chihuahua County water reservoirs (Mexico). *J. Environ. Sci. Health Part A* **2011**, *46*, 1283–1287. [CrossRef]
10. Stong, T.; Osuna, C.A.; Shear, H.; De Anda Sanchez, J.; Ramírez, G.; De Jesús Díaz Torres, J. Mercury concentrations in common carp (*Cyprinus carpio*) in Lake Chapala, Mexico: A lakewide survey. *J. Environ. Sci. Health Part A* **2013**, *48*, 1835–1841. [CrossRef]
11. Torres, Z.; Mora, M.A.; Taylor, R.J.; Alvarez-Bernal, D.; Buelna, H.R.; Hyodo, A. Accumulation and hazard assessment of mercury to waterbirds at Lake Chapala, Mexico. *Environ. Sci. Technol.* **2014**, *48*, 6359–6365. [CrossRef]
12. Nevárez, M.; Leal, L.O.; Moreno, M. Estimation of Seasonal Risk Caused by the Intake of Lead, Mercury, and Cadmium through Freshwater Fish Consumption from Urban Water Reservoirs in Arid Areas of Northern Mexico. *Int. J. Environ. Res.* **2015**, *12*, 1803–1816. [CrossRef]
13. Directive 2002/32/EC of the European Parliament and of the Council of 7 May 2002 on undesirable substances in animal feed. *Off. J. Eur. Commun.* **2006**, *L140*, 9–10.
14. Food and Agriculture Organization of the United Nations. *The State of World Fisheries and Aquaculture 2020: Sustainability in Action*, 1st ed.; FAO: Rome, Italy, 2020; p. 70. [CrossRef]
15. Uribe, C.; Jiménez, A.; Morales, M.; Salazar, C.; Shamah, L. Percepción del peso corporal y de la probabilidad de desarrollar obesidad en adultos mexicanos. *Salud Publ. Mex.* **2018**, *60*, 254–262. Available online: https://scielosp.org/pdf/spm/2018.v60n3/254-262/es (accessed on 12 March 2021). [CrossRef]
16. De Conto Cinier, C.; Petit-ramel, M.; Garin, D.; Bouvet, Y. Kinetics of cadmium accumulation and elimination in carp *Cyprinus carpio* tissues. *Comp. Biochem. Physiol. C Toxicol. Pharmacol.* **1999**, *122*, 345–352. [CrossRef]
17. De Smet, H.; De Wachter, B.; Lobinski, R.; Blust, R. Dynamics of (Cd, Zn) -metallothioneins in gills, liver, and kidney of common carp *Cyprinus carpio* during cadmium exposure. *Aquat. Toxicol.* **2020**, *52*, 269–281. [CrossRef]

18. Kraemer, L.D.; Campbell, P.G.; Hare, L. Dynamics of Cd, Cu and Zn accumulation in organs and sub-cellular fractions in field transplanted juvenile yellow perch (Perca flavescens). *Environ. Pollut.* **2005**, *138*, 324–337. [CrossRef]
19. Calta, M.; Canpolat, O. Calta & Canpolat. Cyprinid heavy metals. *Water Environ. Res.* **2006**, *78*, 548–551.
20. Wagner, A.; Boman, J. Biomonitoring of trace elements in muscle and liver tissue of freshwater fish. *Spectrochim. Acta B* **2003**, *58*, 2215–2226. [CrossRef]
21. Subathra, S.; Karuppasamy, R. Bioaccumulation and Depuration Pattern of Copper in Different Tissues of Mystus vittatus, Related to Various Size Groups. *Arch. Environ. Contam. Toxicol.* **2008**, *54*, 236–244. [CrossRef]
22. Jovičić, K.; Nikolić, D.M.; Jeftic, Z.V.; Đikanović, V.; Skorić, S.; Stefanović, S.M.; Lenhardt, M.; Hegediš, A.; Krpo-Ćetković, J.; Jarić, I. Mapping differential elemental accumulation in fish tissues: Assessment of metal and trace element concentrations in wels catfish (*Silurus glanis*) from the Danube River by ICP-MS. *Environ. Sci. Pollut. Res.* **2015**, *22*, 3820–3827. [CrossRef]
23. Arantes, F.P.; Savassi, L.A.; Santos, H.B.; Gomes, M.V.T.; Bazzoli, N. Bioaccumulation of mercury, cadmium, zinc, chromium, and lead in muscle, liver, and spleen tissues of a large commercially valuable catfish species from Brazil. *An. Acad. Bras. Ciênc.* **2016**, *88*, 137–147. [CrossRef]
24. Has-Schön, E.; Bogut, I.; Vuković, R.; Galović, D.; Bogut, A.; Horvatić, J. Distribution and age-related bioaccumulation of lead (Pb), mercury (Hg), cadmium (Cd), and arsenic (As) in tissues of common carp (*Cyprinus carpio*) and European catfish (Sylurus glanis) from the Buško Blato reservoir (Bosnia and Herzegovina). *Chemosphere* **2015**, *135*, 289–296. [CrossRef]
25. Jia, Y.; Wang, L.; Qu, Z.; Wang, C.; Yang, Z. Effects on heavy metal accumulation in freshwater fishes: Species, tissues, and sizes. *Environ. Sci. Pollut. Res.* **2017**, *24*, 9379–9386. [CrossRef]
26. Watras, C.J.; Back, R.C.; Halvorsen, S.; Hudson, R.J.M.; Morrison, K.A.; Wente, S.P. Bioaccumulation of mercury in pelagic freshwater food webs. *Sci. Total Environ.* **1998**, *219*, 183–208. [CrossRef]
27. Clearwater, S.J.; Farag, A.M.; Meyer, J.S. Bioavailability and toxicity of dietborne copper and zinc to fish. *Comp. Biochem. Physiol. C Toxicol. Pharmacol.* **2002**, *132*, 269–313. [CrossRef]
28. Yi, Y.J.; Zhang, S.H. Heavy metal (Cd, Cr, Cu, Hg, Pb, Zn) concentrations in seven fish species in relation to fish size and location along the Yangtze River. *Environ. Sci. Pollut. Res.* **2002**, *19*, 3989–3996. [CrossRef]
29. Cui, B.; Zhang, Q.; Zhang, K.; Liu, X.; Zhang, H. Analyzing trophic transfer of heavy metals for food webs in the newly-formed wetlands of the Yellow River Delta, China. *Environ. Pollut.* **2011**, *159*, 1297–1306. [CrossRef]
30. Alam, M.G.M.; Tanaka, A.; Allinson, G.; Laurenson, L.J.B.; Stagnitti, F.; Snow, E.T. A comparison of trace element concentrations in cultured and wild carp (*Cyprinus carpio*) of Lake Kasumigaura, Japan. *Ecotoxicol. Environ. Saf.* **2002**, *53*, 348–354. [CrossRef]
31. Sandor, Z.; Csengeri, I.; Oncsik, M.B.; Alexis, M.N.; Zubcova, E. Trace metal levels in freshwater fish, sediment and water. *Environ. Sci. Pollut. Res.* **2001**, *8*, 265–268. [CrossRef]
32. DOF. NOM-127-SSA1-1994. Salud Ambiental. Agua Para Consume Humano. Límites Permisibles de Calidad y Tratamiento a Que Debe Someterse el Agua Para su Potabilización; Diario Oficial de la Federación: Ciudad de México, México, 2000.
33. Bowles, K.C.; Apte, S.C.; Maher, W.A.; Kawei, M.; Smith, R. Bioaccumulation and biomagnification of mercury in Lake Murray, Papua New Guinea. *Can. J. Fish. Aquat. Sci.* **2001**, *58*, 888–897. [CrossRef]
34. Ventura-Lima, J.; Fattorini, D.; Regoli, F.; Monserrat, J.M. Effects of different inorganic arsenic species in *Cyprinus carpio* (*Cyprinidae*) tissues after short-time exposure: Bioaccumulation, biotransformation and biological responses. *Environ. Pollut.* **2009**, *157*, 3479–3484. [CrossRef]
35. Mason, R.P.; Laporte, J.; Andres, S. Environmental Contamination and Toxicology Factors Controlling the Bioaccumulation of Mercury, Methylmercury, Arsenic, Selenium, and Cadmium by Freshwater Invertebrates and Fish. *Arch. Environ. Contam. Toxicol.* **2000**, *297*, 283–297. [CrossRef] [PubMed]
36. Swanson, H.K.; Johnston, T.A.; Leggett, W.C.; Bodaly, R.A.; Doucett, R.R.; Cunjak, R.A. Trophic positions and mercury bioaccumulation in rainbow smelt (Osmerus mordax) and native forage fishes in northwestern Ontario lakes. *Ecosystems* **2003**, *6*, 289–299. [CrossRef]
37. Zhang, J.; Zhu, L.; Li, F.; Liu, C.; Yang, Z.; Qiu, Z.; Xiao, M. Heavy metals and metalloid distribution in different organs and health risk assessment for edible tissues of fish captured from Honghu Lake. *Oncotarget* **2017**, *8*, 101672–101685. [CrossRef] [PubMed]
38. Alipour, H.; Banagar, G.R. Health risk assessment of selected heavy metals in some edible fishes from Gorgan Bay, Iran. *Iran. J. Fish. Sci.* **2018**, *17*, 21–34. [CrossRef]
39. Alam, M.; Tanaka, A.; Stagnitti, F.; Allinson, G.; Maekawa, T. Observations on the effects of caged carp culture on water and sediment metal concentrations in Lake Kasumigaura, Japan. *Ecotoxicol. Environ. Saf.* **2001**, *48*, 107–115. [CrossRef] [PubMed]
40. US Environmental Protection Agency (US EPA). *Mercury Study Report to Congress Vol. IV: An Assessment to Exposure to Mercury in the United States*; EPA 452/R-97-006; US Environmental Protection Agency: Washington, DC, USA, 1997; pp. 7.1–7.3.
41. Trasande, L.; Cortes, J.E.; Landrigan, P.J.; Abercrombie, M.I.; Bopp, R.F.; Cifuentes, E. Methylmercury exposure in a subsistence fishing community in Lake Chapala, Mexico: An ecological approach. *Environ. Health* **2010**, *9*, 1. [CrossRef] [PubMed]
42. Łuczyńska, J.; Paszczyk, B. Health risk assessment of heavy metals and lipid quality indexes in freshwater fish from lakes of Warmia and Mazury Region, Poland. *Int. J. Environ. Res. Public Health* **2019**, *16*, 3780. [CrossRef]
43. Miri, M.; Akbari, E.; Amrane, A.; Jafari, S.J.; Eslami, H.; Hoseinzadeh, E.; Zarrabi, M.; Salimi, J.; Sayyad-Arbabi, M.; Taghavi, M. Health risk assessment of heavy metal intake due to fish consumption in the Sistan region, Iran. *Environ. Monit. Assess.* **2017**, *189*, 583. [CrossRef]

applied sciences

Article

Physico-Chemical Parameters and Health Risk Analysis of Groundwater Quality

Alina Soceanu [1], Simona Dobrinas [1,*], Corina Ionela Dumitrescu [2], Natalia Manea [3], Anca Sirbu [4], Viorica Popescu [1] and Georgiana Vizitiu [1]

[1] Chemistry and Chemical Engineering Department, Faculty of Applied Chemistry and Engineering, "Ovidius" University of Constanta, 900527 Constanta, Romania; asoceanu@univ-ovidius.ro (A.S.); vpopescu@univ-ovidius.ro (V.P.); vizitiugeorgianairina@yahoo.ro (G.V.)

[2] Department of Economics, Faculty of Entrepreneurship, Business Engineering and Management, University POLITEHNICA of Bucharest, 060042 Bucharest, Romania; corina.dumitrescu@upb.ro

[3] Department of Economic Engineering, Faculty of Entrepreneurship, Business Engineering and Management, University POLITEHNICA of Bucharest, 060042 Bucharest, Romania; natalia.manea@upb.ro

[4] Department of Fundamental Sciences and Humanities, Constanta Maritime University, 900663 Constanta, Romania; anca.sirbu@cmu-edu.eu

* Correspondence: sdobrinas@univ-ovidius.ro

Abstract: Groundwater pollution is a very common problem worldwide, as it poses a serious threat to both the environment and the economic and social development and consequently generates several types of costs. The analysis of pollution control involves a permanent comparison between pollution costs and the costs associated with various methods of pollution reduction. An environmental policy based on economic instruments is more effective than an environmental policy focused on command and control tools. In this respect, the present paper provides a case study showing how anthropogenic factors such as wastewater, industrial, agricultural, and natural factors are able to change the physical and chemical parameters of groundwater in the study area, thus endangering their quality. In order to monitor the groundwater quality in the region of Dobrudja, an analysis of physico-chemical parameters was performed. The content of heavy metals was analyzed and the health risk index was taken into account and analyzed, in order to set a better correctness of the metal content from the underground waters. Studies on groundwater quality control have shown that, in many parts of the world, water has different degrees of quality depending on the natural and anthropogenic factors acting on the pertaining environment. This is why more attention should be paid to the prevention of groundwater pollution and the immediate remediation of accidents.

Keywords: groundwater pollution; pollution costs; economic environmental protection instruments; physico-chemical parameters; heavy metals; health risk index; Pearson coefficients

Citation: Soceanu, A.; Dobrinas, S.; Dumitrescu, C.I.; Manea, N.; Sirbu, A.; Popescu, V.; Vizitiu, G. Physico-Chemical Parameters and Health Risk Analysis of Groundwater Quality. *Appl. Sci.* **2021**, *11*, 4775. https://doi.org/10.3390/app11114775

Academic Editor: Elida Nora Ferri

Received: 6 May 2021
Accepted: 19 May 2021
Published: 23 May 2021

1. Introduction

The list of environmental issues that humanity has to face currently includes topics specific to areas (forest degradation) and activities (chemical risk) or global topics (climate change) that bring out three factors that are basic components of the current environmental crisis: overpopulation, depletion of resources, and pollution.

Groundwater occupies a special place in the economy, primarily as the main fresh water resource, as well as for agricultural irrigation, and industrial utilization. Groundwater quality can be affected by hydrogeochemical process and anthropogenic impacts such as rapid urbanization, and industrial and agricultural development [1].

In many countries all over the world, groundwater has become an important part of the drinking water supply process for industrialized areas, for agriculture, and not least for household use [2,3]. Recently, the support of life on earth has been based on the supply of drinking water from groundwater aquifers. They are superior in quality and have a natural protection against microbial activities. Another quality of groundwater is that it can be found all over the globe [4]. Groundwater has recently served as the main source of water supply for the earth's populations, therefore its quality is very important. Both anthropogenic and natural factors are the main sources of water contamination [5].

The loss of the integrity of groundwater occurs due to domestic, public, and industrial wastewater that seeps through various channels in the ground such as defective canals, absorbent boreholes, or even through the ground layer. Another cause is the infiltration of rainwater loaded with pollutants from the atmosphere or even from the soil, due to uninsulated open-air industrial landfills, such as: ash from thermal power plants, wood waste, chloro-sodic landfills, sugar factory sludge, household waste, or metallurgical slag. The suspended materials from vehicular traffic, tire wear, and oil and gas leaks from vehicles are deposited on pavements, and all this material is carried by rainwater and infiltrate the surface of permeable pavements and the other base layers, accumulating over time. Permeable pavement systems, unlike conventional pavement, can reduce the amount of pollutants transported by rainwater and therefore decrease the total quantity of pollutants delivered to the receiving water bodies [6].

Defective tanks containing various fluid substances, losses due to the transport of petroleum products or other fluids, and even loading and unloading stations are other causes of groundwater infestation with pollutants [2]. Some of the factors that destroy the quality of groundwater are soil characteristics, groundwater circulation through different types of rocks, topography, and saline water infusion in coastal areas; but human activities also have effects on groundwater [3]. Of major importance, due to the frequency of infestation by this process, is the spread of fertilizers and phytopharmaceuticals on the soil surface, which then, by means of atmospheric precipitation, infiltrate underground. Of all these aspects, prevention action is the only effective way to protect groundwater quality [2].

Groundwater pollution is a very common problem worldwide [7–11], as it poses a serious threat to both the environment and economic and social development. Groundwater resources are threatened globally due to population growth [12], rapid urbanization [13], and agricultural/industrial development [14].

Nitrogen exists in the soil in the forms of nitrate, nitrite, and ammonium and can easily move into the groundwater by leaching effects. Elevated nitrate is a severe problem affecting regional groundwater quality in Korea [15], Denmark [16], Indonesia [17], China [18], and Italy [19].

Heavy metal pollution of groundwater is a serious problem for human health, as it causes toxicity and carcinogenic effects [3]. Human health, industrialization, as well as the practices of modern agriculture, require increased attention and protection of groundwater quality but also a rational exploitation within the limits of exploitation reserves [1]. In order to assess the groundwater quality, identify sources of contamination, and plan preventive measures to reduce the public health risks, authorities will need to perform routine monitoring of groundwater [20].

In Romania, the main economic sectors regarded as significant sources of water pollution are represented by industrial activities (the extractive industry, mining industry, chemical processing, metallurgical industry), agro-zootechnical activities, as well as unauthorized and inappropriate household and industrial landfills.

Economic Aspects of Pollution.

Pollution is an economic problem primarily because it reduces the value of the resources a society has at its disposal. Secondly, pollution is an economic problem also due to the fact that it involves a process of choice for resolving conflicts of interest (e.g., the polluter–polluted relationship will always be marked by the existence of conflicts of interest). Thirdly, pollution is an economic problem because the means themselves by which this phenomenon can be reduced require the use of resources (which in turn involve certain costs). Therefore, the analysis of pollution control involves a permanent comparison between pollution costs and the costs associated with different methods of pollution reduction.

Pollution generates several types of costs. Pollution reduction costs are the costs generated by reduction of the pollutant amount. The polluter can reduce the amount of pollutant by reducing the production of polluting goods. Yet pollution generates costs that are borne by society as a whole. As soon as pollution has occurred and the polluter fails to prevent it, there are two types of costs that society as a whole face [21].

When analyzing a polluting activity in terms of its economics, one cannot ignore the external cost of said activity. If a company fails to take measures to reduce pollution, the costs will be represented by the reduction of the value of production or consumption activities as affected by pollution. These costs are in the nature of damage, and that is why we have grown accustomed to calling them pollution damage. The variety of the damage (costs) is amazing. They are highlighted, for example, by the measurable loss of profit or utility of those affected by pollution (deterioration of water quality causes increased production costs of the company using this water, and therefore decreased profit).

In general, the decrease in profit is measurable for the simple reason that the additional factors necessary to rectify the situation are assessed on the market. The unfavorable mental consequences of polluting activities are much more difficult to measure, but certainly such effects influence consumption. Leaving aside the theoretical and practical difficulties of measuring these effects, we can say that the decrease in the value of production activities and the decrease in the value of consumption due to pollution are similar from a conceptual point of view. They both reduce the well-being obtained by society by using the resources at its disposal.

In other words, the external cost occurs when a consumption or production activity induces a direct loss of utility or an increase in the cost of production in the sphere of another activity, and these negative effects are not included in the calculations of the producer [22].

1.1. The Efficiency of the Economic Instruments Used in the Environmental Protection Policy

The following are some of the arguments that can be made to prove that an environmental policy based on the use of economic instruments is more effective than an environmental policy that focuses on command and control tools.

1.1.1. Economic Instruments Facilitate the Internalization of Negative Externalities Generated by Polluting Activities

Negative externalities are in the nature of a cost that is not included in the price of the goods produced. Therefore, the main reason for using economic instruments in an environmental protection policy is the inclusion of pollution costs (external costs) in the price of goods generated by a certain economic activity. The internalization of external costs by using economic instruments will place the production of polluting goods at an optimal level from a social point of view.

1.1.2. Economic Instruments Induce Changes in Both the Behavior of Producers and the Behavior of Consumers

For example, if a polluting product is taxed, the producer of that product will try to reduce the pollution. The charged fee will raise the product price, and consumers will buy less of that product. It is true, however, that the change in behavior will be evident if the demand for said goods is elastic, i.e., when the polluting product has many less-polluting potential substitutes.

1.1.3. Economic Instruments Are Often More Cost-Effective Than Other Instruments

According to the regulatory instruments (standards, authorizations, etc.) used for pollution control, all pollutants must reduce pollution according to the same imposed standards, regardless of the costs of achieving this goal. An economic instrument will help each polluter decide whether it is more convenient to pay that fee or to bear the costs of reducing pollution. Those facing low pollution-reduction costs will reduce their emissions more than those who have to bear higher costs. Moreover, given that economic instruments are imposed in terms of unit (kg, t, m^3) of the remaining emissions, there will always be an incentive to reduce pollution.

1.1.4. Economic Instruments Encourage Investments in Green Technology

Economic instruments help the economy move towards an eco-efficient use of limited resources. In the world we live in, there is huge potential for increasing resource productivity via technological solutions that can contribute to sustainable economic growth.

1.1.5. Economic Instruments Generate Income for Environmental Investments

In contrast to command and control instruments, environmental fees generate revenue that is later used for various purposes. In the case of user fees (e.g., wastewater treatment, entrance fees to national parks, etc.), the collected revenues are used by the authorities to provide quality services, without negative effects on the natural environment and allowing full recovery of costs. Revenues from environmental fees can be collected in special environmental funds or in general budgets.

However, there is a paradox that needs to be mentioned: the greater the effectiveness of an environmental tax, the lower the income generated by it. The explanation is as follows: the higher an eco-tax, the more stimulated is the polluter to invest in reducing pollution and therefore the taxing basis is lower. As a result, eco-tax collectors may sometimes be interested in keeping pollution high.

1.1.6. Economic Instruments Support the "Polluter Pays" and "User Pays" Principles

The "polluter pays" principle guarantees that all financial responsibility for reducing pollution lies with the polluter [23–25]. The "user pays" principle states that resources have economic value, and the costs of their usage have to be borne by the user.

2. Materials and Methods

A Case Study: Quality Control of Groundwater in the Region of Dobrudja, Romania.

In this paper, a case study shows how anthropogenic factors such as wastewater, industrial, agricultural, and natural factors, such as soil-water interaction, are able to change the physical and chemical parameters of groundwater in the study area, thus endangering their quality. In the process of evaluating the groundwater quality, the physical and chemical properties represent the most important factors. In this context, the physico-chemical parameters of groundwater such as pH, electrical conductivity, turbidity, alkalinity, hardness, nitrates, nitrites, ammonium, chlorides have been investigated in the groundwater of the region of Dobrudja to generate an assessment of groundwater quality and its use for drinking purposes. The content of heavy metals such as chromium, cadmium, copper, iron, manganese, nickel, lead, and zinc in these samples has been analyzed. In addition, the health risk index was taken into account and analyzed, in order to set a better correctness of the metal content from the underground waters located on Dobrudja's territory.

2.1. The Study Area

The study area was the region of Dobrudja, where the underground catchments have been developed at a fast pace, replacing the surface ones, which are even more expensive and more inadequate in terms of water quality. In the past, Dobrudja was known as a region almost devoid of underground supply possibilities, but more advanced studies carried out on the territory of this area have shown that large areas, such as the basins of the Sava, Telita, and Taita rivers have important resources capable of ensuring a substantial flow of drinking water [26].

The groundwater on the territory of Dobrudja, which is part of the Upper Cretaceous–Jurassic aquifer complex, is bicarbonate-calcium and magnesium, having a mineralization below 500 mg/l, and being drinkable, while the waters from Sarmatian limestones located in the Mangalia area are bicarbonate-calcium and chlorinated sodium with a higher mineralization up to 1000 mg/l, which are also drinkable but with an intense smell of hydrogen sulfide [2].

2.2. Sampling

Six different locations in the region of Dobrudja were established for groundwater sampling from wells (Figure 1) situated in urban areas (Constanta, Palas, Palazu Mare) and in rural areas (Lumina, Corbu, and Casimcea). The collected samples were stored at 4 °C before analysis.

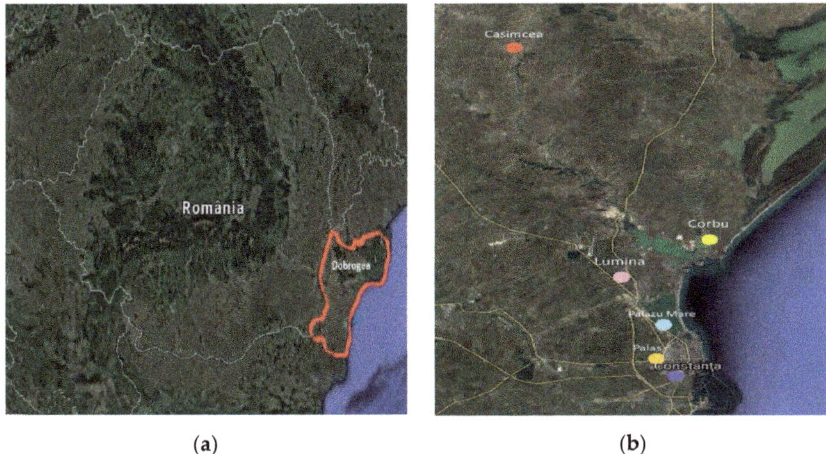

(a)　　　　　(b)

Figure 1. (a) Dobrudja region from Romania; (b) map of groundwater sampling locations in the region of Dobrudja (Constanta, Palas, Palazu Mare, Lumina, Casimcea, Corbu).

2.3. Groundwater Analysis

The potentiometric method was used for pH measurement using a pH meter from Hanna. The conductivity (C) is directly related to the concentration of ions dissolved in water, i.e., the higher the number of ions, the higher the conductivity. The conductivity was determined using a CONSORT K610 conductometer. Water alkalinity is the content of ions that give the water an alkaline character. It is determined against two indicators: "p" alkalinity—compared to phenolphthalein—and "m" alkalinity—compared to methyl orange. The "p" alkalinity is called permanent alkalinity and the "m" alkalinity is called total alkalinity. The alkalinity was determined by titrating a known volume of the sample with an HCl solution 0.1 N in the presence of phenolphthalein up to the discoloration of the solution, followed by titration in the presence of methyl orange to yellow. In the first stage, the anions representing the "p" alkalinity are neutralized, after which the other anions representing the alkalinity of the water are neutralized. The full volume of the HCl solution consumed on titration contains an amount of acid equivalent to the total alkalinity (TA). Nitrites are present in the water due to infiltration into the groundwater in agricultural areas where fertilizers are used intensively, but can also result from improperly stored animal manure. The reaction of the nitrites presents in the sample, with the reagent 4-amino-benzene-sulfonamide in the presence of orthophosphoric acid leading to the formation of a red complex. Nitrates in groundwater come from the washing by rainwater of naturally occurring nitrates in the surface soil, or may have as a source the fertilizers used for soil fertilization. Nitrogen compounds released into the air from combustion processes by automobiles and industry can increase the nitrate content of rainwater and deposited dust, representing another source of nitrate in water. Sulfuric acid and phosphoric acid used in the reaction of nitrates with 2.6-dimethylphenol lead to the formation of 4-nitro-2.6-dimethylphenol. The spectrometric measurement of the absorbance at the wavelength of 324 nm was followed by the reading of the nitrate concentration in the sample using the calibration curve. Determination of total hardness consists of complexing the metal cations that form the hardness, with the disodium salt of ethylene-diamino-tetraacetic acid (EDTA), at pH = 10, in the presence of the indicator eriochrome T. The turbidity of water was determined by measuring the intensity of a diffused light flux when passing through the water sample (variant I) or by measuring the attenuation of the intensity of an incident light flux when passing it through the water sample (variant II). In both cases, the obtained value reads the turbidity on a calibration curve drawn on the basis of the formazine suspension standard, which has the absorbances inscribed on the ordinate, and on the abscissa the corresponding formazine turbidity units. Turbidity is expressed in units of formazine turbidity (UTF) and was read directly on the calibration curve. Determination coefficients (R^2) obtained from linear regression analysis were 0.9999. The equation of the calibration curve is $y = 0.0027x + 0.0002$. Determination of ammonium consists of ammonium's reaction with salicylate and hypochlorite ions in the presence of sodium nitrosopentacyanoferrate (III) (sodium nitroprusside). A blue compound is formed, which will be measured spectrometrically at a wavelength of approximately 650 nm. The reaction of chloramine with sodium salicylate takes place at pH 12.6 in the presence of sodium nitroprusside. The chloramines present in the sample are determined quantitatively, and the role of sodium citrate is to mask the interference given by cations, such as calcium and magnesium. The principle of the method for chlorides determination was chlorides precipitation with silver nitrate solution in the presence of potassium chromate as an indicator. The end point of the titration is highlighted by the appearance of the silver–chromate precipitate, colored brick-red. Metals (Cd, Pb, Cu, Cr, Zn, Mn, Ni, Fe) were determined by atomic absorption spectrometry (AAS). Atomic absorption spectrometer contrAA 700 Analytik Jena was equipped with a graphite furnace with a background correction system and cavity cathode lamp. Table 1 presents the calibration data and detection limits.

Table 1. Calibration data and quantification limits.

Metals	The Range µg/L	R^2	LOQ
Cd	0.4–4	0.9913	0.00041
Cr	2–20	0.9902	0.00206
Cu	3–30	0.9902	0.00310
Fe	3–30	0.9930	0.1183
Mn	1.5–15	0.9978	0.00151
Ni	7–70	0.9940	0.00708
Pb	10–100	0.9946	0.01002
Zn	0.5–5	0.9924	0.1076

Water samples are filtered and preserved by acid treatment. A volume of 20 µL of sample solution was injected into the platform graphite furnace of the atomic absorption spectrometer. The oven is electrically heated. By gradually increasing the temperature the sample is dried, pyrolyzed, and atomized. The light source emits a specific light for each analyte. When the ray of light passes through the cloud of atoms in the heated graphite furnace, light is selectively absorbed by the atoms of the analyte. The decrease in light intensity is measured with a detector at a specific wavelength. The concentration of metals in the sample was determined by comparing the absorbance of the sample with the absorbance of the calibration solutions, according to Lambert Beer's law. Chemical modifiers are used to remove its spectral non-spectral interference from a sample (matrix effects). For Cd, Cr, Cu, Fe, Mn, Zn the chemical modifier $Pd/Mg(NO_3)_2$ 0.1: 0.05% was used, and for Ni and Pb, $Pd(NO_3)_2$ 0.05%.

3. Results

In order to monitor the groundwater quality in the Dobrudja area, an analysis of physico-chemical parameters was performed (Table 2).

Table 2. Values of quality parameters for the studied samples.

Parameters	Constanta	Palazu Mare	Palas	Lumina	Corbu	Casimcea
pH	7.22	7.23	7.5	7.54	7.44	7.91
C (σ) µs/cm	1862	2320	1370	969	1826	3070
TA mmol/L	11.1	13	8.65	11.1	9.4	8
NO_3^- mg/L	175	473	118	0.132	70.1	532
NO_2^- mg/L	0.013	0.02	0.015	0.002	0.043	0.245
d_T German	41.51	44.1	30.3	21.2	22.3	95.9
TU UTF	0.013	0.11	0.2	0.75	0.02	0.15
NH_4^+ mg/L	0.003	0.001	0.001	0.003	0.075	0.047
Cl^- mg/L	112.4	150.7	90.8	89	147.1	312.7

The values of metal concentrations for the studied samples are presented in Table 3.

Table 3. Values of metal concentrations for the studied samples.

Metals µg/L	Constanta	Palazu Mare	Palas	Lumina	Corbu	Casimcea
Cd	<LOQ	<LOQ	<LOQ	<LOQ	<LOQ	<LOQ
Pb	0.47	1.01	<LOQ	0.11	<LOQ	0.11
Cu	<LOQ	<LOQ	<LOQ	0.09	<LOQ	<LOQ
Cr	12.67	15.32	8.25	7.86	6.45	7.82
Zn	9.55	24.89	8.62	37.93	5.05	17.43
Mn	0.67	2.85	1.48	3.66	0.25	0.53
Ni	<LOQ	2.35	<LOQ	<LOQ	<LOQ	<LOQ
Fe	<LOQ	<LOQ	<LOQ	<LOQ	<LOQ	<LOQ

4. Discussion

pH value is an important parameter of water quality and the contamination level in the watershed area. This study has revealed that the pH value of well water is slightly alkaline and varied from 7.22 pH units (Constanta) to 7.91 pH units (Casimcea). The well water samples are within the permissible limit prescribed by Romanian legislation [27] (Table 4).

Table 4. Calibration data and quantification limits.

Parameter	Unit of Measure	Maximum Permissible Limit
Ammonium	mg/L	0.50
Chlorides	mg/L	250
Conductivity	µs·cm^{-1} at 20 °C	2500
Hardness	German degrees	5
Nitrites	mg/L	50
Nitrates	mg/L	0.50
pH	pH units	$\geq 6.5; \leq 9.5$
Turbidity	UTF	≤ 1.0

Regarding the conductivity of this groundwater, the highest value was observed in Casimcea (3070 µs/cm), while Lumina had the lowest value (969 µs/cm). The higher EC values of the well could be attributed to its shallow depth, as ongoing surface activities can affect the conductivity values [28]. Only the sample from Casimcea (3070 µs/cm) exceeded the maximum allowed value of 2500 µs/cm (Table 4) [22].

The permanent alkalinity had the same value for all six groundwater samples in the Dobrudja region (0.000 mmol/L), while the values for total alkalinity fluctuated—the highest value being recorded for the water from Palazu Mare (13.0 mmol/L) and the lowest for the water in the rural district of Casimcea (8.0 mmol/L).

The concentration of nitrate contained in water seems to be from the high agricultural waste and sewage contamination. Such an increase may be due to rapid decomposition of organic matter. When the dead organic matter decomposes in water, it forms complex proteins, which are converted into nitrogenous organic matter and finally to nitrate by bacterial activity. The highest level of nitrates was found in the waters of Casimcea (532 mg/L), and the lowest in Lumina (0.132 mg/L). For nitrites, the highest level was highlighted in the groundwater in Casimcea (0.245 mg/L), and the lowest level in Lumina (0.002 mg/L).

The highest value of hardness (95.9) was found in the waters of Casimcea, while the lowest value was identified in the waters of Lumina (21.2). The studied waters thus fall into the category of hard waters (those from Lumina and Corbu) and very hard waters, with hardness values over 30 degrees German hardness.

Higher turbidity affects life indirectly by cutting the light utilized by phytoplankton growth. It should be noted that the water in Lumina had the highest value of turbidity (0.75 UTF). The lowest value could be observed for the waters in Constanta (0.013 UTF). At all sites turbidity is low (Table 2), because there is no exploitation activity in the area.

Ammonium is an indicator parameter of recent organic pollution. Its sudden increase indicates the intervention of a pollutant of either natural or artificial origin. The ammonium parameter registers the highest value in Corbu (0.075 mg/L), and the lowest value is found simultaneously in two areas: Palazu Mare and Palas (0.001 mg/L).

Chloride is one of the most important parameters in assessing the water quality. The higher concentration of chloride indicates higher degree of organic pollution. The highest value for chloride concentrations was determined in Casimcea (312.7 mg/L) and the lowest value in Lumina (89.0 mg/L).

Following the results obtained, it can be seen that the values of the parameters ammonium, nitrites, pH, and turbidity for all groundwater samples in the region of Dobrudja are in accordance with legal provisions (Table 4). Physico-chemical parameters such as chlorides, conductivity, hardness, and nitrates exceed the maximum permitted levels. The nitrate values from the studied water samples exceed the maximum allowed limit of 50 mg/L, except for the sample of Lumina, which has a value according to the legal provisions of 0.132 mg/L. The hardness parameter presents values above the maximum limits allowed by the Romanian legislation in all groundwater samples in the region of Dobrudja.

Physico-chemical parameters that exceed the permitted limits can cause serious problems for the population and the environment. Values of pH, conductivity, chlorides, nitrates within the maximum permissible limits were obtained in the Chandigarh region [29], while the value for hardness exceeded the maximum permissible limit. Groundwater in the northern China's Yuncheng Basin has reached nitrate levels at different depths [30] and it can be concluded that groundwater in the depths of the earth is less polluted by anthropogenic factors than those at the surface. In the region of Dobrudja, the nitrate level exceeded the maximum limits allowed by the Romanian legislation.

Following the analyses, it was observed that in the samples studied, cadmium and iron were below the detection limit, copper was present only in the water from Lumina, and nickel only in the water sample from Palazu Mare (Table 3). The region of Dobrudja is known as the largest karst area in Romania. The transportation of metals in groundwater can be influenced by hydrogeological and geochemistry properties of the karst area. The precipitation of hydroxide or carbonate phase will inhibit the concentration of metal ions in karstic water to low values [31]. Another reason why metals concentrations are mostly below detection limit is pollutants transport. A karst aquifer has an anisotropic permeability structure that may adsorb substances easily onto the aquifer. Moreover, the increasing pollutant concentrations in a karst aquifer is caused by the larger groundwater velocity of karst areas than other lands [32].

The highest value for chromium was obtained for the sample from Palazu Mare (15.32 µg/L) and the lowest for Corbu (6.452 µg/L). For manganese, the highest value was found in the water samples from Lumina (3.662 µg/L), and the lowest value in the sample from Corbu (0.259 µg/L). The highest value observed for lead in groundwater was in Palazu Mare (1.016 µg/L), while in Palas and in Corbu this metal was below the detection limit. Zinc, the metal found in the largest quantities in all groundwater samples in the Dobrudja region, recorded oscillating values, so the highest value was recorded in Lumina (37.93 µg/L), and the lowest in Corbu (5.036 µg/L).

The concentrations of the studied metals were within the maximum allowed limits established by the Romanian legislation regarding the maximum allowed limits of heavy metals in drinking water [27].

Correlations among metals were determined using the Pearson correlation analysis to provide information on their sources. Table 5 shows the Pearson correlation coefficients for target metals for the six locations in the region of Dobrudja. There are high values for these coefficients for the following pairs of metals: Pb–Cr, Pb–Ni, Cu–Zn, Cu–Mn, Zn–Mn, some of them greater than 0.9 (Pb-Cr, Pb-Ni). These strongly positive correlations between pairs of metals indicate identical behavior of metals in the water column and could suggest the possibility of their common origin. Pb was significantly correlated with Cr ($p < 0.01$) and Ni ($p < 0.05$), and these correlations may reflect similar sources of pollution or similar cumulative characteristics among the corresponding samples. Between Pb–Zn, Pb–Mn, Cr–Mn, Zn–Ni, and Mn–Ni were observed low positive Pearson correlations. However, no statistical correlation was found between these pairs, likely indicating the different origins of these elements. There was an inverse relationship between Pb–Cu, Cu–Cr, and Cu–Ni, which was likely due to their origins from different sources, or antagonism.

Table 5. Pearson correlation coefficients of metals in groundwater samples.

	Cd	Pb	Cu	Cr	Zn	Mn	Ni	Fe
Cd	1							
Pb	NaN	1						
Cu	NaN	−0.351	1					
Cr	NaN	0.963 **	−0.264	1				
Zn	NaN	0.265	0.817 *	0.157	1			
Mn	NaN	0.360	0.736	0.292	0.889 *	1		
Ni	NaN	0.900 *	−0.200	0.791	0.302	0.450	1	
Fe	NaN	NaN	NaN	NaN	NaN	NaN	NaN	1

NaN (not a number) means there is a result, but it cannot be represented in the computer. ** Correlation is significant at the 0.01 level. * Correlation is significant at the 0.05 level.

Positive Pearson coefficients between metals suggest a pollution source (industrial, urban, or agriculture pollution). In addition, to verify if there is variability between groundwater samples from different areas (rural and urban), an analysis of variance (ANOVA) was carried out. According to this test, statistically significant differences were not found between samples from urban and rural areas ($p > 0.05$).

In addition, a Pearson correlation analysis was used to identify the relationship between metals concentrations and water physico-chemical parameters (pH, conductivity, total alkalinity, nitrates, nitrites, hardness, turbidity, ammonium, and chloride)—Table 6.

Pb showed strong positive significant correlation with total alkalinity, and Cu and Zn with turbidity.

In the studied groundwater samples, metals showed small positive correlations with most of the physico-chemical parameters (Pb with C, NO_3^-, d_T; Cu with pH, TA; Cr with C, TA, NO_3^-, d_T; Zn with pH. TA, NO_3^-; Mn with TA, TU and Ni with C, TA, NO_3^-, d_T). For the other studied metals with physico-chemical parameters, a low negative correlation was observed.

Table 6. Pearson's correlation coefficient of metals with physico-chemical parameters in groundwater samples.

	Cd	Pb	Cu	Cr	Zn	Mn	Ni	Fe
pH	NaN	−0.61	0.13	−0.68	0.11	−0.20	−0.47	NaN
C	NaN	0.29	−0.62	0.21	−0.22	−0.48	0.28	NaN
TA	NaN	0.84 *	0.23	0.79	0.48	0.63	0.73	NaN
NO_3^-	NaN	0.53	−0.50	0.45	0.36	−0.14	0.54	NaN
NO_2^-	NaN	−0.23	−0.28	−0.29	−0.08	−0.45	−0.19	NaN
d_T	NaN	0.08	−0.37	0.05	−0.03	−0.36	0.03	NaN
TU	NaN	−0.24	0.96 **	−0.29	0.83 *	0.78	−0.17	NaN
NH_4^+	NaN	−0.46	−0.28	−0.59	−0.44	−0.65	−0.32	NaN
Cl^-	NaN	−0.04	−0.36	−0.14	−0.07	−0.43	0.01	NaN

pH units, C—conductivity, TA—total alkalinity, NO_3^- nitrates, NO_2^- nitrites, d_T hardness, TU turbidity, NH_4^+ ammonium, and Cl^- chloride. NaN (not a number) means there is a result, but it cannot be represented in the computer. ** Correlation is significant at the 0.01 level. * Correlation is significant at the 0.05 level.

Chloride as the representative ion of salinity had negative correlations with all studied metals, except Ni ($r = 0.01$), indicating that terrestrial inputs strongly contributed to the distribution of these metals in groundwater samples.

Thus, in the central area of Bangladesh, located in the south of the Asian continent, values were recorded for Fe from 52 to 19.600 µg/L, for Mn from 0.04 to 4.23 µg/L, for Ni from 1.1 to 18.8 µg/L, for Pb from 0.02 to 28.6 µg/L, and for Zn from 2 to 58 µg/L [33]. Comparing these results of heavy metals in the groundwater of central Bangladesh with those of groundwater in the Dobrudja area, it can be seen that Fe, Mn, and Pb exceed the maximum permissible limits. It can also be seen that in the groundwater of the southern part of the Asian continent there are large quantities of heavy metals compared to those found in the groundwater of the southeastern part of the European continent, more precisely in the Dobrudja area of Romania.

Also in Bangladesh, but this time in the western area, data on the existence of trace metals were monitored, which showed that of the six metals analyzed, namely Cu, Cd, Fe, Mn, Ni and Zn, only Cu and Zn were within the maximum allowed limits, the rest of the metals being found in much larger quantities [34]. This once again demonstrates the increased degree of pollution in the groundwater of Bangladesh, perfectly in contrast to that of groundwater in the Dobrudja area.

Another area where trace metals were monitored was in the Bazman Basin, located in southeastern Iran [35]. From metals analyzed in this area, it was shown that only Cr, Cu, Fe, Mn, Ni, and Zn were within the maximum allowed limits.

Determination of the health risk index.

The health risk index (HRI) was estimated as the ratio of daily metal intake (MRI) to mean reference metal (RfD) values, according to the report [36]:

$$HIR = \frac{DIM}{RfD} = \frac{C_{water} \cdot I}{BM \cdot RfD} \tag{1}$$

where:

C_{water}—metal content in water samples;
I—daily water intake;
BM—average body mass of an adult;
RfD—the value for the average reference dose of metals.

From specialized data [36], it is known that I = 2 L and BM = 70 kg.

The results obtained after determining the health risk index of metals (Cd, Pb, Cu, Cr, Zn, Mn, Ni) are represented in Table 7.

Table 7. Risk index values for the studied samples.

Metals	RfD [mg·kg^{-1}·day^{-1}]	Constanta	Palazu Mare	Palas	Lumina	Corbu	Casimcea
Cd	0.0010	-	-	-	-	-	-
Pb	0.0035	0.329	8.293	-	0.937	-	0.900
Cu	0.0400	-	-	-	0.066	-	-
Cr	1.5000	0.241	0.291	0.157	0.149	0.122	0.148
Zn	0.3000	0.910	2.370	0.821	3.612	0.481	1.66
Mn	0.1400	0.095	0.582	0.302	0.747	0.052	0.108
Ni	0.0200	-	3.355	-	-	-	-

The order of the toxicity of heavy metals according to the average HRI values, calculated for each groundwater sample, was as follows:

- for the groundwater sample from Constanta Zn> Pb> Cr> Mn> Cd, Cu, Ni;
- for the groundwater sample from Palazu Mare Pb> Ni> Zn> Mn> Cr> Cd, Cu;
- for the groundwater sample from Palas Zn> Mn> Cr> Cd, Pb, Cu, Ni;
- for the groundwater sample from Lumina Zn> Pb> Mn> Cr> Cu> Cd, Ni;
- for the groundwater sample from Corbu Zn> Cr> Mn> Cd, Pb, Cu, Ni;
- for the groundwater sample from Casimcea Zn> Pb> Cr> Mn> Cd, Cu, Ni.

Taking into account the fact that the value of the health risk index should be much less than 1 (HRI << 1), it can be concluded that it exceeds this value in the case of water from Palazu Mare for lead, zinc, and nickel. This exceedance is also observed in the case of the waters from Lumina and Casimcea for zinc, thus demonstrating that their use presents a danger for human health.

5. Conclusions

At present, groundwater occupies a special place in nature, primarily as a source of drinking water for the population, and for industry and agriculture. It has better drinking qualities than surface water, which is why its exploitation and capture process has become increasingly important.

Recent studies on groundwater quality control have shown that, in many parts of the world, water has different degrees of quality depending on the natural and anthropogenic factors acting upon them.

- This study has revealed that the pH values of well water are slightly alkaline and varied from 7.22 pH units to 7.91 pH units and are within the permissible limit prescribed by the Romanian legislation.
- Regarding the conductivity of the studied groundwater, the highest value was observed in Casimcea (3070 μs/cm); the higher EC value of the well could be attributed to its shallow depth, as ongoing surface activities can affect the conductivity values.
- The concentration of nitrate contained in water seems to be the high agricultural waste and sewage contamination. The highest level of nitrates was found in the waters of Casimcea (532 mg/L), while for nitrites, the highest level was highlighted also in the groundwater in Casimcea (0.245 mg/L). In the region of Dobrudja, the nitrate level exceeded the maximum limits allowed by the Romanian legislation.
- The studied waters fall into the category of hard waters (those from Lumina and Corbu) and very hard waters, with hardness values over 30 degrees German hardness.
- Higher turbidity affects the life indirectly by cutting the light to be utilized by the phytoplankton growth. At all sites, turbidity is low, because there is no exploitation activity in the area.
- Ammonium is an indicator parameter of recent organic pollution. Its sudden increase indicates the intervention of a pollution that can be both natural and artificial origin. The ammonium parameter registers the highest value in Corbu (0.075 mg/L).

- Chloride is one of the most important parameters in assessing the water quality. The higher concentration of chloride indicates higher degree of organic pollution. The highest value for chloride concentrations was determined in Casimcea (312.7 mg/L) and the lowest value in Lumina (89.0 mg/L).
- The concentrations of the studied metals were within the maximum allowed limits established by the Romanian legislation regarding the maximum allowed limits of heavy metals in drinking water.
- In the case of the water from Palazu Mare the health risk index exceeds the maximum value for lead, zinc, and nickel. This exceedance is also observed in the case of the waters from Lumina and Casimcea for zinc, thus demonstrating that their use presents a danger to human health.

In conclusion, more attention should be paid to the prevention of groundwater pollution and the immediate remediation of accidents. This would lead to lower, cost-effective costs as well as environmental protection.

Author Contributions: Conceptualization, A.S. (Alina Soceanu), N.M., C.I.D., and S.D.; methodology, A.S. (Alina Soceanu), G.V.; software, V.P. and A.S. (Anca Sirbu); validation, S.D.; formal analysis, G.V.; investigation, G.V.; resources, A.S. (Alina Soceanu), S.D., N.M. and C.I.D.; data curation, V.P.; writing—original draft preparation, A.S. (Anca Sirbu); writing—review and editing, A.S. (Alina Soceanu); visualization, V.P.; supervision, S.D., A.S. (Anca Sirbu); project administration, N.M. and C.I.D. All authors have read and agreed to the published version of the manuscript.

Funding: This research received no external funding.

Institutional Review Board Statement: Not applicable.

Informed Consent Statement: Not applicable.

Data Availability Statement: Not applicable.

Conflicts of Interest: The authors declare no conflict of interest.

References

1. Zhang, H.; Cheng, S.; Li, H.; Fu, K.; Xu, Y. Groundwater pollution source identification and apportionment using PMF and PCA-APCA-MLR receptor models in a typical mixed land-use area in Southwestern China. *Sci. Total Environ.* **2020**, *741*, 140383. [CrossRef] [PubMed]
2. Bretotean, M. *Groundwaters, An Important Natural Wealth*; Ceres: Bucharest, Romania, 1981. (In Romanian)
3. Shakerkhatibi, M.; Mosaferi, M.; Pourakbar, M.; Ahmadnejad, M.; Safavi, N.; Banitorab, F. Comprehensive investigation of groundwater quality in the north-west of Iran: Physicochemical and heavy metal analysis. *Groundw. Sustain. Dev.* **2019**, *8*, 156–168. [CrossRef]
4. Xiao, L.; Liu, J.; Ge, J. Dynamic game in agriculture and industry cross-sectoral water pollution governance in developing countries. *Agric. Water Manag.* **2021**, *243*, 106417. [CrossRef]
5. Zacchaeus, O.O.; Balogun Adeyemi, M.; Azeem Adedeji, A.; Adegoke, K.A.; Okehi Anumah, A.; Taiwo, A.M.; Ganiyu, S.A. Effects of industrialization on groundwater quality in Shagamu and Ota industrial areas of Ogun state, Nigeria. *Heliyon* **2020**, *6*, 04353. [CrossRef]
6. Sambito, M.; Severino, A.; Freni, G.; Neduzha, L. A Systematic Review of the Hydrological, Environmental and Durability Performance of Permeable Pavement Systems. *Sustainability* **2021**, *13*, 4509. [CrossRef]
7. Raval, N.P.; Kumar, M. Geogenic arsenic removal through core–shell based functionalized nanoparticles: Groundwater in-situ treatment perspective in the post–COVID anthropocene. *J. Hazard. Mater.* **2021**, *402*, 123466. [CrossRef] [PubMed]
8. Kumar, M. Runoff from firework manufacturing as major perchlorate source in the surface waters around Diwali in Ahmedabad, India. *J. Environ. Manag.* **2020**, *273*, 111091. [CrossRef] [PubMed]
9. Masoud, A.A. Groundwater quality assessment of the shallow aquifers west of the Nile Delta (Egypt) using multivariate statistical and geostatistical techniques. *J. Afr. Earth Sci.* **2014**, *95*, 123–137. [CrossRef]
10. Atwia, M.G.; Abu-Heleika, M.M.; El-Horiny, M.M. Hydrogeochemical and vertical electrical soundings for groundwater investigations, Burg El-Arab area, Northwestern Coast of Egypt. *J. Afr. Earth Sci.* **2013**, *80*, 8–20. [CrossRef]
11. Fathy Abdalla, Ramadan Khalil, Potential effects of groundwater and surface water contamination in an urban area, Qus City, Upper Egypt. *J. Afr. Earth Sci.* **2018**, *141*, 164–178. [CrossRef]
12. He, B.; He, J.; Wang, l.; Zhang, X.; Bi, E. Effect of hydrogeological conditions and surface loads on shallow groundwater nitrate pollution in the Shaying River Basin: Based on least squares surface fitting model. *Water Res.* **2019**, *162*, 114880. [CrossRef] [PubMed]

13. Zhang, F.; Huanga, G.; Houa, Q.; Liua, C.; Zhanga, Y.; Zhang, Q. Groundwater quality in the Pearl River Delta after the rapid expansion of industrialization and urbanization: Distributions, main impact indicators, and driving forces. *J. Hydrol.* **2019**, *577*, 124004. [CrossRef]

14. Jia, X.; O'Connor, D.; Hou, D.; Jin, Y.; Li, G.; Zheng, C.; Ok, Y.S.; Daniel, C.W.; Tsang, D.C.W.; Luo, J. Groundwater depletion and contamination: Spatial distribution of groundwater resources sustainability in China. *Sci. Total Environ.* **2019**, *672*, 551–562. [CrossRef] [PubMed]

15. Koh, E.-H.; Lee, E.; Lee, K.-K. Application of geographically weighted regression models to predict spatial characteristics of nitrate contamination: Implications for an effective groundwater management strategy. *J. Environ. Manag.* **2020**, *268*, 110646. [CrossRef]

16. Hansen, B.; Thorling, L.; Kim, H.; Blicher-Mathiesen, G. Long-term nitrate response in shallow groundwater to agricultural N regulations in Denmark. *J. Environ. Manag.* **2019**, *240*, 66–74. [CrossRef]

17. Taufiq, A.; Effendi, A.J.; Iskandar, I.; Hosono, T.; Hutasoit, L.M. Controlling factors and driving mechanisms of nitrate contamination in groundwater system of Bandung Basin, Indonesia, deduced by combined use of stable isotope ratios, CFC age dating, and socioeconomic parameters. *Water Res.* **2019**, *148*, 292–305. [CrossRef]

18. Huan, H.; Hu, L.; Yang, Y.; Jia, Y.; Lian, X.; Ma, X.; Jiang, Y.; Xi, B. Groundwater nitrate pollution risk assessment of the groundwater source field based on the integrated numerical simulations in the unsaturated zone and saturated aquifer. *Environ. Int.* **2020**, *137*, 105532. [CrossRef]

19. Serio, F.; Miglietta, P.P.; Lamastra, L.; Ficocelli, S.; Intini, F.; Leo, F.D.; Donno, A.D. Groundwater nitrate contamination and agricultural land use: A grey water footprint perspective in Southern Apulia Region (Italy). *Sci. Total Environ.* **2018**, *645*, 1425–1431. [CrossRef]

20. Singh, C.K.; Kumar, A.; Shashtri, S.; Kumar, A.; Kumar, P.; Mallick, J. Multivariate statistical analysis and geochemical modeling for geochemical assessment of groundwater of Delhi, India. *J. Geochem. Explor.* **2017**, *175*, 59–71. [CrossRef]

21. Dumitrescu, I.C. *Sustainable Development and the Natural Environment*; Bren: Bucharest, Romania, 2005. (In Romanian)

22. Molinos-Senante, M.; Maziotis, A.; Sala-Garrido, R. Changes in the total costs of the English and Welsh water and sewerage industry: The decomposed effect of price and quantity inputs on efficiency. *Util. Policy* **2020**, *66*, 101063. [CrossRef]

23. Ruiz-Rosa, I.; García-Rodríguez, F.J.; Antonova, N. Developing a methodology to recover the cost of wastewater reuse: A proposal based on the polluter pays principle. *Util. Policy* **2020**, *65*, 101067. [CrossRef]

24. Luppi, B.; Parisi, F.; Rajagopalan, S. The rise and fall of the polluter-pays principle in developing countries. *Int. Rev. Law Econ.* **2012**, *32*, 135–144. [CrossRef]

25. Tilton, J.E. Global climate policy and the polluter pays principle: A different perspective. *Resour. Policy* **2016**, *50*, 117–118. [CrossRef]

26. Pișota, I.; Buta, I. *Hidrology*; EDP: Bucharest, Romania, 1984. (In Romanian)

27. *Law no. 458/2002 on Drinking Water Quality*; published in the Legal Gazette of Romania, Part I, issue no. 552 of 29 July 2002; Romanian Parliament: Bucharest, Romania, 2002.

28. Abeer, N.; Akbar Khan, S.; Muhammad, S.; Rasool, A.; Ahmad, I. Health risk assessment and provenance of arsenic and heavy metal in drinking water in Islamabad, Pakistan. *Environ. Technol. Innov.* **2020**, *20*, 101171. [CrossRef]

29. Ravindra, K.; Singh Thind, P.; Mor, S.; Singh, T.; Mor, S. Evaluation of groundwater contamination in Chandigarh: Source identification and health risk assessment. *Environ. Pollut.* **2019**, *255*, 113062. [CrossRef]

30. Currell, M.J.; Cartwright, I.; Bradley, D.C.; Han, D. Recharge history and controls on groundwater quality in the Yuncheng Basin, north China. *J. Hydrol.* **2010**, *385*, 216–229. [CrossRef]

31. Astuti, R.D.P.; Mallongi, A.; Amiruddin, R.; Hatta, M.; Rauf, A.U. Risk identification of heavy metals in well water surrounds watershedarea of Pangkajene, Indonesia. *Gac Sanit.* **2021**, *35*, S33–S37. [CrossRef]

32. Pu, J.; Cao, M.; Zhang, Y.; Yuan, D.; Zhao, H. Hydrochemical indications of human impact on karst groundwater in a subtropical karst area, Chongqing, China. *Environ Earth Sci.* **2014**, *72*, 1683–1695. [CrossRef]

33. Bodrud-Doza, M.; Islam, S.M.; Hasan, M.T.; Alam, F.; Haque, M.M.; Rakib, M.A.; Asad, M.A.; Rahman, M.A. Groundwater pollution by trace metals and human health risk assessment in central west part of Bangladesh. *Groundw. Sustain. Dev.* **2019**, *9*, 100219. [CrossRef]

34. Bodrud-Doza, M.; Towfiqul Islam, A.R.M.; Ahmed, F.; Das, S.; Saha, N.; Rahman, M.S. Characterization of groundwater quality using water evaluationindices, multivariate statistics and geostatistics in central Bangladesh. *Water Sci.* **2016**, *30*, 19–40. [CrossRef]

35. Rezaei, A.; Hassani, H.; Hassani, S.; Jabbari, N.; Belgheys, S.; Mousavi, F.; Rezaei, S. Evaluation of groundwater quality and heavy metal pollution indices in Bazman basin, southeastern Iran. *Groundw. Sustain. Dev.* **2019**, *9*, 100245. [CrossRef]

36. Masime, J.O. *Analysis of the Levels of Arsenic, Nitrate, Nitrite and Phosphate in Home Made Brews, Spirits, Tap Water and in Raw Materials in Nairobi County*; MST, Department of Chemistry, Technical University of Kenya: Nairobi, Kenya, 2016.

applied sciences

Article

Evaluation of Multivariate Biomarker Indexes Application in Ecotoxicity Tests with Marine Diatoms Exposed to Emerging Contaminants

Vanessa Leal Pires [1,2], Sara C. Novais [2], Marco F. L. Lemos [2], Vanessa F. Fonseca [1,3] and Bernardo Duarte [1,4,*]

[1] MARE—Marine and Environmental Sciences Centre, Faculdade de Ciências da Universidade de Lisboa, Campo Grande, 1749-016 Lisbon, Portugal; vanessalealpires@gmail.com (V.L.P.); vffonseca@fc.ul.pt (V.F.F.)
[2] MARE—Marine and Environmental Sciences Centre, ESTM, Politécnico de Leiria, 2520-630 Peniche, Portugal; sara.novais@ipleiria.pt (S.C.N.); marco.lemos@ipleiria.pt (M.F.L.L.)
[3] Departamento de Biologia Animal, Faculdade de Ciências da Universidade de Lisboa, Campo Grande, 1749-016 Lisbon, Portugal
[4] Departamento de Biologia Vegetal, Faculdade de Ciências da Universidade de Lisboa, Campo Grande, 1749-016 Lisbon, Portugal
* Correspondence: baduarte@fc.ul.pt

Featured Application: Evaluation of oxidative stress indexes for ecotoxicological evaluation of marine diatoms exposed to emerging contaminants.

Abstract: Worldwide anthropogenic activities result in the production and release of potentially damaging toxic pollutants into ecosystems, thereby jeopardizing their health and continuity. Research studies and biomonitoring programs attend to this emerging problematic by applying and developing statistically relevant indexes that integrate complex biomarker response data to provide a holistic approach, reflecting toxically induced alterations at the organism or population level. Ultimately, indexes allow simple result communications, enhancing policy makers understanding, and contributing to better resource and environmental managing policies. In this study three indexes, the integrated biomarker response index (IBR), the bioeffects assessment index (BAI) and principal components analysis (PCA), were evaluated for their sensitivity in revealing toxically induced stress patterns in cells of the diatom *Phaeodactylum tricornutum* under contaminant exposure. The set of biomarkers selected for index construction comprised the anti-oxidant enzymes APX, CAT and SOD, and the lipid peroxidation marker TBARS. Several significant correlations with the applied concentration gradients were noticed for all indexes, although IBR excelled for its reliability in delivering statistically significant dose-response patterns for four out of the five tested compounds.

Keywords: antioxidant enzymes; pharmaceutical residues; pesticides; detergents; integrative indexes

Citation: Pires, V.L.; Novais, S.C.; Lemos, M.F.L.; Fonseca, V.F.; Duarte, B. Evaluation of Multivariate Biomarker Indexes Application in Ecotoxicity Tests with Marine Diatoms Exposed to Emerging Contaminants. *Appl. Sci.* **2021**, *11*, 3878. https://doi.org/10.3390/app11093878

Academic Editor: Elida Nora Ferri

Received: 14 March 2021
Accepted: 15 April 2021
Published: 25 April 2021

Publisher's Note: MDPI stays neutral with regard to jurisdictional claims in published maps and institutional affiliations.

1. Introduction

Daily worldwide spread of pharmaceuticals, pesticides and detergents that endanger the health of natural aquatic ecosystems are of major concern [1,2]. These emerging contaminants have serious implications for the organisms living in aquatic environments, at different biological levels, leading to biochemical, physiological, metabolic, individual and population-level effects [3,4]. These impacts are often addressed from the biochemical point of view, as a starting point for understanding the early effects of these contaminants in the biota, using a collection of biomarkers [5–7]. These biomarker responses are used as indicators of health status, which are already considered in international environmental policies [8]. The World Health Organization (WHO) defines biomarkers as biological measurements at the biochemical, cellular or molecular level that result from the interaction between an organism and environmental chemical, physical or biological agents [9].

The applicability of subcellular biomarkers to address pollutants in aquatic ecosystems is currently widely studied. Biomarkers of exposure, such as antioxidant systems, are widely used for their sensitivity within the scope of toxicological studies [10,11]. As chemical pollutants induce oxidative damage to biological systems, the activity levels of antioxidant enzymes, like superoxide dismutase (SOD) and catalase (CAT)—which are vitally involved in the detoxification of reactive oxygen species (ROS) and in the reduction of oxidative stress—can be evaluated and understood [7,10]. Likewise, biomarkers of lipid peroxidation (LPO) are vastly investigated [10], especially through the thiobarbituric acid reactive substances (TBARS) test, which measures the levels of secondary lipid oxidation products, such as malondialdehyde (MDA) formation [12]. This set of biomarkers provides a standing of the overall oxidative stress state of cells under a certain compound exposure at a specific dose [13].

While from the scientific point of view, the mode of action of these antioxidant processes and how they are affected by different emerging contaminants is of upmost importance, these biochemical results are often difficult to communicate to non-scientific communities, such as stakeholders and decision makers [14,15]. The pressing need for methodologies and information that can effectively connect researchers and environmental managers, linked to international environmental frameworks, led to the development of a variety of indexes that can easily communicate scientific results to stakeholders, as applicable tools for decision making in environmental managing [16–18]. An array of scoring indexes have emerged in the past few years [16], thus adopting an ecologically transversal approach of great importance in order to develop meaningful community management tools towards addressing contaminants of emerging concern [5,19,20]. Therefore, following biochemical evaluations, new index development must accurately reflect the biological status, follow suitable validation and consider result communication, so that management entities can approach similar new environmental matters within the scope of policy development [16,17]. As mentioned above, several approaches have been developed to translate vast arrays of biochemical data into numeric indexes, such as the integrated biomarker response index (IBR) [5], the bioeffect assessment index (BAI) [19] and the principal component analysis scoring based index (denoted as PCA-index hereafter) [20,21]. These indexes are based on different approaches to the same biochemical data. The IBR index uses the normalization of biochemical data to overcome differences in the magnitude between the different biochemical traits it integrates, and requires expert judgment or previous experience from an operator to define how a certain biochemical trait is altered (enhanced or inhibited) by a given compound [5]. The BAI approach divides each biochemical trait at each exposure level into statistical cohorts, considering quartile distribution, and scoring each biochemical trait accordingly [19]. The PCA index uses the values obtained from a PCA plot of biochemical traits to normalize the values and to weigh each variable according to its importance to the PCA model [20,21]. Thus, these three indexes range from univariate (BAI) to multivariate approaches (IBR and PCA-based indexes), with or without operator intervention (IBR and PCA-based respectively), and a comparative analysis allows to assess which approach is more suitable to be applied in emerging contaminants ecotoxicological studies.

In the present work, the suitability of the three different multivariate indexes (BAI, IBR, and PCA-based indexes) was tested through application of previously obtained oxidative stress biomarker data, using marine diatoms as a study model. The impacts of pharmaceuticals (propranolol [22], fluoxetine [23], ibuprofen [24], detergents (SDS) [25], and herbicides (glyphosate [15])) on primary production and physiological fitness of the marine diatom *Phaeodactylum tricornutum* were used to understand the suitability of multivariate indexes in depicting dose responses and contaminant types, as well as to evaluate the performance and accuracy of each approach for future ecotoxicological trials. In sum, we intended to produce an index that allows to better communicate toxicity results, providing an easy-to-understand framework, which is accessible to managers and the general public.

2. Materials and Methods

2.1. Diatom Exposure Trials and Biomarker Analysis

Data regarding growth and biomarker activity/concentration were collected from previously published works [15,22–25]. *Phaeodactylum tricornutum* Bohlin (Bacillariophyceae; strain IO 108–01, Instituto Português do Mar e da Atmosfera (IPMA)) axenic cell cultures (maintained under asexual reproduction conditions) were placed to grow in f/2 medium [26], under constant aeration in a phytoclimatic chamber, at 18 °C, programmed with a 14/10 h day/night photoperiod (RGB 1:1:1, maximum PAR 80 µmol photons m^{-2} s^{-1}), a sinusoidal function to mimic sunrise and sunset, and light intensity equal to noon, set to replicate a natural light environment [27]. Exposure trials were conducted according to the Organization for Economic Cooperation and Development (OECD) recommendations for algae assays [28], with minor adaptations, and as described in [15,22–25]. Briefly, cultures were exposed to the compounds and target concentrations for 48 h. Target concentrations were selected aiming to cover a concentration gradient reflecting, not only the detected environmental concentrations found in the literature, but also concentrations known to have significant biological effects in *P. tricornutum* [29,30]. Growth inhibition concentration (IC$_{50}$) was calculated according to the OECD guidelines for the algae inhibition test [28]. Biomarker analysis was performed according to standard spectrophotometric procedures as described in [15,22–25]. Briefly, catalase (CAT), ascorbate peroxidase (APx) and superoxide dismutase (SOD) activities were assayed by spectrophotometric means using specific substrates as previously described [31–33]. Lipid peroxidation products were analyzed spectrophotometrically [34], using trichloroacetic acid extraction before the reaction with thiobarbituric acid.

2.2. Multivariate Index Calculation

2.2.1. Integrated Biomarker Response (IBR) Index

The IBR index was calculated for each tested compound according to Beliaeff and Burgeot (2002) [5], posteriorly adapted by Broeg and Lehtonen (2006) [35]. Briefly, it was calculated by summing up triangular star plot areas calculated for each two neighbouring variables. To calculate IBR integrating all biomarkers, data of each biomarker were standardized to obtain Y as:

$$Y = \frac{X - m}{s}$$

where X is the value for each biomarker replicate within a given concentration, m is the general mean of all data (including concentrations) regarding a given biomarker, and s is the standard deviation of the biomarker data for a given treatment/concentration. Then Z was calculated using $Z = -Y$ or $Z = Y$, in the case of a biological effect corresponding respectively to an inhibition or a stimulation and it was determined using the slope of the biomarker activity/concentration towards the applied exogenous concentration. Positive slopes corresponded to a stimulation, while negative slopes corresponded to an inhibition. Regarding the biological effect, biomarkers can either increase or decrease depending on the type and concentration of the compound exposure, but they can also vary among organisms.

Subsequently, the score (S) was calculated as

$$S = Z + |Min|$$

where, $S \geq 0$ and $|Min|$ is the absolute value for the minimum value for all calculated Y for a given biomarker at all the made measurements (again all concentrations considered). Star plots were then used to display score results (S) and to calculate the IBR index as:

$$IBR = \sum_{t=15}^{n} A_i$$

A_i is the area between two consecutive clockwise scores in a given star plot:

$$A_i = \frac{S_i}{2} \sin \beta (S_i \cos \beta + S_{i+1} \sin \beta)$$

$$\beta = \tan^{-1} \frac{S_{i+1} \sin \alpha}{S_i - S_{i+1} \cos \alpha}$$

where S_i and S_{i+1} are two consecutive clockwise scores of a given star plot; n the number of biomarkers used and α:

$$\alpha = \frac{2\pi}{n}$$

2.2.2. Principal Component Analysis (PCA)-Based Index

To assess which variables were more suitable for the index elaboration and implementation, a previously successfully tested statistical approach was used [20,21,36]. This was preformed independently for each compound. Principal component analysis (PCA) was performed in order to select the appropriate measured parameters to integrate the index. For this selection, the five variables with higher weighing factors from the factor axis with higher explanatory value percentages were chosen [21]. PCA was performed after data normalization and transformation using PRIMER 6 [37]. The PCA-based index is defined as:

$$PCA - based\ index = \sum W_i E_i$$

where W is the PCA weighing factor of the PCA selected variable and E its respective score. The scores were normalized using a sigmoidal equation limited from 1 to 0, as follows [21,36,38]:

$$E = \frac{a}{1 + \frac{x}{x_0}^b}$$

where x is the variable verified value, a is the maximum score of the variable (in this case 0.535 so that the final index has 1 as its maximum value), x_0 is the average value of the variable and b is the value of the slope of the equation. The slope was defined as -2.5 for the better-adjusted curve tending to 1 for all the variables proposed.

2.2.3. Bioeffect Assessment Index (BAI)

Numerical values were assigned to the index variables, based on the degree or severity of damage to an organ or tissue caused by environmental stressors [19]. The BAI was designed as an index for the assessment of multifactorial contamination situations of coastal areas. Thus, it only includes biomarkers of general toxicity [19]. To assess the individual values of each biomarker within the BAI stages, biomarker population quartils were determined. The index value of biomarker (Biomarker$_{index}$) was attributed, depending on the quartil interval of each value:

- Values > 3rd quartile: 40
- Values between 3rd and 2nd quartiles: 30
- Values between 2nd and 1st quartiles: 20
- Values < 1st quartile: 10

The BAI value results from:

$$BAI = \sum Biomarker_{index}{}^{1/n}$$

where n is the number of biomarkers used for index calculation.

2.3. Statistical Analysis

Owing to the absence of normality and homogeneity of variances in our data, pairwise comparisons between different sample groups were assessed through non-parametric Kruskal–Wallis tests. Spearman correlation tests were performed to evaluate if there was a dose–response behavior between the exogenous concentrations tested and the biomarker variables. Both Kruskal–Wallis and Spearman tests were performed using Statistica software (StataSoft, version 12.5.192.7). Statistical significance was considered at $p < 0.05$. A multivariate approach was employed to test for variations in the complete oxidative stress data package [23,38,39]. Canonical analysis of principle (CAP) coordinates, using Euclidean distances, were preformed to plot in a canonical space the dissimilarities regarding biomarker studied variables while preforming a cross-validation step and determining the allocation efficiency into the different treatment groups. This multivariate methodology is unaffected by heterogeneous data and is frequently used to compare different sample assemblies using the inherent features of each assembly (metabolic traits) [39–42]. Multivariate statistical analyses were performed using Primer 6 software (version 6.1.13, Plymouth, UK) [37].

3. Results

3.1. Diatom Growth and Ecotoxicological Parameters (IC$_{50}$)

Data on the exposure concentration and toxicity of five contaminants that inhibit the growth of the marine diatom *P. tricornutum* were obtained from previous studies [15,22–25] and are presented in Table 1. According to the IC$_{50}$ values, which refer to the exposure concentration capable to inhibit half of the maximum growth in this marine diatom, SDS and fluoxetine are the most toxic compounds tested in *P. tricornutum*, with the lowest IC$_{50}$ values of the tested cultures (<50 µg L^{-1}). In opposition, ibuprofen appears as the least toxic compound tested, with the highest dose (ca. 350 µg L^{-1}) needed to reduce to half the growth of *P. tricornutum*.

Table 1. Exposure compounds concentrations, *Phaeodactylum tricornutum* relative growth inhibition and calculated IC$_{50}$.

Compound	Concentrations Tested (µg L^{-1})	Inhibition (%)	IC$_{50}$ (µg L^{-1})	Reference
Propranolol	0.3	0.15		[22]
	8	4.11		
	80	41.11	194.6	
	150	77.08		
	300	154.16		
Fluoxetine	0.3	0.63		[23]
	0.6	1.27		
	20	42.26	47.3	
	40	84.52		
	80	169.04		
Ibuprofen	0.8	0.23		
	3	0.86		[24]
	40	11.41	350.6	
	100	28.52		
	300	85.57		
Glyphosate	10	4.43		
	50	22.13		[15]
	100	44.27	225.9	
	250	110.67		
	500	221.34		
SDS	0.1	15.5		
	1	21.1		[25]
	3	24.6	11.4	
	10	43.8		

3.2. Oxidative Stress Biomarkers

To evaluate potential cell damage due to oxidative stress conditions induced by contaminant exposure, several oxidative stress biomarkers were evaluated in the diatom cells exposed to different pollutant levels. Oxidative stress biomarker values (average $\pm$ std deviation), linear regression slope and the statistical relation between each biomarker and the external concentration of the different contaminants applied are presented in Supplementary Material Table S1. Further discussion on the biological considerations of these results can be found in the studies contributing to the present work [15,22–25]. To study the dose-response relationship of each of the analyzed biomarkers, correlations between the biomarker response and the exogenous dose applied were also determined (Supplementary Material Table S1). Propranolol gradient concentrations showed a positive dose relationship for SOD antioxidant enzyme ($r^2 = 0.87$, $p < 0.05$) and TBARS ($r^2 = 0.72$, $p < 0.05$), whereas CAT and APX showed no significant relation to this compound. A significant rise in lipid peroxidation products (TBARS) was observed in *P. tricornutum* cells exposed to fluoxetine concentration gradient, revealing a positive and significant correlation ($r^2 = 0.73$, $p < 0.05$). Among the tested antioxidant enzymes, SOD and APX followed a similar a trend similar to the lipid peroxidation, though not significant, but showing the highest activity levels under the highest applied concentration (80 µg L^{-1}), whereas CAT activity did not respond to any of the exogenous fluoxetine concentrations. A positive correlation with the applied exogenous ibuprofen dose is shown for TBARS ($r^2 = 0.77$, $p < 0.05$) and APX activities ($r^2 = 0.62$, $p < 0.05$). On the other hand, CAT and SOD antioxidant enzymes were irresponsive to this contaminant concentration variation. Glyphosate exposure revealed a positive correlation to APX ($r^2 = 0.60$, $p < 0.05$) and TBARS ($r^2 = 0.92$, $p < 0.05$) activities, where TBARS activity levels showed an increasing trend up to the maximum glyphosate exogenous concentration (500 µg L^{-1}). CAT antioxidant enzyme revealed no correlation to glyphosate exposure but a great increase in activity under the applied highest concentration (500 µg L^{-1}). SOD activity decreased as the compound concentration increased, revealing a negative but significant dose–response correlation ($r^2 = 0.64$, $p < 0.05$). The applied set of oxidative stress biomarkers showed no correlation to the exogenous SDS exposure concentrations applied to the *P. tricornutum* cell culture.

To determine the efficiency of the considered biomarkers in classifying the sample exposures, a multivariate canonical analysis was conducted for each of the studied compounds (Figure 1). Fluoxetine (Figure 1B) and glyphosate (Figure 1D) multivariate analysis revealed the lowest classification efficiency of the samples (38.9%) using the considered biomarkers. This was attributed to a high degree of misclassification of the samples exposed to intermediate concentrations, indicative of a low impact of these compounds at these concentrations on the considered biomarkers. On the other hand, the assessed biomarkers in the cells exposed to propranolol (Figure 1A) and SDS (Figure 1E) were able to provide a correct classification of more than half of the evaluated samples (55.6% and 53.3%, respectively). Once again, in intermediate concentrations the considered biomarkers had lower efficiency in classifying the samples, due to possible reduced impacts of these compounds in intermediate concentrations.

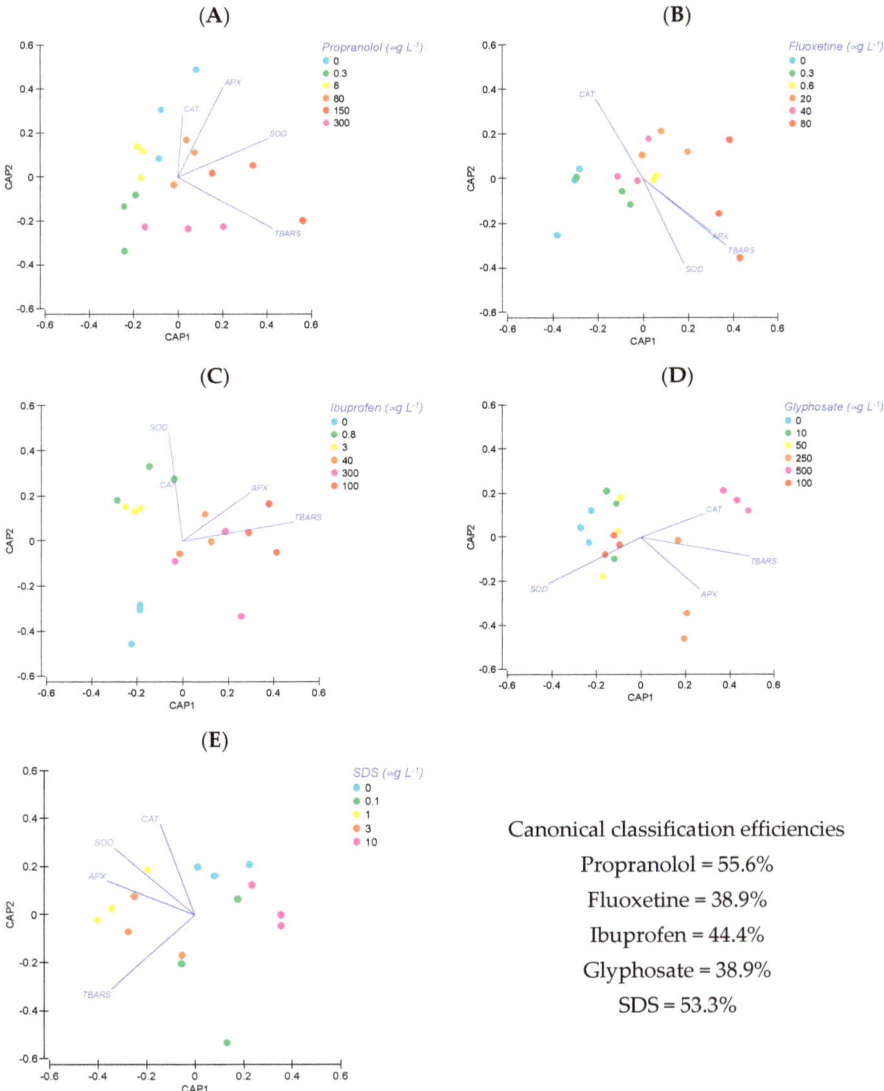

Figure 1. Canonical analysis of principal components plots of the diatom cells exposed to propranolol (**A**), fluoxetine (**B**), Ibuprofen (**C**), glyphosate (**D**), and SDS (**E**) at different exogenous concentrations, having the biomarker values as classifying variables. Data sources are available in Table 1.

3.3. Oxidative Stress Biomarkers Indexes

To evaluate the applicability of these biomarkers in multivariate indexes, towards a better communication of the overall effect of each tested compound, three index approaches were undertaken: the Integrated biomarker response (IBR) index; the principal component analysis (PCA)-based index; and the bioeffect assessment index (BAI).

The IBR index applied to biomarker results on propranolol and SDS contaminant exposure gradient, showed no significant variance between the exogenous concentrations tested (Figure 2A,D respectively). Nevertheless, a trend can be noticed for both pollutants, as the highest index values reflected higher concentrations. For the glyphosate exogenous concentration gradient, IBR scores showed significant increases on index values at 250 µg L^{-1} and 500 µg L^{-1} exposure concentrations (Figure 2B), akin to fluoxetine (Figure 2C) delivering a significantly higher index value at 80 µg L^{-1} concentration, suggesting a significant dose–response tendency for both pollutants. Concerning exposure to ibuprofen (Figure 2E), the IBR index revealed a significantly higher score at the maximum tested concentration of 300 µg L^{-1}, and in fact, this increase reflects a high significant positive trend alongside the exogenous ibuprofen gradient.

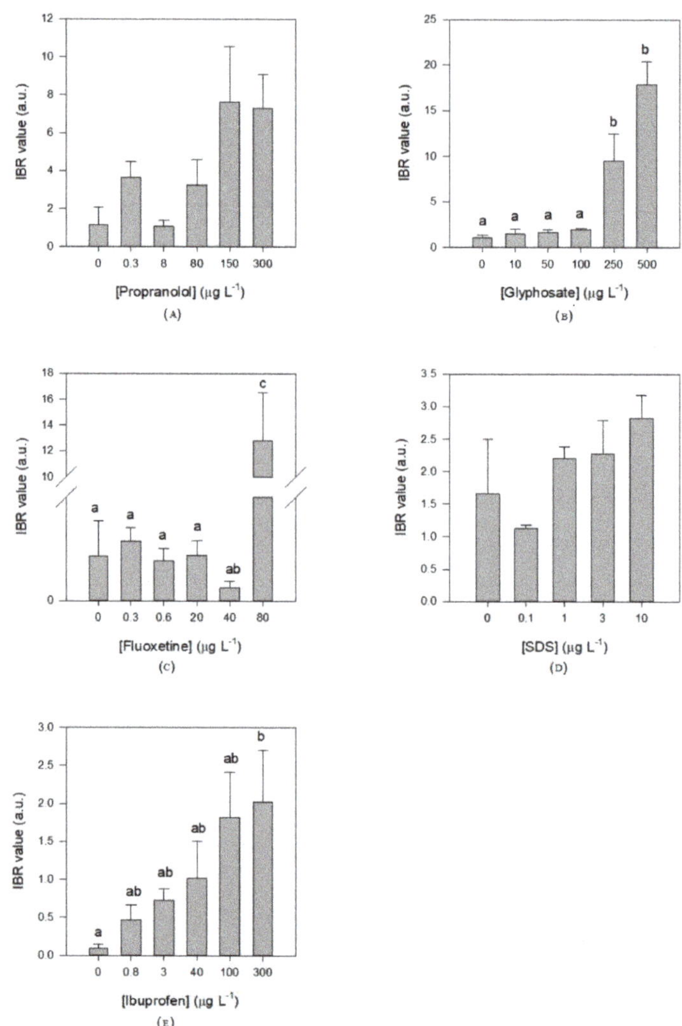

Figure 2. Integrated biomarker response (IBR) index of *Phaeodactylum tricornutum* cells exposed to different contaminants and concentrations. (**A**) Propranolol. (**B**) Glyphosate. (**C**) Fluoxetine. (**D**) SDS. (**E**) Ibuprofen. Average ± standard deviation, n = 3, different letters indicate significant differences at $p < 0.05$.

Regarding PCA-based index results for propranolol exposure, a significantly lower value was found in cultures exposed to 0.3 µg L^{-1}, whilst the highest significant value was verified at 150 µg L^{-1} of propranolol concentration (Figure 3A). For glyphosate and fluoxetine exposure (Figure 3B,C respectively), the PCA-based index attributes a significantly higher score at the maximum concentrations of 500 μg L^{-1} and 80 μg L^{-1}, respectively, evidencing a positive correlation with both exogenous concentrations of these contaminants. Concerning SDS detergent (Figure 3D), PCA-based index scores did not show any significant variance under the increasing contaminant exposure gradient. On the other hand, in relation to ibuprofen (Figure 3E), significant and approximate high index values are presented for cultures exposed to 0.8 μg L^{-1} ibuprofen and all the concentrations above.

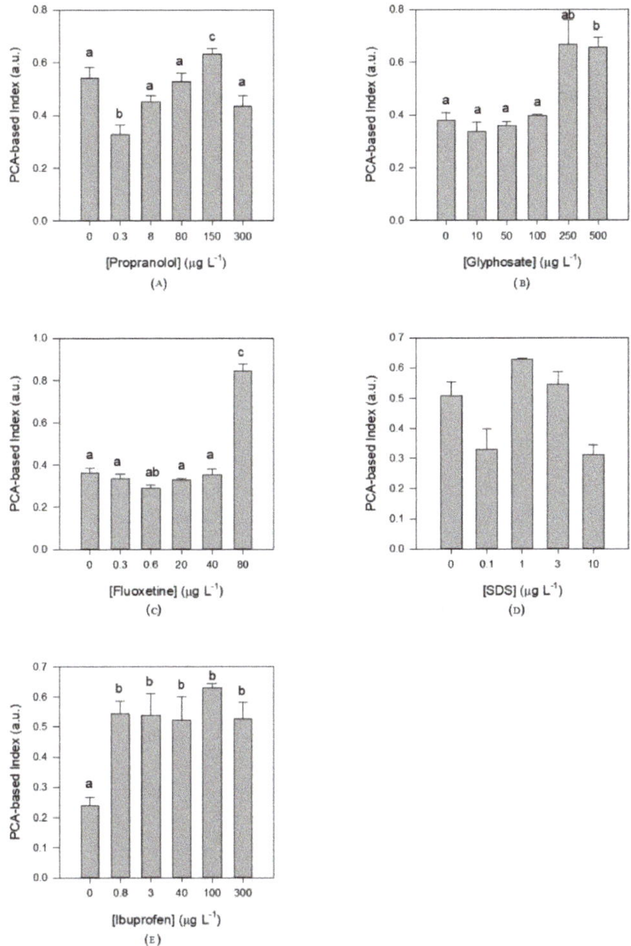

Figure 3. Principal component analysis (PCA)-based index of *Phaeodactylum tricornutum* cells exposed to different contaminants and concentrations. (**A**) Propranolol. (**B**) Glyphosate. (**C**) Fluoxetine. (**D**) SDS. (**E**) Ibuprofen. Average ± standard deviation, n = 3, different letters indicate significant differences at $p < 0.05$.

For the propranolol exposure gradient, the BAI index provided similar results as the abovementioned PCA-based index, where a significant low index value is attributed to the minimum exogenous concentration of 0.3 μg L^{-1}, and a significant index value increase was registered under the application of 150 μg L^{-1} of propranolol, underlining a positive trend for dose–response (Figure 4A). Regarding glyphosate, BAI was unsuccessful to provide significant values for the evaluation of antioxidant stress response under this contaminant exposure gradient (Figure 4B). However, a slight increase in the index values was followed by an increase in the contaminant concentration. Considering fluoxetine, BAI index values were only found to be considerably higher in the cells exposed to 80 µg L^{-1}, the maximum contaminant concentration applied (Figure 4C). For the SDS detergent, a significantly higher index value is obtained in the cells exposed to 1 µg L^{-1} SDS, whereas an inverse trend is noticed for the maximum exogenous concentration, with significantly lower index scores, suggesting a noteworthy inhibition of the activity levels of the antioxidant enzymes in the cells exposed to 10 µg L^{-1} SDS (Figure 4D). Applied to ibuprofen exposure data, the BAI index classifies the response to contaminant concentration gradient similarly to the PCA-based index, attributing substantial and approximately high scores for cells exposed to 0.8 µg L^{-1} ibuprofen and all the concentrations above, without the suggestion of a dose–response trend (Figure 4E).

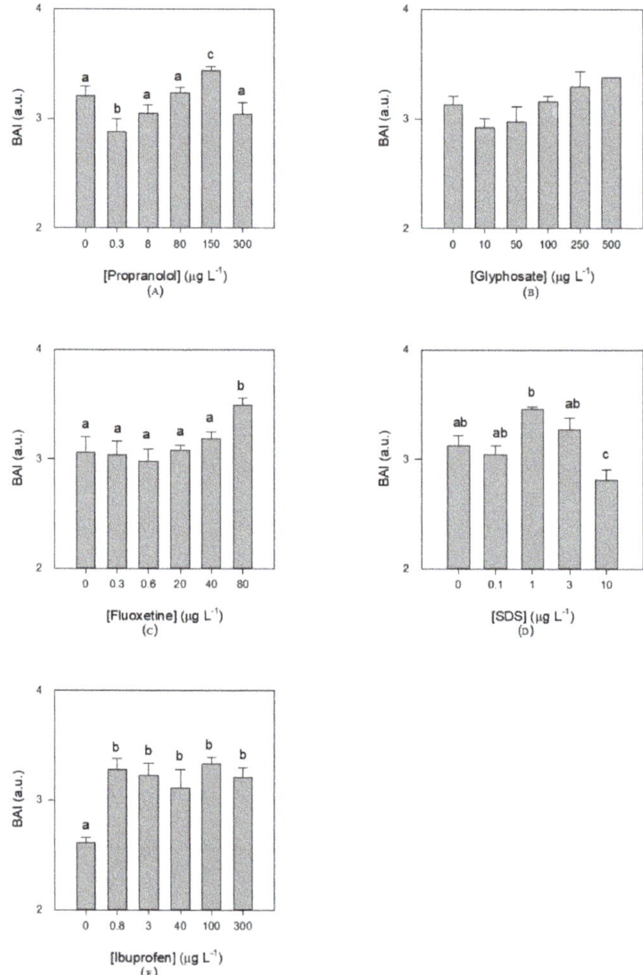

Figure 4. Bioeffect assessment index (BAI) of *Phaeodactylum tricornutum* cells exposed to different contaminants and concentrations. (**A**) Propranolol. (**B**) Glyphosate. (**C**) Fluoxetine. (**D**) SDS. (**E**) Ibuprofen. Average ± standard deviation, n = 3, different letters indicate significant differences at $p < 0.05$.

In order to provide a comparison of the obtained index results, a visual exploration of the correlations between the exogenous contaminant concentrations and the suitability of the subjected indexes is presented in Figure 5, shown as cluster heat maps. These clusters display the statistical significances of the associations between the considered variables and allow the evaluation of dataset quality. When comparing the IBR values with the exogenous concentration a significant positive dose–response correlation was found in the cells exposed to propranolol, ibuprofen, glyphosate and SDS, while for fluoxetine the inverse trend could be observed (Figure 5). If the same approach is performed for the PCA-based index it is possible to observe the values of this index in the cells exposed to ibuprofen and glyphosate, present an inverse significant dose–response relationship with the exogenous dose for these compounds. As for the BAI values, these showed only significant correlations with the exogenous dose applied in the cells exposed to fluoxetine and glyphoshate. It is also noticeable that, in most cases, the majority of the indexes also present significant correlations between them.

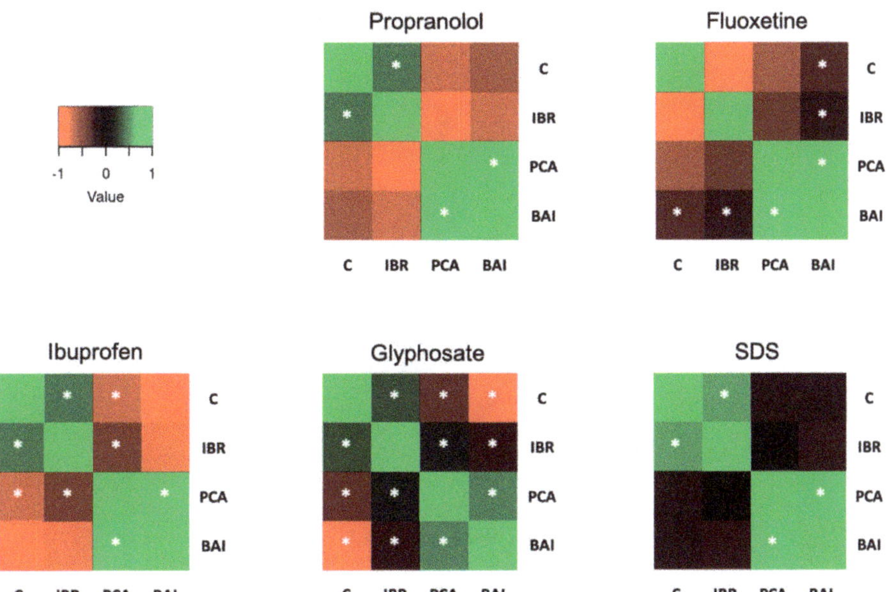

Figure 5. Spearman correlations heatmaps between the calculated index values and the applied exogenous concentrations for each applied compound. (*) denote significant differences at $p < 0.05$ for all tested compounds. Indexes include the integrated biomarker response (IBR) index; the principal component analysis (PCA)-based index; and the bioeffect assessment index (BAI), whilst (C) refers to the gradient of exposure concentration.

4. Discussion

In the present work, selected biomarker data from five ecotoxicological contributions on emerging contaminants (three pharmaceuticals, one detergent and one herbicide), were subjected to integrated response analyses using three methods, the integrated biomarker response (IBR) [5], the principal components analysis (PCA)-based index and the bioeffect assessment index (BAI) [19], to devise which could be more suitable in delivering an accurate measurement on the physiological stress response of diatom cell cultures under pollutant exposure.

Greater emphasis on evaluating the suitability of the already existing indexes prior to developing new ones is essential [16]. The set of biomarkers should be selected with flexibility considering cell function, biological level and pollutant affinity, in order to calculate relevant integrated responses of organisms to different contamination compounds, given that certain marker responses could emphasize certain functions or systems, while other levels of sensitivity could be at stake. Additionally, attention to the proper interpretation of biomarker responses integrating the index construction as parameters, and later, interpreting the derived index scores, is paramount [35]. In sum, a representative index is able to expose toxically induced response trends in natural systems which, therefore, are relevant to policy resolutions.

Previous contributions concerning aquatic environmental contamination, based their analysis on biomarker responses, proving that the adverse effects resulting from a wide variety of pollutants can be evaluated based on an array of specific and general exposure biomarkers, such as oxidative stress biomarkers, exhibiting sensitivity and accuracy in developing valid results [6,43], which are essential for the correct computation of the indexes [35]. Here, the efficiency of the applied set of biomarkers was determined by CAP (canonical analysis of principal coordinates) multivariate statistical approach. The antioxidant enzymes CAT, SOD and APX, and the non-enzymatic TBARS oxidative stress biomarkers, involved in the stress response physiological processes of a cosmopolitan marine diatom species, *Phaeodactylum tricornutum*, revealed an overall suitability for characterizing the different exogenous contaminants and the applied concentration gradients. This set of biomarkers produced statistically significant dose–response relations for four out of the five pollutants, fitting the experimental purpose, as determined by CAP efficiency results.

The IBR index is one of the most widely used and proficient indexes for evaluating good ecological status and contamination levels of sites and in fact, available literature on biological effects following chemical exposure integrate the IBR index as an important tool for contamination assessment [5–7,35,44–47]. IBR appears as a useful tool to mark pollution in field assessments, as it reflects contaminant levels at sites, despite the variability of biomarkers used for its calculation. Field and biomonitoring studies, concerning environmental pollution, rely on IBR for result support [48–52]. Fewer laboratory studies testify to IBR's applicability within the scope of treatment comparisons [47,53].

One of the premises of the IBR index is the "user judgment step", as one should decide the weight of each biomarker for the index construction a priori, as it may positively or negatively influence the index, depending on an induction or an inhibition in terms of physiological response to contamination. In the present work, the decision was based on the statistical relation of each biomarker alongside a gradient of contaminant concentrations [5]. This is the only empirical step of the IBR index, and can also be its weakness, putting at stake the index's reliability as it is highly dependent on a user's expert judgement and on data availability. As was previously observed, the same biomarker can have different responses depending on the species and on the tested compound and thus this attribution of a positive or negative influence has to be greatly based on experimental data [54,55]. As the IBR index integrates multiple biomarker scores through standardization of diverse biological responses, it enables direct comparisons among contaminants and performs suitable scores for the set of selected biomarkers. In the end, the complex collected data are presented within a scoring framework, where high values represent environmental stress and low values represent a lack of stress or inhibition. The results of IBR data analysis in this study were consistent and reflected the overall oxidative stress conditions induced in *P. tricornutum* cells by the exogenous contaminants, translating the expected response of the cell physiological mechanisms to overcome stressful contaminated conditions. Higher index values were attributed to higher pollutant concentrations, which conforms with the general assumption of IBR, revealing statistically significant dose-response trends for four out of the five harmful compounds examined. Our considerations on IBR's advantages and disadvantages must consider some factors: (i) a relatively large set of biomarkers is recommended as the higher the number of biomarkers, the more significant the index becomes [35,44]; (ii) it is recommended to classify which markers can be combined for IBR, differentiating effect and defense responses [35]; (iii) to compare spatial and temporal series of datasets, new data should be normalized and incorporated together with previously obtained numerical values, since the numerical values of IBR are comparable only within each dataset, [35]; (iv) IBR values must be interpreted in combination with individual biomarker scores [7,47]; and (v) IBR is appropriate for qualitative assessments [50]. In this context, IBR constitutes a robust and practical tool to assess the susceptibility of *P. tricornutum* cells to contaminants using multiple biomarker responses.

The BAI index came into existence as a modification of the "health assessment index" (HAI) [56], intending to assess contamination levels for coastal areas. It integrates only biomarkers of general toxicity, leading to a more holistic approach since the use of biomarkers that are non-specific to toxic effects allow the index to respond to a wider set of chemicals [19]. This is a valuable characteristic for an environmental health index, as the anthropogenic sources of contamination comprise a complex mixture of pollutants and tend to show an exponential worldwide increase [16]. The BAI index has confirmed its suitability to address alterations at different bio-organizational levels, reflecting general toxicity patterns and revealing health status under polluted conditions, capturing degradative and restorative information on routine monitoring programs and allowing comparison between geographically distinct areas with differing contamination conditions [6,19,34,35]. The BAI index, a univariate statistical approach, transforms complex alterations of biomarkers into single classes of values, losing some of the original data variability, so the combination of BAI analysis with other integrative index approaches is frequently applied for assessment protocols [6,45]. The BAI index revealed low sensitivity against our data set, confirming its scope of applicability as more adequate for general toxicity biomarkers and parameters of non-specific stress, as suggested in the literature [35], while present results were constructed from specific exposure oxidative stress biomarkers. The BAI index exhibited a significant correlation to the PCA-based index, both indexes suggesting similar stress patterns under the application of all contaminant gradients. On the other hand, comparisons between the BAI index and the IBR index reflected the dissimilarities of these index productions, as, generally, IBR will be based on parameters of general health biomarkers or complemented with specific biomarkers, whereas the BAI takes only general toxicity markers [35]. Still, the BAI results support an overall significantly negative relation between index scores and pollutant concentration gradients, confirmed particularly for fluoxetine and glyphosate.

The PCA is a multivariate statistical technique, familiar in terms of the identification of environmental pollution patterns and in the discrimination of the main variables responsible for the variance of chemical accumulation in organisms and consequential biological effects and, thus, is largely applied for the development of interpretative models of pollution data for decision makers [48,57–59]. PCA is an effective tool when applied to high-dimensional data sets, reducing their dimensionality while identifying the patterns between the inter-related variables, retaining as much as possible the variability of the original data set [60,61]. Additionally, PCA demonstrates applicability for index construction, selecting the appropriate measured parameters to integrate the index and identifying the main axes of variance within a dataset [60,62,63]. Improved interpretation on other indexes results, such as the IBR index, may be supported by the application of PCA analysis [48]. A correct PCA procedure assumes: (i) large complex datasets; (ii) the normalization and/or standardization of the original data should be considered; (iii) biomarker weigh and influence are set solely based on statistical results; and (iv) patterns are suggested without reference to prior knowledge about the samples; thus, it is also recommended for scaling to not be adjusted to match prior knowledge of the data. Nevertheless, limitations on PCA applicability should be considered on output interpretation, which are, according to Lever et al. (2017) [63]: (a) the assumption of a linear underlying structure of the data, (b) highly correlated patterns may be undetermined due to the resultant uncorrelated principal components (axes of variance), and (c) maximizing variance is the objective rather than clustering results. When applied to our dataset, PCA satisfactorily delivered a significant negative correlation for the overall experiment relative to compound concentration and index scores, particularly for ibuprofen and glyphosate gradients. In fact, the observed trend was identical to using either the PCA or BAI indexes, which were statistically positively correlated. PCA relies on data variance to create next step assumptions, and a large dataset delivers better results with this statistical method.

5. Conclusions

This study provides information about the suitability and reliability of three biological indexes in pollution monitoring research. An adequate index provides an integrative approach, constructed with the integration of multi-biomarker response data, to better clarify the links between contaminant exposure and biological effects in aquatic systems. The potential use of stress biomarkers in biological effects studies is highlighted.

Overall, the three indexes addressed in the present study (IBR, PCA and BAI) could be successfully applied for the evaluation of environmental quality, given that all of them depicted the response triggered by the tested pollutants under varying concentration gradients, reflecting disruption in the cell physiological processes by contamination, and providing a simplified output of the biomarker data despite their varying sensitivities and resolutions. IBR performed the best, successfully reflecting the expected level of exposure, thus revealing the stress patterns exposed by the set of specific oxidative stress markers here selected, under a contaminant gradient, corroborating it as a suitable tool for the assessment of the biological effects of pollution in marine organisms. IBR combines multi-biomarker responses, delivers a qualitative approach and provides temporal assessment of contamination patterns.

Open access publications examining the applicability of the already available tools for pollution assessment, such as the present, improve scientific communication and efficiency within the research community, sustaining result communication to the general public and bringing great advantages for management deliberations.

Supplementary Materials: Supplementary materials can be found at https://www.mdpi.com/article/10.3390/app11093878/s1.

Author Contributions: Conceptualization, B.D. and V.F.F.; formal analysis, B.D. and V.F.F.; writing—original draft preparation, V.L.P.; writing—review and editing, B.D., V.F.F., S.C.N. and M.F.L.L.; project administration, B.D.; funding acquisition, B.D. All authors have read and agreed to the published version of the manuscript.

Funding: The authors would like to thank Fundação para a Ciência e a Tecnologia (FCT) for funding the research via project grants PTDC/MAR-EST/3048/2014 (BIOPHARMA), PTDC/CTA-AMB/30056/2017 (OPTOX), UIDB/04292/2020 and UID/MULTI/04046/2019. Work was also funded by the Integrated Programme of SR&TD SmartBioR (reference Centro-01-0145-FEDER-000018), co-funded by Centro 2020 program, Portugal 2020, European Union, through the European Regional Development Fund. B. Duarte and V. F. Fonseca were supported by researcher contracts (CEECIND/00511/2017 and DL57/2016/CP1479/CT0024).

Institutional Review Board Statement: Not applicable.

Informed Consent Statement: Not applicable.

Data Availability Statement: Data available upon request.

Conflicts of Interest: The authors declare no conflict of interest. The funders had no role in the design of the study; in the collection, analyses, or interpretation of data; in the writing of the manuscript, or in the decision to publish the results.

References

1. Asimakopoulos, A.; Bletsou, A.; Thomaidis, N. Emerging contaminants: A tutorial mini-review. *Glob. Nest J.* **2012**, *14*, 72–79.
2. EUROSTAT. EUROSTAT. Available online: https://ec.europa.eu/eurostat/statisticsexplained/index.php?title=Archive:Chemicals_production_statistics&oldid=199190 (accessed on 11 February 2021).
3. Duarte, B.; Caçador, I. *Ecotoxicology of Marine Organisms*, 1st ed.; Duarte, B., Caçador, I., Eds.; CRC Press: Boca Raton, FL, USA; Taylor & Francis Group: Abingdon-on-Thames, UK, 2019; ISBN 1138035491.
4. Lopes, D.G.; Duarte, I.A.; Antunes, M.; Fonseca, V.F. Effects of antidepressants in the reproduction of aquatic organisms: A meta-analysis. *Aquat. Toxicol.* **2020**, *227*, 105569. [CrossRef]

5. Beliaeff, B.; Burgeot, T. Integrated biomarker response: A useful tool for ecological risk assessment. In *Proceedings of the Environmental Toxicology and Chemistry*; SETAC Press: Pensacola, FL, USA, 2002; Volume 21, pp. 1316–1322.

6. Schettino, T.; Caricato, R.; Calisi, A.; Giordano, M.E.; Lionetto, M.G. Biomarker Approach in Marine Monitoring and Assessment: New Insights and Perspectives. *Open Environ. Sci.* **2012**, *12*. [CrossRef]

7. Duarte, I.A.; Reis-Santos, P.; França, S.; Cabral, H.; Fonseca, V.F. Biomarker responses to environmental contamination in estuaries: A comparative multi-taxa approach. *Aquat. Toxicol.* **2017**, *189*, 31–41. [CrossRef]

8. European Comission. *European Comission Marine Strategy Framework Directive (MSFD)*; Europen Comision: Luxembourg, 2017.

9. WHO. *Biomarkers and Risk Assessment: Concepts and Principles, Environmental Health Criteria*; WHO: Geneva, Switzerland, 1993.

10. Valavanidis, A.; Vlahogianni, T.; Dassenakis, M.; Scoullos, M. Molecular biomarkers of oxidative stress in aquatic organisms in relation to toxic environmental pollutants. *Ecotoxicol. Environ. Saf.* **2006**, *64*, 178–189. [CrossRef]

11. Livingstone, D.R. Contaminant-stimulated reactive oxygen species production and oxidative damage in aquatic organisms. *Mar. Pollut. Bull.* **2001**, *42*, 656–666.

12. Draper, H.H.; Squires, E.J.; Mahmoodi, H.; Wu, J.; Agarwal, S.; Hadley, M. A comparative evaluation of thiobarbituric acid methods for the determination of malondialdehyde in biological materials. *Free Radic. Biol. Med.* **1993**, *15*, 353–363. [CrossRef]

13. Anjum, N.A.; Duarte, B.; Caçador, I.; Sleimi, N.; Duarte, A.C.; Pereira, E. Biophysical and biochemical markers of metal/metalloid-impacts in salt marsh halophytes and their implications. *Front. Environ. Sci.* **2016**, *4*, 24.

14. Borja, A.; Bricker, S.B.; Dauer, D.M.; Demetriades, N.T.; Ferreira, J.G.; Forbes, A.T.; Hutchings, P.; Jia, X.; Kenchington, R.; Marques, J.C.; et al. Overview of integrative tools and methods in assessing ecological integrity in estuarine and coastal systems worldwide. *Mar. Pollut. Bull.* **2008**, *56*, 1519–1537. [CrossRef]

15. De Carvalho, R.C.; Feijão, E.; Matos, A.R.; Cabrita, M.T.; Novais, S.C.; Lemos, M.F.L.; Caçador, I.; Marques, J.C.; Reis-Santos, P.; Fonseca, V.F.; et al. Glyphosate-based herbicide toxicophenomics in marine diatoms: Impacts on primary production and physiological fitness. *Appl. Sci.* **2020**, *10*, 7391. [CrossRef]

16. Borja, A.; Dauer, D.M. Assessing the environmental quality status in estuarine and coastal systems: Comparing methodologies and indices. *Ecol. Indic.* **2008**, *8*, 331–337. [CrossRef]

17. Beck, M.W.; O'Hara, C.; Lowndes, J.S.S.; Ma-Zor, R.D.; Theroux, S.; Gillett, D.J.; Lane, B.; Gearheart, G. The importance of open science for biological assessment of aquatic environments. *PeerJ* **2020**, *8*, 1–27. [CrossRef]

18. Cabral, H.N.; Fonseca, V.F.; Gamito, R.; Gonalves, C.I.; Costa, J.L.; Erzini, K.; Gonalves, J.; Martins, J.; Leite, L.; Andrade, J.P.; et al. Ecological quality assessment of transitional waters based on fish assemblages in Portuguese estuaries: The Estuarine Fish Assessment Index (EFAI). *Ecol. Indic.* **2012**, *19*, 144–153. [CrossRef]

19. Broeg, K.; Westernhagen, H.V.; Zander, S.; Körting, W.; Koehler, A. The "bioeffect assessment index" (BAI): A concept for the quantification of effects of marine pollution by an integrated biomarker approach. *Mar. Pollut. Bull.* **2005**, *50*, 495–503. [CrossRef]

20. Duarte, B.; Freitas, J.; Caçador, I. Sediment microbial activities and physic-chemistry as progress indicators of salt marsh restoration processes. *Ecol. Indic.* **2012**, *19*, 231–239. [CrossRef]

21. Bastida, F.; Luis Moreno, J.; Teresa, H.; García, C. Microbiological degradation index of soils in a semiarid climate. *Soil Biol. Biochem.* **2006**, *38*, 3463–3473. [CrossRef]

22. Duarte, B.; Feijão, E.; de Carvalho, R.C.; Duarte, I.A.; Silva, M.; Matos, A.R.; Cabrita, M.T.; Novais, S.C.; Lemos, M.F.L.; Marques, J.C.; et al. Effects of propranolol on growth, lipids and energy metabolism and oxidative stress response of *Phaeodactylum tricornutum*. *Biology* **2020**, *9*, 478. [CrossRef] [PubMed]

23. Feijão, E.; Cruz de Carvalho, R.; Duarte, I.A.; Matos, A.R.; Cabrita, M.T.; Novais, S.C.; Lemos, M.F.L.; Caçador, I.; Marques, J.C.; Reis-Santos, P.; et al. Fluoxetine Arrests Growth of the Model Diatom *Phaeodactylum tricornutum* by Increasing Oxidative Stress and Altering Energetic and Lipid Metabolism. *Front. Microbiol.* **2020**, *11*, 1803. [CrossRef]

24. Silva, M.; Feijão, E.; da Cruz de Carvalho, R.; Duarte, I.A.; Matos, A.R.; Cabrita, M.T.; Barreiro, A.; Lemos, M.F.L.; Novais, S.C.; Marques, J.C.; et al. Comfortably numb: Ecotoxity of the non-steroidal anti-inflammatory drug ibuprofen on *Phaeodactylum tricornutum*. *Mar. Environ. Res.* **2020**, *161*. [CrossRef]

25. Silva, M.; Feijão, E.; Carvalho, R.C.; Matos, A.R.; Cabrita, M.T.; Lemos, M.F.L.; Novais, S.C.; Marques, J.C.; Caçador, I.; Reis-Santos, P.; et al. Ecotoxity of the detergent SDS on *Phaeodactylum tricornutum* primary productivity, fatty acid metabolism and oxidative stress biomarkers. 2021; preprint.

26. Guillard, R.R.; Ryther, J.H. Studies of marine planktonic diatoms. I. Cyclotella nana Hustedt, and Detonula confervacea (cleve) Gran. *Can. J. Microbiol.* **1962**, *8*, 229–239. [CrossRef] [PubMed]

27. Feijão, E.; Gameiro, C.; Franzitta, M.; Duarte, B.; Caçador, I.; Cabrita, M.T.; Matos, A.R. Heat wave impacts on the model diatom *Phaeodactylum tricornutum*: Searching for photochemical and fatty acid biomarkers of thermal stress. *Ecol. Indic.* **2018**, *95*, 1026–1037. [CrossRef]

28. OECD. OECD guidelines for the testing of chemicals. Freshwater alga and cyanobacteria, growth inhibition test. *Organ. Econ. Coop. Dev.* **2011**, 1–25. [CrossRef]

29. Franzellitti, S.; Buratti, S.; Du, B.; Haddad, S.P.; Chambliss, C.K.; Brooks, B.W.; Fabbri, E. A multibiomarker approach to explore interactive effects of propranolol and fluoxetine in marine mussels. *Environ. Pollut.* **2015**, *205*, 60–69. [CrossRef] [PubMed]

30. Claessens, M.; Vanhaecke, L.; Wille, K.; Janssen, C.R. Emerging contaminants in Belgian marine waters: Single toxicant and mixture risks of pharmaceuticals. *Mar. Pollut. Bull.* **2013**, *71*, 41–50. [CrossRef] [PubMed]

31. Teranishi, Y.; Tanaka, A.; Osumi, M.; Fukui, S. Catalase activities of hydrocarbon-utilizing candida yeasts. *Agric. Biol. Chem.* **1974**, *38*, 1213–1220. [CrossRef]

32. Tiryakioglu, M.; Eker, S.; Ozkutlu, F.; Husted, S.; Cakmak, I. Antioxidant defense system and cadmium uptake in barley genotypes differing in cadmium tolerance. *J. Trace Elem. Med. Biol.* **2006**, *20*, 181–189. [CrossRef]

33. Marklund, S.; Marklund, G. Involvement of the Superoxide Anion Radical in the Autoxidation of Pyrogallol and a Convenient Assay for Superoxide Dismutase. *Eur. J. Biochem.* **1974**, *47*, 469–474. [CrossRef]

34. Heath, R.L.; Packer, L. Photoperoxidation in isolated chloroplasts. I. Kinetics and stoichiometry of fatty acid peroxidation. *Arch. Biochem. Biophys.* **1968**, *125*, 189–198. [CrossRef]

35. Broeg, K.; Lehtonen, K.K. Indices for the assessment of environmental pollution of the Baltic Sea coasts: Integrated assessment of a multi-biomarker approach. *Mar. Pollut. Bull.* **2006**, *53*, 508–522. [CrossRef]

36. Sinha, S.; Masto, R.E.; Ram, L.C.; Selvi, V.A.; Srivastava, N.K.; Tripathi, R.C.; George, J. Rhizosphere soil microbial index of tree species in a coal mining ecosystem. *Soil Biol. Biochem.* **2009**, *41*, 1824–1832. [CrossRef]

37. Clarke, K.; RN, G. Primer v6: User Manual/Tutorial. 2006. Available online: https://www.scienceopen.com/document?vid=87ec9d6d-cae1-4f49-9204-0fbf723dcdbc (accessed on 3 March 2021).

38. Masto, R.E.; Chhonkar, P.K.; Singh, D.; Patra, A.K. Alternative soil quality indices for evaluating the effect of intensive cropping, fertilisation and manuring for 31 years in the semi-arid soils of India. *Environ. Monit. Assess.* **2008**, *136*, 419–435. [CrossRef] [PubMed]

39. Cabrita, M.T.; Duarte, B.; Gameiro, C.; Godinho, R.M.; Caçador, I. Photochemical features and trace element substituted chlorophylls as early detection biomarkers of metal exposure in the model diatom *Phaeodactylum tricornutum. Ecol. Indic.* **2018**, *95*, 1038–1052. [CrossRef]

40. Duarte, B.; Cabrita, M.T.; Vidal, T.; Pereira, J.L.; Pacheco, M.; Pereira, P.; Canário, J.; Gonçalves, F.J.M.; Matos, A.R.; Rosa, R.; et al. Phytoplankton community-level bio-optical assessment in a naturally mercury contaminated Antarctic ecosystem (Deception Island). *Mar. Environ. Res.* **2018**, *140*, 412–421. [CrossRef] [PubMed]

41. Duarte, B.; Prata, D.; Matos, A.R.; Cabrita, M.T.; Caçador, I.; Marques, J.C.; Cabral, H.N.; Reis-Santos, P.; Fonseca, V.F. Ecotoxicity of the lipid-lowering drug bezafibrate on the bioenergetics and lipid metabolism of the diatom *Phaeodactylum tricornutum. Sci. Total Environ.* **2019**, *650*, 2085–2094. [CrossRef] [PubMed]

42. Duarte, B.; Pedro, S.; Marques, J.C.; Adão, H.; Caçador, I. *Zostera noltii* development probing using chlorophyll a transient analysis (JIP-test) under field conditions: Integrating physiological insights into a photochemical stress index. *Ecol. Indic.* **2017**, *76*, 219–229. [CrossRef]

43. Kaviraj, A.; Unlu, E.; Gupta, A.; El Nemr, A. Biomarkers of environmental pollutants. *Biomed Res. Int.* **2014**, *2014*, 806598. [CrossRef]

44. Devin, S.; Burgeot, T.; Giambérini, L.; Minguez, L.; Pain-Devin, S. The integrated biomarker response revisited: Optimization to avoid misuse. *Environ. Sci. Pollut. Res.* **2014**, *21*, 2448–2454. [CrossRef]

45. Meng, F.-P.; Yang, F.-F.; Cheng, F.-L. Marine environmental assessment approaches based on biomarker index: A review. *J. Appl. Ecol.* **2012**, *23*, 1128–1136.

46. Marigómez, I.; Garmendia, L.; Soto, M.; Orbea, A.; Izagirre, U.; Cajaraville, M.P. Marine ecosystem health status assessment through integrative biomarker indices: A comparative study after the Prestige oil spill "mussel Watch.". *Ecotoxicology* **2013**, *22*, 486–505. [CrossRef]

47. Novais, S.C.; Gomes, N.C.; Soares, A.M.V.M.; Amorim, M.J.B. Antioxidant and neurotoxicity markers in the model organism *Enchytraeus albidus* (Oligochaeta): Mechanisms of response to atrazine, dimethoate and carbendazim. *Ecotoxicology* **2014**, *23*, 1220–1233. [CrossRef]

48. Brooks, S.J.; Escudero-Oñate, C.; Gomes, T.; Ferrando-Climent, L. An integrative biological effects assessment of a mine discharge into a Norwegian fjord using field transplanted mussels. *Sci. Total Environ.* **2018**, *644*, 1056–1069. [CrossRef] [PubMed]

49. Serafim, A.; Company, R.; Lopes, B.; Fonseca, V.F.; Frana, S.; Vasconcelos, R.P.; Bebianno, M.J.; Cabral, H.N. Application of an integrated biomarker response index (IBR) to assess temporal variation of environmental quality in two Portuguese aquatic systems. *Ecol. Indic.* **2012**, *19*, 215–225. [CrossRef]

50. Raftopoulou, E.K.; Dimitriadis, V.K. Assessment of the health status of mussels *Mytilus galloprovincialis* along Thermaikos Gulf (Northern Greece): An integrative biomarker approach using ecosystem health indices. *Ecotoxicol. Environ. Saf.* **2010**, *73*, 1580–1587. [CrossRef]

51. Felício, A.A.; Freitas, J.S.; Scarin, J.B.; de Souza Ondei, L.; Teresa, F.B.; Schlenk, D.; de Almeida, E.A. Isolated and mixed effects of diuron and its metabolites on biotransformation enzymes and oxidative stress response of Nile tilapia (*Oreochromis niloticus*). *Ecotoxicol. Environ. Saf.* **2018**, *149*, 248–256. [CrossRef]

52. Rodrigues, A.P.; OlivaeTeles, T.; Mesquita, S.R.; DelerueeMatos, C.; Guimarães, L. Integrated biomarker responses of an estuarine invertebrate to high abiotic stress and decreased metal contamination. *Mar. Environ. Res.* **2014**, *101*, 101–114. [CrossRef] [PubMed]

53. Silva, C.S.E.; Novais, S.C.; Simões, T.; Caramalho, M.; Gravato, C.; Rodrigues, M.J.; Maranhão, P.; Lemos, M.F.L. Using biomarkers to address the impacts of pollution on limpets (*Patella depressa*) and their mechanisms to cope with stress. *Ecol. Indic.* **2018**, *95*, 1077–1086. [CrossRef]

54. Duarte, B.; Carreiras, J.; Pérez-Romero, J.A.; Mateos-Naranjo, E.; Redondo-Gómez, S.; Matos, A.R.; Marques, J.C.; Caçador, I. Halophyte fatty acids as biomarkers of anthropogenic-driven contamination in Mediterranean marshes: Sentinel species survey and development of an integrated biomarker response (IBR) index. *Ecol. Indic.* **2018**, *87*, 86–96. [CrossRef]

55. Duarte, B.; Durante, L.; Marques, J.C.; Reis-Santos, P.; Fonseca, V.F.; Caçador, I. Development of a toxicophenomic index for trace element ecotoxicity tests using the halophyte *Juncus acutus*: Juncus-TOX. *Ecol. Indic.* **2021**, *121*, 107097. [CrossRef]

56. Adams, S.M.; Brown, A.M.; Goede, R.W. A Quantitative Health Assessment Index for Rapid Evaluation of Fish Condition in the Field. *Trans. Am. Fish. Soc.* **1993**, *122*. [CrossRef]

57. Clarke, K.R. Nonmetric multivariate analysis in community-level ecotoxicology. *Environ. Toxicol. Chem.* **1999**, *18*, 118–127. [CrossRef]

58. Jollands, N.; Lermit, J.; Patterson, M. Aggregate eco-efficiency indices for New Zealand—A principal components analysis. *J. Environ. Manag.* **2004**, *73*, 293–305. [CrossRef]

59. Isibor, P.O.; Imoobe, T.O.T.; Enuneku, A.A.; Akinduti, P.A.; Dedeke, G.A.; Adagunodo, T.A.; Obafemi, D.Y. Principal Components and Hierarchical Cluster Analyses of Trace Metals and Total Hydrocarbons in Gills, Intestines and Muscles of Clarias gariepinus (Burchell, 1822). *Sci. Rep.* **2020**, *10*, 1–15. [CrossRef] [PubMed]

60. Shin, P.K.S.; Lam, W.K.C. Development of a Marine Sediment Pollution Index. *Environ. Pollut.* **2001**, *113*, 281–291. [CrossRef]

61. Reid, M.K.; Spencer, K.L. Use of principal components analysis (PCA) on estuarine sediment datasets: The effect of data pre-treatment. *Environ. Pollut.* **2009**, *157*, 2275–2281. [CrossRef] [PubMed]

62. Caçador, I.; Neto, J.M.; Duarte, B.; Barroso, D.V.; Pinto, M.; Marques, J.C. Development of an Angiosperm Quality Assessment Index (AQuA-Index) for ecological quality evaluation of Portuguese water bodies—A multi-metric approach. *Ecol. Indic.* **2013**, *25*, 141–148. [CrossRef]

63. Lever, J.; Krzywinski, M.; Altman, N. Principal component analysis. *Nat. Methods* **2017**, *14*, 641–642. [CrossRef]

applied sciences

Article

Human Health Impact Analysis of Contaminant in IoT-Enabled Water Distributed Networks

Essa Q. Shahra [1,†], Wenyan Wu [1,*,†] and Roberto Gomez [2]

[1] Engineering Department, Faculty of Computing, Engineering and Built Environment, Birmingham City University, Birmingham B4 7XG, UK; Essa.Shahra@bcu.ac.uk

[2] Urban Water Networks Department, TPF Getinsa Eurostudios, 28028 Madrid, Spain; Roberto.Gomez@tpfingenieria.com

* Correspondence: Wenyan.Wu@bcu.ac.uk; Tel.: +44-121-3004291

† These authors contributed equally to this work.

Abstract: This paper aims to assess and analyze the health impact of consuming contaminated drinking water in a water distributed system (WDS). The analysis was based on qualitative simulation performed in two different models named hydraulic and water quality in a WDS. The computation focuses on quantitative analysis for chemically contaminated water impacts by analyzing the dose level in various locations in the water network and the mass of the substance that entered the human body. Several numerical experiments have been applied to evaluate the impact of water pollution on human life. They analyzed the impact on human life according to various factors, including the location of the injected node (pollution occurrence) and the ingested dose level. The results show a significant impact of water contaminant on human life in multiple areas in the water network, and the level of this impact changed from one location to another in WDSs based on several factors such as the location of the pollution occurrence, the contaminant concentration, and the dose level. In order to reduce the impact of this contaminant, water quality sensors have been used and deployed on the water network to help detect this contaminant. The sensors were optimally deployed based on the time-detection of water contamination and the volume of polluted water consumed. Numerical experiments were carried out to compare water pollution's impact with and without using water quality sensors. The results show that the health impact was reduced by up to 98.37% by using water quality sensors.

Keywords: human health analysis; water contamination; water quality model; hydraulic model; WDS; water quality sensors

Citation: Shahra, E.Q.; Wu, W.; Gomez, R. Human Health Impact Analysis of Contaminant in IoT-Enabled Water Distributed Networks. *Appl. Sci.* **2021**, *11*, 3394. https://doi.org/10.3390/app11083394

Academic Editor: Elida Nora Ferri

Received: 23 February 2021
Accepted: 6 April 2021
Published: 10 April 2021

Publisher's Note: MDPI stays neutral with regard to jurisdictional claims in published maps and institutional affiliations.

1. Introduction

One of the fundamental human rights of all human beings, regardless of their race, nationality, religion, and beliefs, is to have pure and safe drinking water [1]. At a fundamental level, daily consumption requires a minimum amount of water for survival, and thus access to a particular form of water is necessary for life [2]. Water has a much more significant impact on health and well-being. However, concerns in many subjects, such as the quality and quantity of water supplied, are essential in determining the health of individuals and entire communities [3]. Water utilities' fundamental duties are to provide and distribute water to customers of an acceptable quality, quantity, and pressure through the entire water distributed system (WDS) [4]. The usable water is at first treated to meet the drinking water quality standards before delivery [5]. Because of the spatial scale and the high complexity of water distributed networks, adverse accidental incidents commonly occur. For this unwanted event, the growth of infectious diseases begins with biological and chemical contaminants in drinking water [6]. The use of contaminated drinking water and poor sanitation is directly related to the transmission of diseases and epidemics, for example, diarrhea, polio, dysentery, cholera, hepatitis, gastric ulcers, pneumonia, and pulmonary disease [7]. Water supplies may contain various non-biological contaminants, including silica, sodium, sulfur, ammonia, and chlorine. In addition to these contaminants, microbiological contaminants in drinking water at the stage of consumption represent another problem of water quality. Poor drinking water quality significantly affects consumers' health [8]. In recent years, many developing countries have set public health goals to reduce waterborne diseases and develop safe water resources, and the situation has improved slightly. This somewhat improved situation can be further exacerbated by increased water demand and lower water availability due to population growth and economic growth. Understanding the factors that influence the quality of drinking water is also a necessary step in making the right decisions about protecting and managing the quality of drinking water [9]. In general, the quality of drinking water is affected by the variety of water sources, the treatment in predistribution water treatment plants, the water distribution system, and the containers or tanks used for water storage and household filters. However, in rural areas and small municipalities, drinking water is generally pumped directly from wells or rivers, lakes, and reservoirs without adequate treatment [10]. Therefore, the quality of water sources plays a fundamental role in determining the quality of drinking water.

Protecting water source quality, which plays a crucial role in providing safe drinking water, usually requires the combined efforts of many partners, such as water managers, industrial plants, water supply, communities, and the public [11]. Evaluation of individual water quality parameters may not give a good understanding of water quality between partners with different scientific backgrounds. On the other hand, the traditional method, even by water quality experts, can also cause subjective judgments and prejudices. To solve the problems associated with characterizing and interpreting the state of water quality, various indicators of water quality have been developed to convert the levels of water quality parameters into a numerical index integrated using mathematical tools [12]. Many factors are taken into account when calculating these water quality indicators, and the two most important factors are the time from the incidence of contamination to the water delivery to the customer and the load and concentration of pollution taken from the tap water by the user.

The main contributions of this paper are as follows: (1) Propose a health impact analysis model based on hydraulic and water quality models. (2) Analyze the impact of water contamination based on its occurrence locations and ingested dose level. (3) Reduce the contamination impact using water quality sensors that were optimally deployed based on the time of detection and volume of contaminated water.

The paper is organized as follows. Section 2 presents the selective related literature review. Section 3 explains the methodology and the case study. Section 4 shows the results and the discussion. Section 5 concludes the work.

2. Related Work

The use of a feasible and effective method of assessing the quality of drinking water is essential to achieve reliable results and to facilitate informed decision-making. Since the 1960s, when Horton developed the first water quality index (WQI), many methods of assessing water quality have been proposed [13–15]. The rapid development of water quality assessment methods improves our understanding of water quality either by a single numeric number or by a more sophisticated interpretation of water quality characteristics [16].

Abtahi et al. [11,17] proposed Drinking Water Quality Indicators (DWQIs) to understand the general status of drinking water quality in water networks. The DWQI method works to classify the variety of drinking water sources based on a comparison of the inputs measured values of the water quality parameters with the standard fundamental values (benchmark). The DWQI scores classify the quality of drinking water into five different categories: excellent, good, fair, marginal, and weak. The results showed that the DWQI is a stable, flexible, simple, and reliable indicator, and it could be used as an essential method for characterizing the quality of the drinking water supply. Likewise, Akter et al. [18] used the weighted arithmetic WQI approach to provide details on water quality in Bangladesh. The WQI method converts massive numbers of water quality variables into digital numbers, and it is useful for understanding the water quality, making it one the most common methods for measuring water quality with some limitations. Furthermore, Ba and McKenna [19] developed a time series monitoring and event detection method for water quality. This method composites an affine projection algorithm and an autoregressive (AR) model to predict the time series of water quality. In order to assess the existence of contamination events or not, they used the online change point detection methods for estimated residuals. Most recently, Braun et al. [20] studied the impact of demand uncertainties on the water age in a network model and applied polynomial chaos expansion (PCE) on two real networks. The results show the PCE performs considerably better for a fixed number of simulations, while perturbation approaches are not appropriate for this task. Moreover, the scale of the network has been shown to have a significant impact on the maximum water age.

Nonetheless, these approaches to evaluating water quality rely on water quality criteria for classifying water quality. Therefore, what matters is establishing water quality standards while taking into account specific regional local water quality conditions, the context of water quality indicators, and the health problem [21]. It would be difficult to fully understand the water quality without incorporating these issues into the water quality standards. In contrast, the rapid growth of methods to evaluate water quality is also very promising [22].

Contrary to the rapid development of water quality assessment strategies, assessing human health risks relies mainly on the model suggested by the United States Environment Protection Agency (USEPA) or updated USEPA models. However, the slow development of the evaluation model does not impede health risk assessment activities. Over the past three decades, increasingly more health risk assessment studies have been performed. Researchers slowly recognized the vital role that public health risk assessment plays in effective water quality control and consumer safety, instead of waterborne diseases. For example, Chen et al. [23] used the triangular Fuzzy Number Approach to consider inconsistencies in the health risk assessments provided by the USEPA. The results showed that the health risk assessment model is successful in quantifying risk uncertainty with lower complexity. Moreover, the findings in this paper help managers mitigate possible safety risks and provide new insight to address water uncertainties. Another study by Zhang et al. [24] where strontium (Sr) is detected in drinking water was studied through a monthly sampling strategy in twelve locations in Xi'an, Northwest China. A Monte Carlo simulation was carried out in a safety risk assessment for various age groups and exposure routes. The results of this study could be useful in developing possible risk control. In the study by Ali et al. [25], organophosphate pesticide concentrations (OPPs) were determined in drinking water samples obtained in the Peshawar Basin, Pakistan. The findings showed the highest contamination for methamidophos was detected among OPPs, while the lowest was found for dichlorvos. This study, therefore, recommends stopping the use of shallow drinking water and using deep groundwater at depths >98 m, which showed relatively less pollution. Joshi et al. [26] assessed the combination of drinking water quality and meteorological factors with the incidence of enteroviral diseases among children in seven Korean metropolitan provinces. They concluded that drinking water quality was one of the main determinants of enteroviral diseases.

Full awareness of the water quality status and the possible impacts on human health is the first step towards sustainable governance of water quality [27]. Adimalla and Li [28] assessed the quality of groundwater for consumption and irrigation and estimated the possible adverse impact on health for consumers in the rapidly urbanizing areas in Telangana State, India. Results showed that, due to the use of highly contaminated drinking water with nitrate and fluoride in the study region, children were more vulnerable to health risk. Kumar et al. [29] proposed a method using sensitivity analysis (SSA) to assess the relative impact of inputs in estimating the dose to safety. This method is useful in evaluating the ambiguity and vulnerability of parameters in the input health hazard framework. Shahra and Wu [30] proposed a sensor placement using an evaluations algorithm (EA). The proposed method uses multiple objectives—the volume of the contaminated water (VCW) and time detection (TD)—to find the optimal sensor placements. Two real case studies of WDNs were used to evaluate the effectiveness of the proposed method. The results showed the ability of the proposed method to maximize the coverage of the deployed sensors and reduce the time detection of the water contaminant. Ung et al. [31] used Monte Carlo and identification methods for sensor placement. It calculates a ranked list of potential contamination location nodes and contamination times using binary responses from sensors. The backtracking algorithm is used to determine the source identification. The locations of sensors that are best suited for allocating the source of contamination are determined using a greedy algorithm and a local search algorithm, all of which are combined with a Monte Carlo approach. The proposed method was tested using a real French water network with 2500 nodes. In another study, Hooshmand et al. [32] proved the quality of the two expectation-based models (IBM and IPM) to identify the appropriate configuration of the sensors. These expectation models are extended to include the value at risk (VaR), worse case, and conditional VaR measures (CVaR). Numerical experiments on a real network of WDNs have shown that the optimal sensor placement of CVaRIBM causes an increase of approximately 18.5% in the expectation values compared with the optimal sensor of WorstIBM. The optimal sensor placement of CVarIBM causes increases of 36% in the value of the worst case compared with the values of IBM and VaRIBM. Santonastaso et al. [33] used three different methods to find the detection quality point in WDS; these three methods are based on topology, optimization, and empiricism. The comparisons showed that the widely used empirical approach of the water utility is unsatisfactory. Although the optimization-based approach performs significantly better, it is difficult to implement because it necessitates a calibrated hydra. Due to its simplicity compared to the optimization-based approach, the topological approach proves to be successful, and it can be easily adopted by water utilities to determine the location for quality detection points. Hu et al. [34] conducted a theoretical analysis of an optimal scheduling problem before demonstrating that the valve and hydrant scheduling are NP-complete. Following that, they propose a multi-objective optimization model to respond to the contaminant. The proposed model focuses on two opposing goals: first, minimizing the volume of contaminated water exposed to the public, and second, minimizing the hydrant and valves' operational costs. The validity of the proposed model is demonstrated by two different water networks.

3. Materials and Methods

3.1. Materials

3.1.1. Case Study

The real case study represents the drinking water network distributed in a district of Madrid City, specifically the city center, Spain. It has dimensions of X = 1.5 Km, Y = 1.03 Km, as shown in Figure 1. It includes one reservoir, two pumps, fifty-five valves, three-hundred-and-ten pipes, and three-hundred-and-twelve junctions. This study was carefully chosen to include different cases in terms of water demand. The three cases were divided based on the water demand, and they are labeled in the case study by a circle, square, and node_3654. Whereas the water demand is considered high in the center and the areas near the source of drinking water (labeled by the circle in Figure 1), in the isolated regions (not connected to the rest of the network) the water demand is minimal, as at Nod_3654 shown in Figure 1. There are also other areas related to the primary system but with a medium need for drinking water, as labeled by a square in the case study presented in Figure 1. Based on these three different cases, in this study only chemical water contaminant is used, so it is generated and spread in this network for 8 h, see Section 3.2.1 for more details. Through this, the impact of water contaminants on human health will be calculated and analyzed in all the previous cases that have been mentioned above.

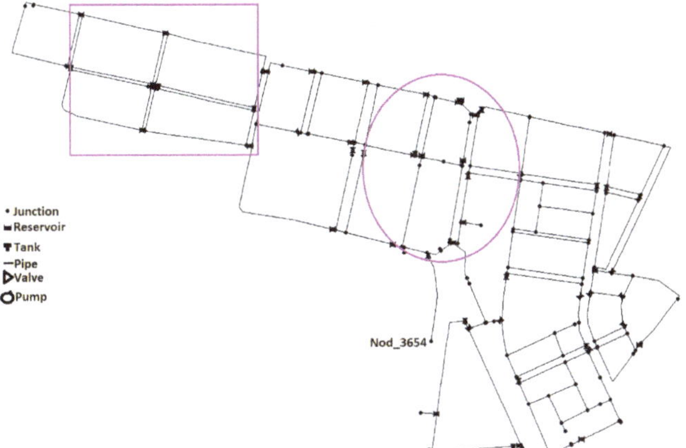

Figure 1. Water distributed networks from Madrid (city center).

3.1.2. Tools

The analysis of this paper was performed with TEVA-SPOT [35], which uses the toolkit of an EPANET programmer (2.00.12) for hydraulic and water quality modeling in a WDS. TEVA-SPOT was created by the National Homeland Security Research Center of the USEPA to include the ability to determine the impact of deliberate and accidental releases of a contaminant into a WDS. It is one of the tools that can be conveniently and effectively used in a water network model to determine the impacts of injections at any or all nodes. The TEVA-SPOT included the ability to monitor contaminant mass using the concentration results given by EPANET to allow a better understanding of the distribution of a contaminant in a network after injection and to improve quality control for simulations [36].

EPANET is a software program used to model water delivery networks around the globe. It was developed as a method to understand the movement and fate of components of drinking water within distribution systems. It can be used in distribution system analysis for many different types of applications. It can also be used to model risks of pollution and determine resistance to threats of safety or natural disasters [37].

3.2. Methods

The risk assessment approach relies on the determination of the safety and life-threatening consequences of water supply involving contaminants. The proposed model of health impact analysis encompasses various phases: contaminants generation, dose level calculation, and health impact analysis as shown in Figure 2, more details of these phases are given in the following subsections.

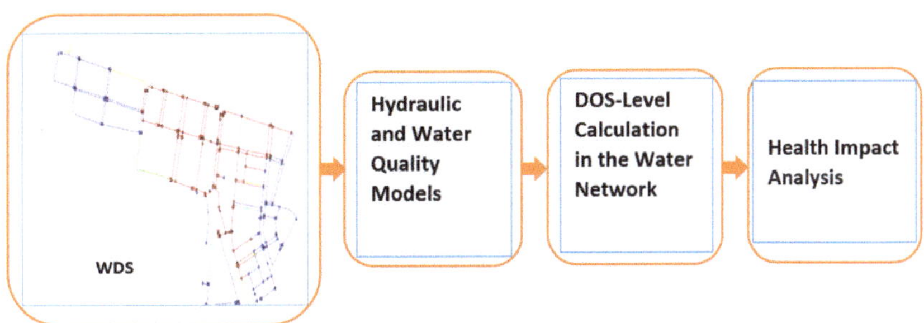

Figure 2. The proposed model of health analysis

3.2.1. Contaminant Generation

In the first phase, the water contaminants are generated and start to propagate in the system (see Figure 3). Chemical/toxin contaminants are used to represent the events in this study. For these contaminants, the health impacts are based on the lethal dose (LD50). Therefore, some parameters need to be set for this kind of contaminant, such as LD50, resulting in death to 50% of the population; body weight; and others. All parameter values required for the contaminant have been set up based on the EPA's guidance [35], and they are summarized in Table 1. Another point that needs clarification is how the water contaminant spread in the water network for eight hours. Figure 3 depicts all nodes in the water network model that have been color-coded based on the concentration level of the chemical contaminant that would result across the network given an injection node at the water source (reservoir). The legend on the left hand side of the map provides the scale of the contaminant for both links and nodes, where the red color represents the high level of the contaminant concentration and the blue shows the low level of the same contaminant.

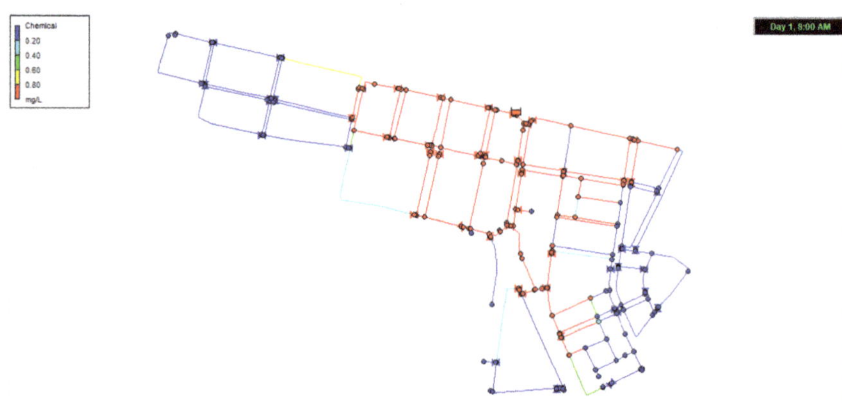

Figure 3. The concentration of the contaminant after 8 h for the injected node (reservoir).

Table 1. Water contaminant parameters.

Parameters	Values
Contaminant type	Chemical/Toxin
Average Body Mass (KG)	70.0
Water Ingestion Rate (Liters per day)	1.41
Gallons per Person per Day (GPD)	200
LD50 (mg/kg)	0.001
Simulation Start time	00:00 A.M.
Simulation Length	24 h
Injection Start Time	00:00 A.M.
Injection Duration	8 h
Injection Node	Reservoir

3.2.2. Dose Level Calculation

Hydraulics and water quality models are performed in a water pipeline network for predicting spatial and temporal changes in the concentration of contaminants in the water network. Dose delivered to a subsection of the water network which is supplying N people needs to be calculated, and its level in each branch needs to be determined. For this, Equation (1) was used and applied [38].

$$L_i = \sum_{t=0}^{T} C_i t \times Q_{it} \times \Delta t \tag{1}$$

where
L_i = load of substance delivered to people from i sub-network (pipe),
$C_i t$ = concentration of contamination in water at time (t),
$Q_i t$ = Water demand in section i at a time (t), and
Δt = Exposure time.

Based on Davis et al. [39], a procedure for evaluating a simple contaminant concentration linked to an adverse outcome was illustrated. Equation (2) was used to determine the concentration of the contaminant ingested [36]:

$$CC = \frac{q \times M}{Q} \tag{2}$$

where CC is the contaminant concentration, q is a constant for the water ingestion rate, M is the mass of the contaminant, and Q is the volume of the water usage rate. The target dose of a person for a specific contaminant can be calculated from the dose–response data available for the contaminant as well as the individual's body mass. The dose values are typically expressed in units of mg per kg of body mass. The maximum number of people (N) who can receive an ingested dose (d) equals the mass ingested divided by the dose as shown in Equation (3) [36]:

$$N = \frac{q \times M}{Q \times d} \tag{3}$$

where N represents the population served as represented by the model of the water distribution system, and d is the dose of the target contaminant resulting from the ingestion of tap water. The target dose of a person for a specific contaminant can be calculated from the available dose response data for the contaminant along with the body mass of the individual. The lethal dose ($LD50$), for example, is the dose at which 50 percent of the population that receives such a dose dies, this defines the $LD50$ for a contaminant. Using the contaminant-specific $LD50$ information and the mean body mass of the population, a target dose for a contaminant (TDC) can be determined as shown in Equation (4):

$$TDC = LD50 \, \frac{mg}{kg} \times \text{BodyMass} \, (Kg) \tag{4}$$

3.2.3. Health Impact Analysis

Health impacts are measured in terms of the number of people exposed to or affected by the contaminated water. Potential public health impacts on contaminants can differ dramatically across networks. Public health impacts increase with the network population for contaminants with high toxicities, particularly for the scenarios with injection locations that have consequences near the upper end of the distribution. On the other hand, for less toxic contaminants potential public health impacts have not been shown to be sensitive to the network population, even those across the network that differ by more than a factor of population size. This is because contaminants with low toxicity appear to create localized impacts that are constrained by the mass of the available contaminant rather than the contaminants with high toxicity. The public health impact $I_e i$ is analyzed and evaluated by calculating the mass of the substance to enter the body using Equation (5):

$$I_e i = \frac{q \sum_{t=0}^{t} C_i \times \Delta t}{G \times t} \tag{5}$$

where q is the volume of consumed water and C_i is the contamination concentration. G is the average weight of the human body during the exposure, and t is the duration of exposure. The risk will be ignored if the value of I_{ei} is too small relative to the tolerable daily intake. The volume of consumed water is represented using Equation (6):

$$q_{ne}(t) = \begin{cases} C_{ne} \times d_i(t) \times \Delta t & if \quad C_{ne}(t) \geq C_{min} \\ 0 & otherwise \end{cases} \tag{6}$$

where q_{ne} is the volume of the water at node n for event e, C_{min} is the threshold hazard concentration, and C_{ne} is the concentration at node n for event e. C_{ne} is the contaminant concentration, d_i (t) denotes the water demand at node i at a time step, and Δt represents the time step interval. The volume of contaminant water for all event scenarios is calculated based on Algorithm 1.

Algorithm 1 Calculating the volume of the contaminant water

1: Generate Contamination Scenarios (S).
2: Select one scenario (C) from S.
3: WHILE ($C \leq S$)
4: Simulate Hydraulic and Quality models.
5: If (Contamination Concentration(CC)>>Threshold)
6: Contaminate water volume (q) $\leftarrow$ cc $\times \Delta t \times$ demand

---

The research methodology represented in Figure 2 can be summarized as follows:

- Generate and simulate water contamination propagation on the pipeline networks (WDS).
- Develop the hydraulic and water quality models of the pipeline networks.
- Calculate dose level delivered to one section of the water network, which is supplying N people using Equation (1).
- Evaluate the health impact based on the calculation of the mass of the substance entering the body using Equation (5).

4. Results and Discussion

4.1. Health Analysis Results

To show the results from the Madrid case study described in the Section 3.1.1, we start explaining the distribution of contaminants based on the water demand of the network. In this work, we used the demand-based method to define the number of water gallons used by each person per day (GPD) to define the distribution of the population. Figure 4 depicts that high demand is focused on the center of the network, while low demand is focused on the top left of the water network. Various demand levels are represented by different colors in the map, where blue represents the deficient water demand and dark red represents the high water demand.

Figure 4. Population distribution based on water demand.

In this study, public health impacts have been described as the size of the population receiving a contaminant intake dose above a certain level. To calculate this, we simulate the impact of water contaminants in different locations in the case study used in this work, as shown in Figure 5, and then investigate the impact of the contaminant for each location.

Table 2 presents the impacts of the contaminant based on where the contaminant has occurred (injected node location) in the case study. This table shows the essential properties of the water contaminants, such as the location, max concentration, max individual dose, and the number of estimated infected people. Referring to Table 2, the locations of the contaminant are selected carefully to include all cases, where Loc.6 represents that the contaminants occurs at the source of the water (Reservoir). This kind of contaminant has a high impact on the number of water nodes in WDS, which leads to a high number of people becoming infected from this contaminant. While Loc. 1 means that a contaminant is present in one of the isolated areas in the case study (see Figure 5), these areas labeled by *nod_3654*; in these areas, the impact of the contaminant on water nodes is small, which leads to a small number of people being affected by this contaminant.

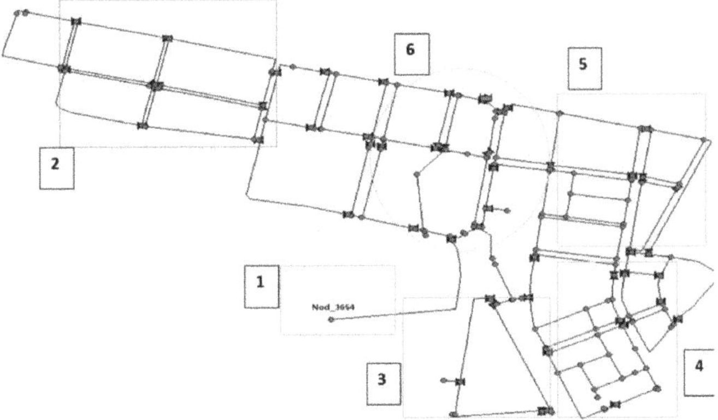

Figure 5. Areas of the contaminant locations.

From Table 2, we note that the location of the injected node is the essential property of contamination as it recorded a high number of estimated infected people (146,729) when the contaminant occurred in the water source (Loc. 6), even though the contaminant was carrying a lower concentration. While very few infected people, ranging from zero to 2750 people, were recorded when the contaminant occurred in isolated locations (Loc. 2) that have a low population, the contamination was recorded at a higher concentration. On the other hand, the remaining locations—Loc. 3, Loc. 4, and Loc. 5—recorded a number of different infected people with a variable contaminant concentration as well. To examine and investigate the impact of water contaminant based on its occurrence in more details, two different cases were chosen (Loc. 2 and Loc. 6), and the extent of the impact is explained in detail as follows.

Table 2. Infected people based on contamination locations.

Contamination Locations	Maximum Concentration (Mg/L)	Maximum Individual Dose (mg)	Number of Estimated Infected People
Loc. 1	2993	1	0
Loc. 2	30	2	2750
Loc. 3	44	5	7217
Loc. 4	224	11	29,017
Loc. 5	27	16	43,996
Loc. 6	2	91	146,729
Mean	217	8	18,568

The first case (Loc. 6, Figure 6) has a contaminant that occurs at the source of the water, which means that the injected node is the reservoir, and this contaminant has a massive impact on the water nodes. From Figure 6, the impact of this contaminant appears at most of the nodes in the network, where the impact denoted by different colors, in which the colors indicate the number of people, has been affected by this contaminant (fatality). The different colors used in the map represent different levels of contaminant impact, in which red represents the high impact with a number of population is close to eight-thousand-and-seventy-nine people (8079 people), and the blue color represents the low impact with a number of people in the range between zero and two-hundred. Another way this impact was shown is in terms of fatalities, infection, and diseased people. These categories of impact were done based on the dose threshold values. If the people receive a dose greater than the threshold value, it is reported to one of these categories based on the dose level received. Figure 7 shows the estimated number of people impacted in these three categories, and it depicts that more than six hours after the contaminant spread in the water network, the impact began to appear in the form of infection in the seventh hour to reach the highest value by affecting more than one-hundred-thousand people. The impact appears as diseases in the next hour, recording the same number of impacted people after eight hours, the actual emergence of the fatality began to appear after nine hours, and it reaches its highest value after 7–10 hours with nearly one-hundred-and-fifty-thousand registered infected people.

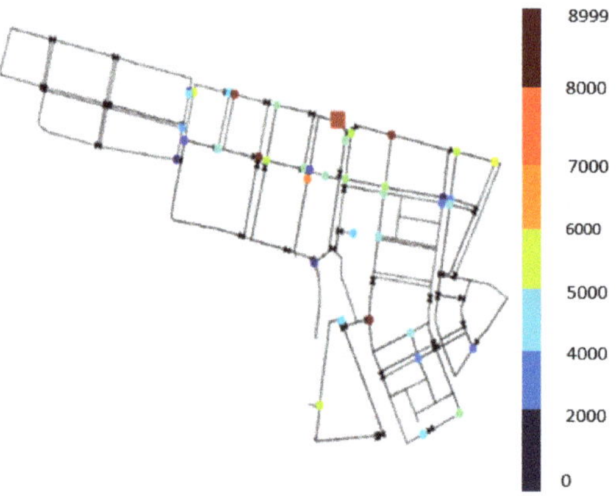

Figure 6. Estimated impact for contaminant occurred at water source (Reservoir).

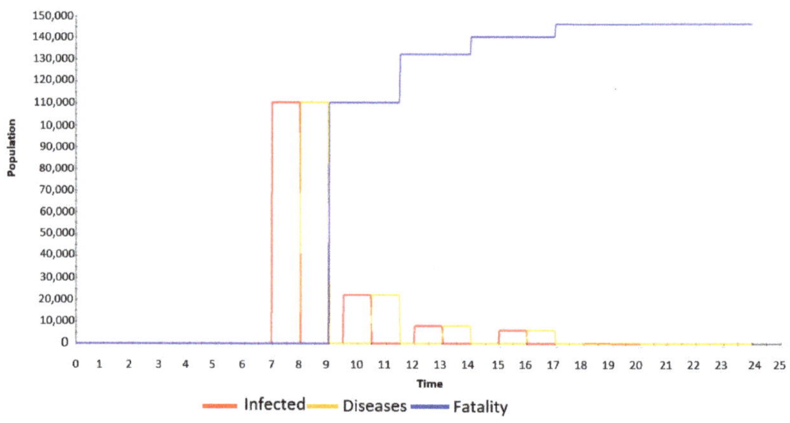

Figure 7. Estimated infection, disease, and fatalities (contamination at source).

In the second case (Loc. 1), the contaminant occurred in one of the smallest areas of water network (isolated area). Figure 8 shows the special impacts on the node directly related to the contaminated source; the impact is minimal and strictly confined. Comprehensive analysis of the impact of this contaminant source is shown in Figure 9, and we notice that the impact appeared consecutively in the form of infection in the seventh hour, followed by the impact in the way of diseases in the next hour. Finally, the emergence of the fatality rate expected at nine o'clock and the same percentage of impact for people at a rate equal to two-thousand-two-hundred infected people was recorded.

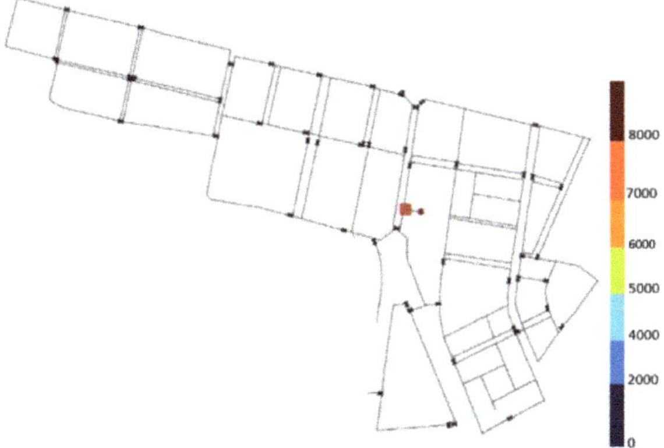

Figure 8. Estimated impact for contaminant at one of the small isolated networks.

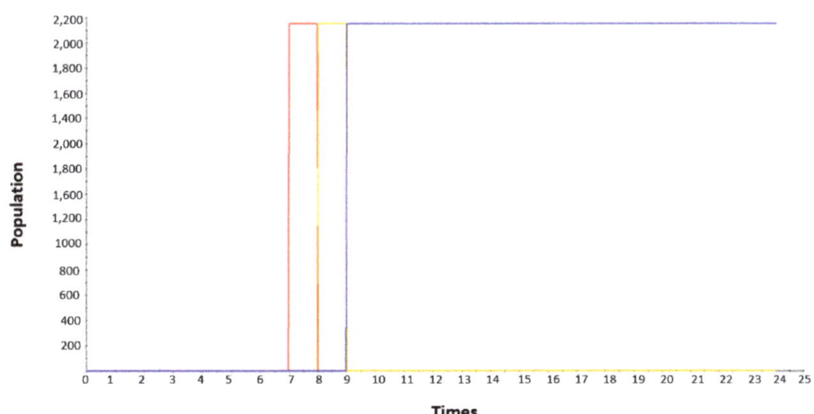

Figure 9. Estimated infection, disease, and fatalities (contaminant at the isolated branch).

4.2. Reducing the Impact of the Contaminant using Water Quality Sensors

The main goal of deploying water quality sensors in WDS is to minimize the consequences associated with water contamination. The water quality sensor detects the contaminant based on the contaminant concentration that would change one or more parameters of water quality (i.e., chlorine, TOC, turbidity, and conductivity) enough that an event detection device (sensor) or a water utility operator can detect the change. For instance, if the residual chlorine decreased by 1 mg/L, the conductivity increased by 150 Sm/cm, or the TOC increased by 1 mg/L.

In this section, the impact of water contamination on the lives of consumers is conducted, where the impact of the contaminated water used by people will be recorded before the water quality sensors are deployed. Then, the extent of the contaminant water impact is recorded after the water quality sensors have been deployed based on our proposed approach [30] reviewed in Section 2. Three different numerical experiments have been conducted in this section:

1. The first experiment investigated the impact of the contaminant without using water quality sensors.
2. The second experiment shows the impact using the water quality sensors optimally deployed on WDS utilizing the volume of the contamination water (VCW) as an objective function for deploying the sensors.
3. The third experiment analyzes the impact using water quality sensors that are optimally deployed on WDS using time detection (TD) as an objective function to deploy the sensors.

In the first experiment, the health impact was analyzed without using water quality sensors to detect the water contaminant. Table 3 shows the impact of water contaminant on the people and contains the essential properties to represent the impact of the contaminant, such as the contaminants' injections nodes, detection time, and the impact as infected people. In Table 3, we have to select thirty injected nodes (as samples) from more than one-thousand nodes included in the Madrid case study, and these samples will be used in all the numerical experiments. As we present this experiment without using deployed sensors (water quality sensors), the detection time is recorded as 1440 min, which means there is no detection until the end of the simulation (24 h). The impact is computed using the proposed model presented in Section 3.2, and it computes the number of people got affected. The impact of the contaminated water without sensors is represented clearly in Figure 10. The x-axis represents the water nodes that are infected with contamination, and the y-axis represents the impact of this contamination concerning the number of people that are affected by this contamination. From Figure 10, we notice that the results were significantly recorded in this experiment, as it was estimated 1that 14,721 cases were recorded at injected node 17_No_1195. At the same time, the rest of the injected water nodes recorded significant numbers of people.

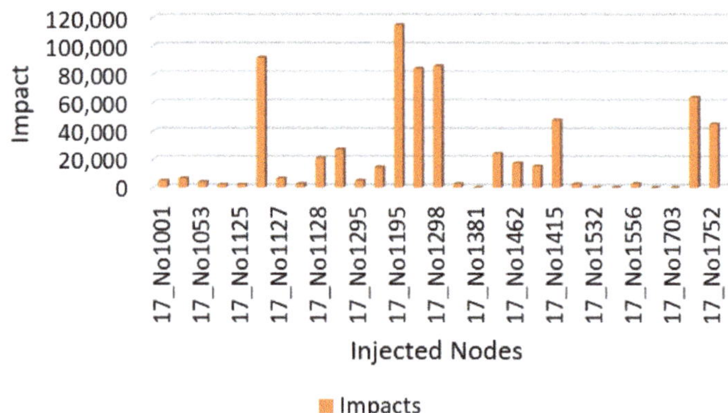

Figure 10. Health impact without using water quality sensors.

In the second experiment, we have deployed ten sensors as monitoring stations to detect the water contaminant, the deployment of sensors used the volume of contaminant water (VCW) as its optimization objective [30]. Table 4 shows the impact of the contaminant after deploying the ten monitoring sensors, and the table contains the same properties presented in the previous experiment, while the detected stations represent the sensor station that detects the contaminant. The time detection column recorded the detection time of the contaminant by the monitoring stations. From Table 4, despite using monitoring stations (sensors), we still find that some injected nodes are not directly detected by any monitoring stations such as 17_No_1381, 17_No1532, 17_No_1702, and 17_No_1703. These nodes kept the contamination until the end of the simulation (1440 min). The main reason for this is the optimization method used to deploy the sensor in this experiment is still unable to cover all injected nodes. The estimated impact of using ten sensor stations (based on the volume of contaminated consume water) is represented clearly in Figure 11. The x-axis represents the water nodes that are infected with contamination, and the y-axis represents the impact of this contamination concerning the number of people that got impact from this contamination. From Figure 11, we notice that the impact decreased significantly compared to what was presented in the previous experiment. The maximum number of cases recorded in this experiment was estimated at 2211 cases, recorded at injected node 17_No_1462.

Table 3. Health impact without using any water quality monitor stations.

Injection Nodes in WDS	Detection Time (Minutes)	Infected People
17_No1001	1440	4858
17_No1002	1440	6499
17_No1053	1440	3899
17_No1124	1440	2071
17_No1125	1440	2042
17_No1113	1440	91,710
17_No1127	1440	6364
17_No1126	1440	2512
17_No1128	1440	20,742
17_No1196	1440	27,215
17_No1295	1440	4821
17_No1296	1440	14,360
17_No1195	**1440**	**114,721**
17_No1297	1440	84,217
17_No1298	1440	85,784
17_No1365	1440	2492
17_No1381	1440	23
17_No1414	1440	24,377
17_No1462	1440	17,485
17_No1463	1440	14,949
17_No1415	1440	47,880
17_No1483	1440	2493
17_No1532	1440	186
17_No1533	1440	183
17_No1556	1440	2666
17_No1702	1440	64
17_No1703	1440	64
17_No1654	1440	63,882
17_No1752	1440	45,190

In the third experiment, we deployed ten sensors as monitoring stations to detect water contamination and used time detection (TD) as optimization functions for deployed sensors. Using the same samples used in the previous numerical experiments, Table 5 includes the same properties used in the previous numerical experiments. However, the results show that the number of injected nodes that were not detected by any of the monitor stations (sensors) decreased significantly compared to the previous numerical experiments. Only one injected node (17_No_1532) was not directly covered by any of the monitoring stations, and this injected node recorded the maximum impact of the water contamination. The estimated impacts of pollution are represented clearly in Figure 12. Figure 12 shows the extent of the impact of water pollution, and it has decreased significantly compared to previous experiences in this experiment. The highest number of infected cases was recorded as 1464 cases, recorded at 17_No_1532.

Table 4. Health impact with using 10 water quality sensors using (VCW).

Injection Nodes in WDS	Detection Sensors	Detection Time	Infected People
17_No1001	17_No1891	106	3
17_No1002	17_No1891	66	24
17_No1053	17_No1483	382	982
17_No1124	17_No1483	1410	1829
17_No1125	17_No1483	1166	25
17_No1113	17_No1297	8	30
17_No1127	17_No525	612	485
17_No1126	17_No1483	32	16
17_No1128	17_No3856	104	28
17_No1196	17_No3707	24	35
17_No1295	17_No525	494	10
17_No1296	17_No525	496	10
17_No1195	17_No3707	26	378
17_No1297	17_No1297	0	0
17_No1298	17_No1415	48	32
17_No1365	17_No1483	4	17
17_No1381		1440	23
17_No1414	17_No3856	512	140
17_No1462	**17_No4702**	**232**	**2211**
17_No1463	17_No2087	310	2111
17_No1415	17_No1415	0	0
17_No1483	17_No1483	0	0
17_No1532		1440	186
17_No1533		1440	183
17_No1556	17_No1483	354	11
17_No1702		1440	64
17_No1703		1440	64
17_No1654	17_No1415	54	29
17_No1752	17_No1483	342	12

Finally, the comparison of water contaminant impact before deploying the sensors and after the deployment of the sensors is conducted and presented in Figure 13. The results in Figure 13 show a significant impact on the number of people impacted by the contaminant water before the water quality sensors were deployed and recorded a massive amount of more than one-hundred-thousand people. On the other hand, this number decreased with the use of water quality sensors to monitor the quality of the water. These sensors were distributed in the water network based on the previously described method, and different numbers were recorded for the number of people affected by water contaminant. Using the contaminated water volume as an objective function to determine the location of the sensors, the number of people affected was close to two-thousand. On the other hand, the number of people in the situation that time detection is used as a target to locate the sensors is close to one-thousand-and-four-hundred. In summary, the use of water sensors to monitor water quality has shown a big difference in reducing the number of people exposed to the risk of using contaminated water, which shows the effectiveness of the proposed method used to distribute the water quality sensors.

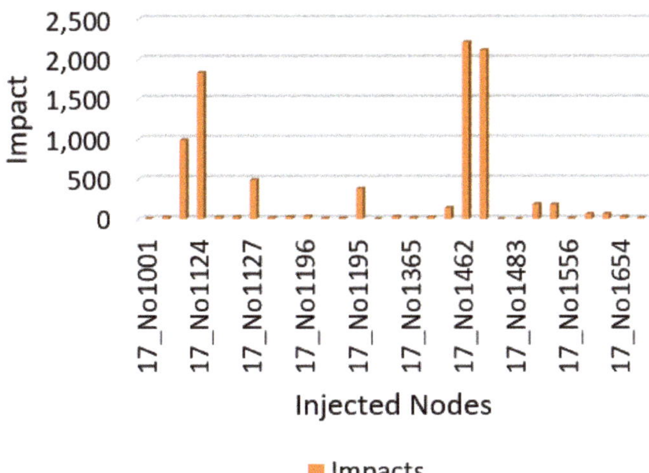

Figure 11. Health impact with 10 sensors deployed using volume of contaminated water (VCW).

Table 5. Health impact with using 10 water quality sensors using time detection (TD).

Injection Nodes In WDS	Detection Sensors	Detection Time	Infected People (People)
17_No1001	17_No1891	106	108
17_No1002	17_No2830	62	65
17_No1053	17_No1483	382	386
17_No1124	17_No1483	1410	1415
17_No1125	17_No1483	1166	1172
17_No1113	17_No5633	120	127
17_No1127	17_No4089	460	468
17_No1126	17_No1483	32	41
17_No1128	17_No4089	102	112
17_No1196	17_No525	320	331
17_No1295	17_No525	494	506
17_No1296	17_No525	496	509
17_No1195	17_No525	86	100
17_No1297	17_No5633	148	163
17_No1298	17_No5633	60	76
17_No1365	17_No1483	4	21
17_No1381	17_No348	50	68
17_No1414	17_No4089	510	529
17_No1462	17_No4702	232	252
17_No1463	17_No2087	310	331
17_No1415	17_No1483	334	356
17_No1483	17_No1483	2	23
17_No1532		**1440**	**1464**
17_No1533	17_No1483	1200	1225
17_No1556	17_No_1483	354	380
17_No1702	17_No961	38	65
17_No1703	17_No961	22	50
17_No1654	17_No1483	142	171
17_No1752	17_No1483	342	372

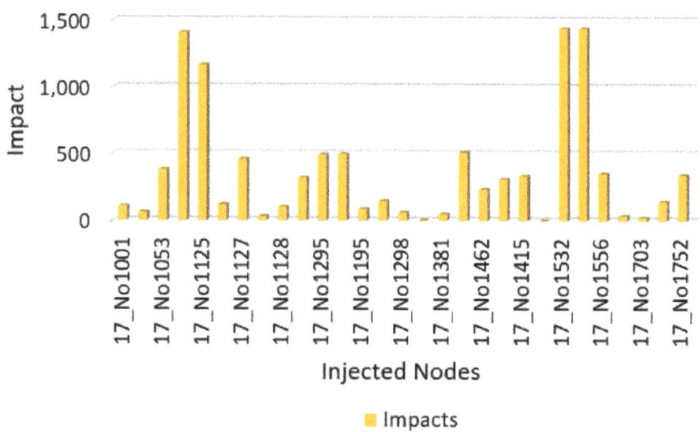

Figure 12. Health impact with 10 sensors deployed using time detection (TD).

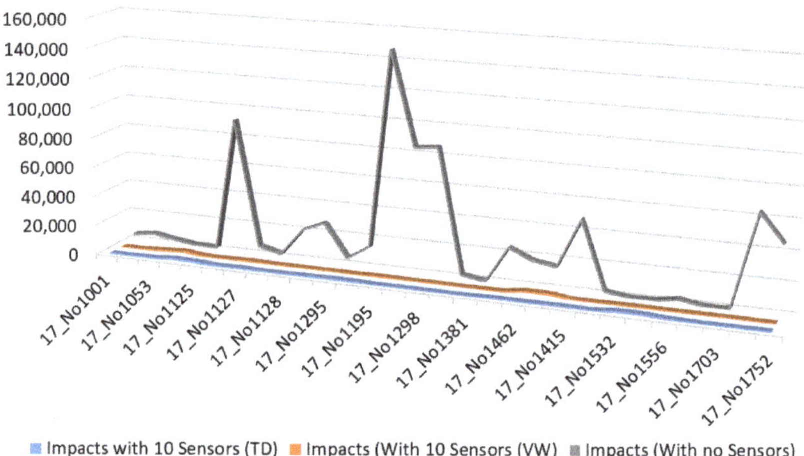

Figure 13. Comparison of the health impact for three scenarios.

5. Conclusions

In this paper, a health impact analysis model of WDS is proposed to assess and analyze the health impact of consuming contaminated drinking water in a water distributed system (WDS). The health impact analysis process mainly depends on the hydraulic and water quality models in WDS. Water contaminants are injected in various locations and spread in the water network for 8 h before we start the analysis. The health impact for the contaminated scenarios in WDS is examined and simulated using a real case study from a district of Madrid, Spain. The impact of water pollution has been analyzed in different locations with multiple critical threshold values of the dose level. The number of people that were infected, diseased, or died from pollution was recorded in each simulated experiment. Finally, the extent of pollution's impact was compared without using water quality stations and using ten water quality sensors. These sensors are optimally deployed on WD based on the time detection and volume of the contaminated water. The results showed a very significant reduction in the impact using the deployed sensors, where the health impact was reduced by up to 98.37%.

Author Contributions: Conceptualization E.Q.S. and W.W.; methodology, E.Q.S.; software, E.Q.S. and R.G.; formal analysis, E.Q.S. and W.W.; data , R.G.; writing—original draft preparation, E.Q.S.;

writing—review and editing, E.Q.S. and W.W. All authors have read and agreed to the published version of the manuscript.

Funding: European Union's Horizon 2020 research and innovation programme Under the Marie Skłodowska-Curie–Innovative Training Networks (ITN)-IoT4Win-Internet of Things for Smart Water Innovative Network (765921).

Institutional Review Board Statement: Not applicable.

Informed Consent Statement: Not applicable.

Data Availability Statement: The data and the case study used in this article have been provided by TPF Getinsa Eurostudios, Madrid, Spain. These data are not available on the Internet and have been used in this article only.

Acknowledgments: This research is supported by European Union's Horizon 2020 research and innovation programme Under the Marie Skłodowska-Curie–Innovative Training Networks (ITN)-IoT4Win-Internet of Things for Smart Water Innovative Network (765921).

Conflicts of Interest: The authors declare no conflict of interest.

References

1. Li, P.; Wu, J. Drinking water quality and public health. *Expo. Health* **2019**, *11*, 73–79. [CrossRef]
2. Padulano, R.; Del Giudice, G. A nonparametric framework for water consumption data cleansing: An application to a smart water network in Naples (Italy). *J. Hydroinform.* **2020**, *22*, 666–680. [CrossRef]
3. Damiani, M.; Núñez, M.; Roux, P.; Loiseau, E.; Rosenbaum, R.K. Addressing water needs of freshwater ecosystems in life cycle impact assessment of water consumption: State of the art and applicability of ecohydrological approaches to ecosystem quality characterization. *Int. J. Life Cycle Assess.* **2018**, *23*, 2071–2088. [CrossRef]
4. Liu, G.; Zhang, Y.; Knibbe, W.J.; Feng, C.; Liu, W.; Medema, G.; van der Meer, W. Potential impacts of changing supply-water quality on drinking water distribution: A review. *Water Res.* **2017**, *116*, 135–148. [CrossRef] [PubMed]
5. Kara, S.; Karadirek, I.E.; Muhammetoglu, A.; Muhammetoglu, H. Real time monitoring and control in water distribution systems for improving operational efficiency. *Desalin. Water Treat.* **2016**, *57*, 11506–11519. [CrossRef]
6. Frisbie, S.H.; Mitchell, E.J.; Dustin, H.; Maynard, D.M.; Sarkar, B. World Health Organization discontinues its drinking-water guideline for manganese. *Environ. Health Perspect.* **2012**, *120*, 775–778. [CrossRef]
7. Lavoie, R.; Joerin, F.; Rodriguez, M. ATES: A geo-informatics decision aid tool for the integration of groundwater into land planning. *J. Hydroinform.* **2015**, *17*, 771–788. [CrossRef]
8. De Vera, G.A.; Wert, E.C. Using discrete and online ATP measurements to evaluate regrowth potential following ozonation and (non) biological drinking water treatment. *Water Res.* **2019**, *154*, 377–386. [CrossRef]
9. Long, D.T.; Pearson, A.L.; Voice, T.C.; Polanco-Rodríguez, A.G.; Sanchez-Rodríguez, E.C.; Xagoraraki, I.; Concha-Valdez, F.G.; Puc-Franco, M.; Lopez-Cetz, R.; Rzotkiewicz, A.T. Influence of rainy season and land use on drinking water quality in a karst landscape, State of Yucatán, Mexico. *Appl. Geochem.* **2018**, *98*, 265–277. [CrossRef]
10. Scheili, A.; Rodriguez, M.; Sadiq, R. Seasonal and spatial variations of source and drinking water quality in small municipal systems of two Canadian regions. *Sci. Total Environ.* **2015**, *508*, 514–524. [CrossRef]
11. Abtahi, M.; Golchinpour, N.; Yaghmaeian, K.; Rafiee, M.; Jahangiri-rad, M.; Keyani, A.; Saeedi, R. A modified drinking water quality index (DWQI) for assessing drinking source water quality in rural communities of Khuzestan Province, Iran. *Ecol. Indic.* **2015**, *53*, 283–291. [CrossRef]
12. Rodriguez-Narvaez, O.M.; Peralta-Hernandez, J.M.; Goonetilleke, A.; Bandala, E.R. Treatment technologies for emerging contaminants in water: A review. *Chem. Eng. J.* **2017**, *323*, 361–380. [CrossRef]
13. Chang, K.; Gao, J.L.; Wu, W.Y.; Yuan, Y.X. Water quality comprehensive evaluation method for large water distribution network based on clustering analysis. *J. Hydroinform.* **2011**, *13*, 390–400. [CrossRef]
14. Soltani, A.A.; Oukil, A.; Boutaghane, H.; Bermad, A.; Boulassel, M.R. A new methodology for assessing water quality, based on data envelopment analysis: Application to Algerian dams. *Ecol. Indic.* **2021**, *121*, 106952. [CrossRef]
15. Baştürk, E. Assessing Water Quality of Mamasın Dam, Turkey: Using Water Quality Index Method, Ecological and Health Risk Assessments. *CLEAN-Air Water* **2019**, *47*, 1900251. [CrossRef]
16. Wu, Z.; Zhang, D.; Cai, Y.; Wang, X.; Zhang, L.; Chen, Y. Water quality assessment based on the water quality index method in Lake Poyang: The largest freshwater lake in China. *Sci. Rep.* **2017**, *7*, 1–10. [CrossRef]
17. Abtahi, M.; Yaghmaeian, K.; Mohebbi, M.R.; Koulivand, A.; Rafiee, M.; Jahangiri-rad, M.; Jorfi, S.; Saeedi, R.; Oktaie, S. An innovative drinking water nutritional quality index (DWNQI) for assessing drinking water contribution to intakes of dietary elements: A national and sub-national study in Iran. *Ecol. Indic.* **2016**, *60*, 367–376. [CrossRef]
18. Akter, T.; Jhohura, F.T.; Akter, F.; Chowdhury, T.R.; Mistry, S.K.; Dey, D.; Barua, M.K.; Islam, M.A.; Rahman, M. Water Quality Index for measuring drinking water quality in rural Bangladesh: A cross-sectional study. *J. Health Popul. Nutr.* **2016**, *35*, 1–12. [CrossRef]

19. Ba, A.; McKenna, S.A. Water quality monitoring with online change-point detection methods. *J. Hydroinform.* **2015**, *17*, 7–19. [CrossRef]
20. Braun, M.; Piller, O.; Deuerlein, J.; Mortazavi, I.; Iollo, A. Uncertainty quantification of water age in water supply systems by use of spectral propagation. *J. Hydroinform.* **2020**, *22*, 111–120. [CrossRef]
21. Banda, T.D.; Kumarasamy, M.V. Development of Water Quality Indices (WQIs): A Review. *Pol. J. Environ. Stud.* **2020**, *29*. [CrossRef]
22. Gara, T.; Fengting, L.; Nhapi, I.; Makate, C.; Gumindoga, W. Health safety of drinking water supplied in Africa: A closer look using applicable water-quality standards as a measure. *Expo. Health* **2018**, *10*, 117–128. [CrossRef]
23. Chen, J.; Qian, H.; Gao, Y.; Li, X. Human health risk assessment of contaminants in drinking water based on triangular fuzzy numbers approach in Yinchuan City, Northwest China. *Expo. Health* **2018**, *10*, 155–166. [CrossRef]
24. Zhang, H.; Zhou, X.; Wang, L.; Wang, W.; Xu, J. Concentrations and potential health risks of strontium in drinking water from Xi'an, Northwest China. *Ecotoxicol. Environ. Saf.* **2018**, *164*, 181–188. [CrossRef] [PubMed]
25. Ali, N.; Khan, S.; ur Rahman, I.; Muhammad, S. Human health risk assessment through consumption of organophosphate pesticide-contaminated water of Peshawar basin, Pakistan. *Expo. Health* **2018**, *10*, 259–272. [CrossRef]
26. Joshi, Y.P.; Kim, J.H.; Kim, H.; Cheong, H.K. Impact of drinking water quality on the development of enteroviral diseases in Korea. *Int. J. Environ. Res. Public Health* **2018**, *15*, 2551. [CrossRef]
27. Zhang, Q.; Xu, P.; Qian, H. Groundwater quality assessment using improved water quality index (WQI) and human health risk (HHR) evaluation in a semi-arid region of northwest China. *Expo. Health* **2020**, *12*, 487–500. [CrossRef]
28. Adimalla, N.; Li, P. Occurrence, health risks, and geochemical mechanisms of fluoride and nitrate in groundwater of the rock-dominant semi-arid region, Telangana State, India. *Hum. Ecol. Risk Assess. Int. J.* **2019**, *25*, 81–103. [CrossRef]
29. Kumar, D.; Singh, A.; Jha, R.K.; Sahoo, S.K.; Jha, V. A variance decomposition approach for risk assessment of groundwater quality. *Expo. Health* **2019**, *11*, 139–151. [CrossRef]
30. Shahra, E.Q.; Wu, W. Water contaminants detection using sensor placement approach in smart water networks. *J. Ambient. Intell. Humaniz. Comput.* **2020**, 1–16. [CrossRef]
31. Ung, H.; Piller, O.; Gilbert, D.; Mortazavi, I. Accurate and optimal sensor placement for source identification of water distribution networks. *J. Water Resour. Plan. Manag.* **2017**, *143*, 04017032. [CrossRef]
32. Hooshmand, F.; Amerehi, F.; MirHassani, S. Risk-Based Models for Optimal Sensor Location Problems in Water Networks. *J. Water Resour. Plan. Manag.* **2020**, *146*, 04020086. [CrossRef]
33. Santonastaso, G.F.; Di Nardo, A.; Creaco, E.; Musmarra, D.; Greco, R. Comparison of topological, empirical and optimization-based approaches for locating quality detection points in water distribution networks. *Environ. Sci. Pollut. Res.* **2020**, 1–10. [CrossRef]
34. Hu, C.; Yan, X.; Gong, W.; Liu, X.; Wang, L.; Gao, L. Multi-objective based scheduling algorithm for sudden drinking water contamination incident. *Swarm Evol. Comput.* **2020**, *55*, 100674. [CrossRef]
35. Janke, R.; Murray, R.; Haxton, T.; Taxon, T.; Bahadur, R.; Samuels, W.; Berry, J.; Boman, E.; Hart, W.; Riesen, L.; et al. *Threat Ensemble Vulnerability Assessment-Sensor Placement Optimization Tool (TEVA-SPOT) Graphical User Interface User'S Manual*; Researchgate: Berlin, Germany, 2012; pp. 1–109.
36. Bałut, A.; Urbaniak, A. Application of the TEVA-SPOT in designing the monitoring of water networks. *E3S Web Conf. EDP Sci.* **2018**, *59*, 00009. [CrossRef]
37. Davis, M.J.; Janke, R.; Taxon, T.N. Mass imbalances in EPANET water-quality simulations. *Drink. Water Eng. Sci.* **2018**, *11*, 25–47. [CrossRef] [PubMed]
38. Studziński, A.; Pietrucha-Urbanik, K. Simulation model of contamination threat assessment in water network using the EPANET software. *Ecol. Chem. Eng. S* **2016**, *23*, 425–433. [CrossRef]
39. Davis, M.J.; Janke, R.; Magnuson, M.L. A framework for estimating the adverse health effects of contamination events in water distribution systems and its application. *Risk Anal.* **2014**, *34*, 498–513. [CrossRef]

MDPI

Article

In Situ Bioremediation of a Chlorinated Hydrocarbon Plume: A Superfund Site Field Pilot Test

Peter Guerra [1], Akemi Bauer [2,3,*], Rebecca A. Reiss [4] and Jim McCord [5]

[1] Wood Environment & Infrastructure Solutions, Albuquerque, NM 87113, USA; peter.guerra@woodplc.com
[2] University of Arkansas, Fayetteville, AR 72701, USA
[3] ECCI, Little Rock, AR 72223, USA
[4] Biology Department, New Mexico Institute of Mining and Technology, Socorro, NM 87801, USA; rebecca.reiss@nmt.edu
[5] Lynker Technologies, LLC, Leesburg, VA 20175, USA; jtmccord@lynker.com
* Correspondence: abauer@ecci.com; Tel.: +1-501-975-8100

Abstract: The North Railroad Avenue Plume, discovered in 1989, contained chlorinated solvent groundwater plumes extending over 23.5 hectares (58 acres) and three hydrostratigraphic units. The source contaminant, tetrachloroethene, stemmed from release at a dry cleaner/laundromat business. The anaerobic biodegradation byproducts trichloroethene, isomers of dichloroethene (DCE), and vinyl chloride were detected in groundwater samples collected prior to remedial action. The impacted aquifers are the sole source drinking water aquifers for the communities near the site. Following the remedial investigation and feasibility study, the selected alternative for full-scale remedial action at the site was enhanced reductive dichlorination (ERD) focused on four treatment areas: the shallow source zone, the shallow hotspot area, the shallow downgradient area, and the deep zone. Pilot testing, which was conducted in the source zone and hotspot areas, is the subject of this paper. The primary objectives of the pilot test were to obtain the necessary information to select an ERD treatment formulation, dose, and frequency of dosing for use during full-scale remedial action, as well as to refine the site's hydrogeologic conceptual site model and design parameters. Four (4) test cells, each of which contained well pairs of injection and downgradient extraction wells, were used to test ERD bio-amendment formulations: ethyl lactate, dairy whey, emulsified vegetable oil (EVO), and a combination of EVO and a hydrogen gas infusion. A conservative tracer, bromide, was added to the recirculation flow to record tracer breakthrough, peak, and dissipation at extraction wells. The results of these dipole tracer tests were used to reassess the hydraulic conductivity and hydrodynamic dispersity used in the remedial design. In addition to water quality analyses of contaminants and substrates, groundwater samples were also analyzed for biological analyses before, during, and after the addition of bioamendment. Analyses of phospholipid fatty acids and deoxyribonucleic acid (DNA) extracts from fresh groundwater samples informed decisions on the capacity for complete ERD without DCE stalling and tracked the shifts in the bacterial and archaeal taxonomy and phylogeny stemming from the addition of bioamendments. The pilot test concluded that EVO was the most suitable, considering (1) support of the native microbial consortia for ERD, (2) mechanics and hydraulics of the remediation system, and (3) sustainability/retention of the substrate in the subsurface. Along with EVO, the addition of a nutrient broth derived from brewery waste accelerated and sustained the desired conditions and microbial diversity and population levels. The pilot test results were also used to assess the utilization kinetics of the injected substrates based on total organic carbon (TOC) concentrations measured in the groundwater. After determining that substrate utilization followed Monod kinetics, a TOC threshold at 300 milligrams per liter, equivalent to approximately twice its half-saturation constant was established. Full scale treatment dosing and dose frequency were designed around this threshold, assuming the maximum substrate utilization would yield optimum ERD.

Citation: Guerra, P.; Bauer, A.; Reiss, R.A.; McCord, J. In Situ Bioremediation of a Chlorinated Hydrocarbon Plume: A Superfund Site Field Pilot Test. *Appl. Sci.* **2021**, *11*, 10005. https://doi.org/10.3390/app112110005

Academic Editor: Simone Morais

Received: 10 September 2021
Accepted: 22 October 2021
Published: 26 October 2021

Publisher's Note: MDPI stays neutral with regard to jurisdictional claims in published maps and institutional affiliations.

Keywords: superfund; bioremediation; enhanced reductive dechlorination (ERD); dense nonaqueous-phase liquid (DNAPL); chlorinated hydrocarbons; tetrachloroethene (PCE); trichloroethene (TCE); dichloroethene (DCE); anaerobic reduction; metagenomics

1. Introduction

The North Railroad Avenue Plume (NRAP) Site was discovered in 1989. The 23.5 hectares (58 acres) of groundwater were contaminated with tetrachloroethene (PCE), trichloroethene (TCE), dichloroethene (DCE) isomers (1,2-cis DCE and 1,2-trans DCE), and vinyl chloride (VC) from a release at a former dry cleaner and laundromat business. The three known groundwater units underlying the site are defined by differences in water pressures and geology. All three units contain chlorinated solvent impacts from the NRAP source zone. The contaminated groundwater aquifers are the sole source drinking water aquifers for the City of Española, Santa Clara Pueblo, and individual water supply wells near the site. Figure 1 shows the site area with groundwater plume. In 1999, the site was added to the National Priority List (NPL) and recognized as a superfund site. A study to determine a remediation plan was conducted from 1999 through 2001 [1,2]. Based on the findings, the original field test plan (FTP) was prepared in 2003 proposing bioremediation, specifically enhanced reductive dechlorination (ERD) with surfactant flushing for the source zone [3,4].

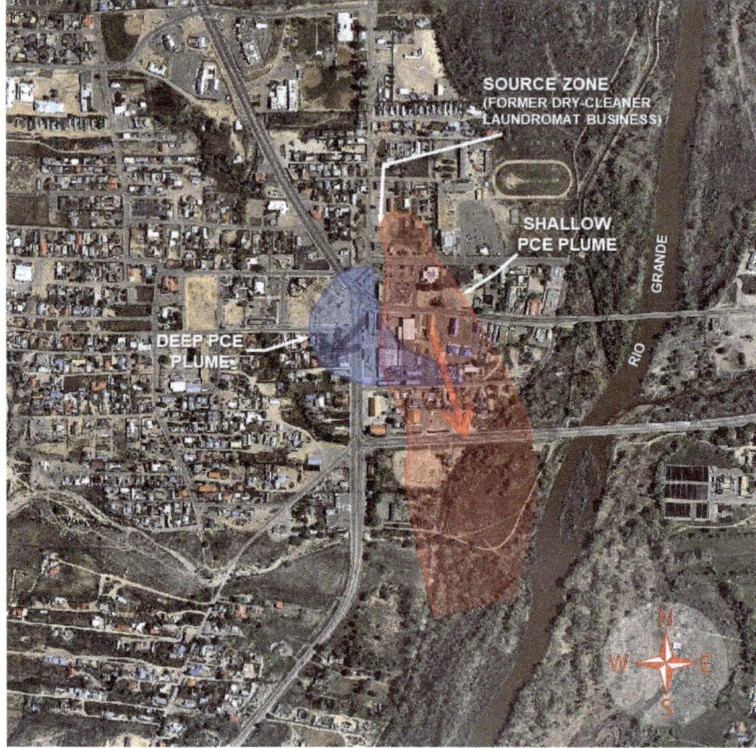

Figure 1. North Railroad Avenue Plume Superfund Site. Approximate groundwater plumes prior to remediation and the groundwater flow directions in the shallow aquifer (red plume with red arrow) and deep aquifer (blue plume and blue arrow) are shown. The location of the source of contamination, a former dry cleaner/laundromat business, is also shown.

Approximately 100 wells (injection/extraction/monitoring) were installed during the summer and fall of 2005. Intensive soil/sediment and groundwater sampling, screening, and analyses were conducted during the well installation operations. Results improved the understanding of the contamination distribution and provided an enhanced picture of the lithology, especially concerning the shallow aquifer and aquitard. Based on the findings stemming from the well installation and soil/groundwater sampling and analysis campaigns, surfactant flushing was eliminated, and the 2003 FTP was revised. The primary changes to the 2003 FTP included new amendment selection, new test cell well pairs, and additional test cell to evaluate source area remediation. Construction of the remediation system was completed in 2006, and pilot test treatment was initiated in 2007 through 2008. This article presents the result of the Revised FTP.

Remedial action at the site was divided into four treatment areas: the source zone, a hotspot, and downgradient treatment areas located in the shallow groundwater aquifer and the deep zone area. Based on high-concentration dissolved-phase PCE detected in the source zone area, ganglia-phase dense non-aqueous phase liquid (DNAPL) contamination may have been present. Alternatively, the high-concentration dissolved-phase PCE may have been the result of increased solubility due to surfactants from the laundry wastewater discharge that mixed with the dry-cleaner chemicals at the release site, an inline lint trap. The hotspot area emanates downgradient from source zone in the shallow aquifer. Pilot testing was conducted in the source zone and hotspot areas. Although pilot test activities were not conducted at the downgradient or deep-zone areas, the results were used to inform remedial action approach for these areas.

Shifts in microbial consortia diversity, including changes in microbes that harbor the key enzyme-producing genes involved in the biodegradation of chlorinated hydrocarbons, were monitored during the pilot test by analyses of the microbiome, using molecular microbiological methods (MMMs). Changes in biomass measured by microscopic examination of concentrated consortia samples [5] was employed to assess the overall response of the microbiome to changing aquifer conditions. Phospholipid fatty acids, the main component of the cell membrane of all microbes were separated from fresh groundwater samples and analyzed to estimate biomass and reveal the signature and relative population of bacteria, such as *Firmicutes*, *Proteobacteria*, metal reducers, and sulfur-reducing bacteria (SRB), as well as the presence of eukaryotes in microbial communities [6]. Quantitative polymerase chain reaction (qPCR) was used to detect the presence and concentration of bacterial taxa such as *Dehalococcoides* (DHC) and *Dehalobacter*, and genes for important functions harbored by these taxa, the reductive dehalogenases (RDases) [7–9]. The MMMs implemented for the pilot test were performed by Microbial Insights, including their Census® qPCR analysis, which quantitates these microbes and genes in groundwater samples. When normalized with biomass calculation, qPCR has a detection limit of 100 cells per sample. The benefit of the MMMs implemented for the pilot test is that the analyses provide a snapshot that can be compared to the current knowledge base. This is the case with the services provided Microbial Insights, which, for example, were used to study bioremediation potential at a 1,4-dioxane- and TCE-contaminated site [10].

Metagenomic analysis using shotgun sequencing of DNA does not require prior amplification. Sequences are identified by comparison to microbial genomic databases and the increased throughput provides a more detailed picture of the consortia diversity. With adequate sampling, it is even possible to detect and assemble genomes of unculturable, and hence undiscovered, microbes. The benefit of applying high-throughput DNA sequencing techniques is exemplified by an extensive project focused on a groundwater site undergoing in situ injections [11]. As a result of this research, 47 new candidate phyla are proposed and extensive insights to changes in biogeochemical cycles are revealed by gene annotations.

The primary objective of the pilot testing was to obtain the necessary information to select a treatment formulation for use during the full-scale treatment operations. The tested treatment formulations were electron donors and nutrients. Bio-augmentation was not tested because the site conditions did not indicate it would be necessary (e.g., detectable VC in portions of the NRAP groundwater plume overlapped with petroleum contamination from a nearby leaking retail fuel business). Secondary objectives for the pilot test were to inform designers on the dosage and frequency of the selected bio-amendment and to obtain more representative information on the subsurface hydrogeology and as-built well and piping hydraulics for use in optimizing the full-scale operations. The field tests results consisted of biorecirculation tests to select amendments for full-scale treatment and the development of site-specific understanding of PCE contaminant breakdown pathways and rates.

2. Materials and Methods

2.1. Hydrogeologic Setting and Plume Characteristics of the Pilot Test

The source zone and hotspot areas, where the tests described in this paper were conducted, is underlain by interbedded gravel, sand, clay, and silt, typical of Rio Grande Basin alluvial deposits in northern New Mexico. The water table aquifer, referred to as the shallow zone, contains the source zone and hotspot areas [4]. The water table in these areas is encountered at approximately 1.5 m (5 feet) below ground surface (bgs) and extends to approximately 9.1 m (30 feet) bgs before encountering a clay aquitard. This upper section is comprised primarily of gravelly and sandy material with intermittent lenses of clay and silt. The arithmetic mean of the shallow-zone hydraulic conductivity is approximately 18.2 m per day (60 feet per day), ranging from approximately 1.5 m/day (5 ft/day) to 30.5 m/day (100 ft/day), based on in-well slug testing and numerical modeling conducted as part of the remedial investigation and design. Historical water level measurements from wells screened within this unit indicate that the groundwater flows south to southeast (slightly towards the Rio Grande) at a representative horizontal hydraulic gradient of 0.002 m per meter (m/m). Assuming an effective porosity of 0.15 (based on typical values for soils of the types found at this site) a representative lateral groundwater flow velocity is estimated at 0.2 m/day (0.8 ft/day). Considering the length of the treatment zone along the groundwater flow path at 67.1 m (220 feet) long, the time for groundwater to travel from the source zone to the periphery of the hotspot area is estimated at 275 days.

The maximum anticipated depth for treatment varies spatially across the hotspot area and is equal to the depth to the lower surface of the shallow groundwater plume. The vertical extent of the groundwater plume has been limited by extensive but discontinuous silt and clay layers, with contaminated groundwater penetrating deeper into the shallow zone as the plume migrated downgradient of the source zone, to the south and southeast. The maximum depth of the plume penetration beneath the source area, using a PCE concentration cutoff value of approximately 5 micrograms per liter (µg/L), is between 6.4 and 7.0 m (21 and 23 feet) bgs. At the south end of the hotspot area the maximum depth approaches 9.1 m (30 feet) bgs. The representative pore volume for the source zone and hotspot areas (or its volume), assuming the total porosity is 0.25, is estimated to be 4842.2 cubic meter (171,000 cubic feet) or 4.85 million liter (1.28 million gallons).

PCE concentrations within the source zone/hotspot portion of the groundwater plume (excluding the source area), along the plume centerline and prior to remedial action, range from approximately 10,000 µg/L to approximately 3000 µg/L, installed during phase II of remedial design. Evidence of natural reductive dehalogenation of PCE appears to occur within the hotspot area, as 830 µg/L TCE has been detected since March 2006. 1,2-cis DCE has also been detected in the hotspot area in groundwater and soil samples. The groundwater concentration ranges were developed from only a few sampling points. Thus, there is considerable uncertainty in the structure of the plume in this area of the Site

The source area within the larger hotspot area, encompassing approximately 83.6 square meters (900 square feet) of the 3344.5-square-meter (36,000-square-foot) treatment area, is characterized by high concentrations of PCE in the groundwater (dissolved phase) and on the sediment (adsorbed phase). Groundwater PCE concentrations in the source zone groundwater have been detected above 60,000 µg/L, with sediment PCE concentrations ranging up to 800,000 micrograms per kilogram (at SB-04 at a sampling depth of 3.7 m (12 feet) bgs).

2.2. Pilot Testing Design and Operation

Two key guidance documents [12,13], each of which were developed based on a review of research and real-world applications of ERD, were used in the development of the sampling and analysis plan and operational procedures of the pilot test. The infrastructure utilized in conducting the pilot test included the groundwater extraction and injection wells, bio-amendment mixing tanks and chemical metering pumps, flow-control manifolds, and monitoring and control systems including monitoring wells and a Supervisory Control and Data Acquisition (SCADA) system for continuous monitoring and control of the remediation system.

2.2.1. Injection and Extraction Operations

The extraction pumping and injection flow rate scheme as designed was used in operating the pilot test, whereby extracted groundwater was fed manifolds and reinjected into the injection wells. By this method, injection and extraction rates were balanced and a hydraulic recirculation cell was established for each of the four pilot tests. The pilot test also included a shut-in period, during which recirculation was stopped and the test cell was allowed to incubate under ambient conditions. Based on the ambient hydraulic gradient in the upper aquifer and the shut-in duration, the injectate plume was not anticipated to migrate during the shut-in period.

Hydrogeologic modeling conducted during the remedial design indicated that the average time of travel between adjacent injectors and extractor would be approximately 26 days at the 9.5 L per minute (lpm) (2.5 gallons per minute) recirculation flow rate [3]. As described earlier, the subsurface is heterogeneous, with interbedded layers and lenses of higher- and lower-permeability materials. These heterogeneities give rise to flow field anisotropy and preferential pathways between injector/extractor well sets. The conservative tracer (bromide) added along with the injectate indicated an earlier arrival travel time (presumably along preferential flow pathways) that was significantly less than 26 days (see Section 3.1 below). This result was taken into consideration in subsequent result interpretation for ERD for each of the pilot tests.

This pilot test design allowed us to determine if this flow rate was appropriate for full-scale implementation. In addition, pilot test results were employed to indicate whether modification of the design flow rate schedule was necessary to achieve one or more of the following objectives: (i) adequate hydraulic control against plume spreading; (ii) improved efficiency of the treatment zone sweep (i.e., distribution, mixing and contact); (iii) improved in-situ treatment efficiency; and/or (iv) improved equipment and system operation and maintenance efficiency.

2.2.2. Pilot Test Cells

The pilot test was operated in four test cells consisting of extraction wells (designated HSE and EW) and injection wells (designated as HSI and IW) as follows: HSI-14/HSE-11 which was initially dosed with ethyl lactate (EL); HSI-8/HSE-6 which was initially dosed with dairy whey; HSI-19/HSE-14 which was initially dosed with emulsified vegetable oil (EVO); and, the source zone test cell, IW-2S, IW-2D/EW-3, which was dosed with EVO and groundwater infused with hydrogen gas. A second dose of EVO was added to each of the test cell at the approximate midpoint of the pilot test. The locations of these field test cell well pairs are shown in Figure 2. Figure 3 provides a generalized geologic cross section of a typical test cell.

2.3. Pilot Test Procedures

The revised, final FTP scope consisted of a sequence of activities involving recirculation of groundwater through the remediation system wells and periodic addition of bio-amendment formulation. In chronological order, these activities were: pre-pumping baseline sampling, biorecirculation testing, and verification sampling. Figure 4 below provides a flow chart of the FTP procedure and chronology.

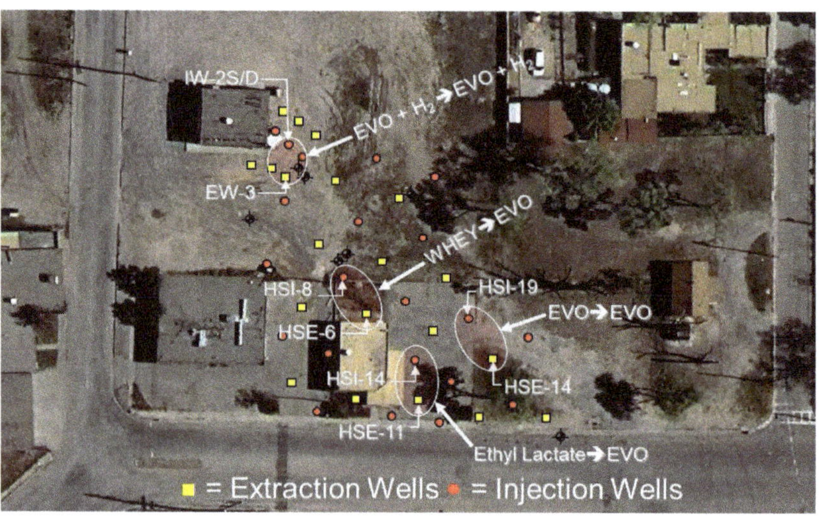

Figure 2. Pilot test cell well pairs. The substrate formulations added to each test cell at the beginning and midpoint of the pilot testing are shown.

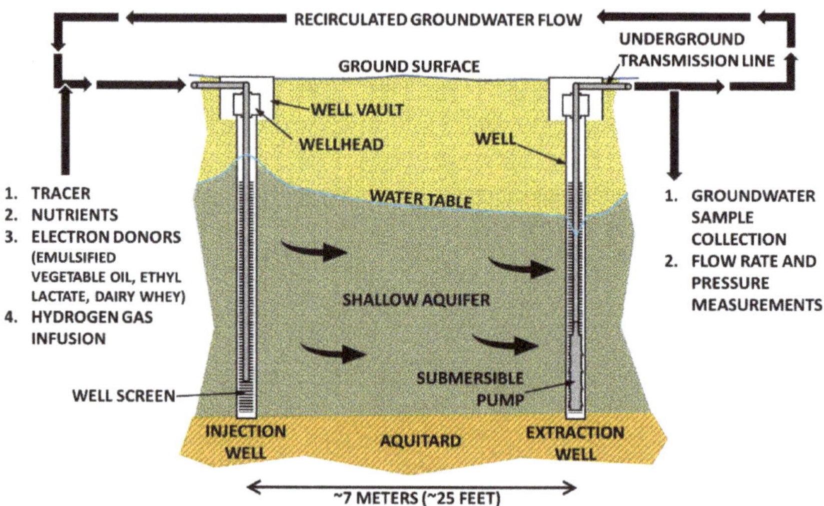

Figure 3. Cross-section of a typical test cell. Induced flow between the injection and extraction wells is shown.

Groundwater pressures during testing were measured using two methods. At the start of testing, water table elevations were measured using a water level meter to gauge the depth to groundwater from an established reference elevation at each wellhead. Wellhead reference elevations were measured by a Professional Land Surveyor licensed in the State of New Mexico, USA. Following the establishment of the water table elevations, submersible pressure transducers were installed in test cell and surrounding wells to provide continuous monitoring of water level changes. Extracted and injected water flow rates were measured using paddle-wheel flow indicating transmitters installed in manifolds housed within a remediation building centrally located in the hotspot area. Extracted and injected liquids from and to wells was transmitted via underground piping from the wells to the extraction and injection manifolds, respectively. The manifolds were used to isolate test cell flows, regulate flow and pressures, and inject/add tracers, bio-amendments, and nutrient mixtures.

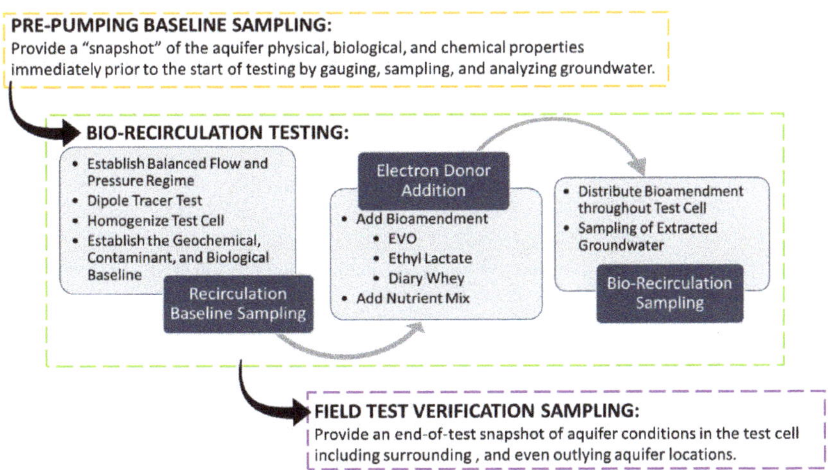

Figure 4. Field test plan procedure flow chart.

Groundwater field screening for general water quality parameters were analyzed in the filed using handheld digital meters fitted with electrochemical and solid-state sensors. These digital meters were calibrated prior to deployment in the field and checked with reference solutions prior to use.

Groundwater and groundwater filtrate samples were collected and analyzed pursuant to the materials, methods, and procedures prescribed in the Revised FTP [4]. The groundwater sample collection methods prescribed in the Revised FTP include containerization and preservation approaches that meet US EPA methodology. Similarly, analyses of groundwater samples and groundwater filtrate were performed, reported, and validated at US EPA accredited laboratories and third-party data validation services. The methods for groundwater and groundwater filtrate sample analysis results presented in this paper follow:

- PCE, TCE, DCE isomers, and VC analysis: Gas chromatograph with mass spectroscopy detector (GC/MS) via US EPA Method 8260 [14].
- TOC: Carbonaceous Analyzer via US EPA Method 9060A [15].
- Groundwater microbial consortia sampling: Collected on a 0.40-micron sterilized filter.
- Phospholipid fatty acid analysis (PLFA): quantification of total viable biomass and general profile of the microbial community.
- The CENSUS qPCR: DNA-based method to accurately quantify specific microorganisms and functional genes responsible for biodegradation of contaminants.

PLFA and CENSUS qPCR analyses were developed and performed by Microbial Insights. Details of the materials and methods employed for these analyses are available on their webpage at https://microbe.com. (accessed on 25 October 2021)

2.3.1. Pre-Pumping Baseline Sampling

It is important that the plume in the source zone and hotspot area be characterized prior to the start of the pilot tests to be able to update the contaminant, geochemical, and biological baseline for the aquifer where the FTP was performed. Each well within the source and hotspot areas was gauged to determine the water table elevations under ambient flow, screened in the field for general water quality parameters (temperature, dissolved oxygen, pH, reduction-oxidation (redox) potential and specific conductance), and sampled to establish a geochemical, biological, and contaminant baseline immediately prior to the start of induced flow within the test cells.

2.3.2. Biorecirculation Testing

This phase of the FTP was performed to gather data for the assessment of the in situ performance of three candidate electron donors and the addition of hydrogen gas injection at the source zone. Biorecirculation testing consisted of four tasks: tracer testing, recirculation baseline sampling, and electron donor addition and biorecirculation sampling. These tasks were performed on all four test cells.

Tracer Testing

The tracer study incorporated into the recirculation testing was a critical step in understanding the travel time between injection and extraction wells and the expected flow and dispersion of the bio-amendments in situ. Bromide (in the form of sodium bromide salts) was used as the conservative (non-adsorbing) tracer. The tracer was added to the injection well at the manifold using chemical metering pumps. Monitoring of the tracers was accomplished using two methods: bromide probes installed in each of the test cell extraction wells (EW-3, HSE-11, and HSE-14) and sampling of the extraction well manifold. Extraction manifold samples were submitted for bromide laboratory analysis. Bromide probes were programmed to record bromide concentration at a frequency of once every half hour and bromide results were logged in the probe's on-board datalogger. The datalogger was accessed and data downloaded weekly. Furthermore, samples were collected for laboratory bromide analyses during the recirculation testing. In addition, each batch of stock tracer solution was sampled and submitted for bromide analysis so that the injection concentration calculations could be checked for accuracy.

Baseline Recirculation Sampling and Analysis

Immediately prior to the additional bio-amendments to the test cell recirculation, the flow stream from the extraction well was sampled to assess the homogenization of the groundwater within the test cell. Similar to the pre-pumping baseline sampling, baseline recirculation samples were field screened (temperature, dissolved oxygen, pH, redox potential and specific conductance) and submitted for a full suite of laboratory analyses.

Electron Donor Addition

Upon completion of the baseline sampling tasks, the electron donor/nutrient test began. The electron donor volumes and recirculation requirements were established using the tracer test results. An injection of a nutrient mix was also included with each injection of electron donor. The nutrient mix was a yeast fermentation metabolite provided by JRW Bioremediation, LLC. It was added to accelerate the growth of the microbial populations. All the field test cells received supplemental injection of EVO. EVO was chosen as the electron donor for all four test cells primarily because vinyl chloride had begun to form in the EVO test cells but not in the EL or dairy whey test cells. Table 1 summarizes electron donor used at each test cell.

EVO Test Cell (HIS-19/HSE-14): The EVO used was a soybean oil-based product with a total organic carbon concentration of approximately 510 mg/L. Approximately 1381.7 L (365 gallons) of EVO and 283.9 L (75 gallons) of nutrient mix were fed as a stock solution into the injection well recirculation flow stream at an approximate feed rate of 15.1 L per hour (4 gallons per hour). Dosing of the EVO and nutrient mix required over five days. During and after the injection, recirculation was continued until the end of the 5-month testing period. Due to decreasing total organic carbon (TOC) concentrations measured in the extraction well discharge, supplemental electron donor addition was performed in the middle of the testing period. HSI-19 received approximately 946.4 L (250 gallons) of EVO solution over a 6h period and 227.1 L (60 gallons) of nutrient mix injected over a 2h period as a supplemental EVO dose.

Table 1. Summary of electron donor addition, sampling analysis, and method details by test cell.

Test Cell Name	Ethyl Lactate Test	EVO Test	Dairy Whey Test	Source Zone Test
Well Pair	HIS-14/HSE-11	HIS-19/HSE-14	HIS-8/HSE-6	IW-2S/IW-2D/EW-3
Target Zone	Hot Spot	Hot Spot	Hot Spot	Source Zone
Electron Donor	Ethyl Lactate	EVO	Dairy Whey	EVO
Date	6/12/2007	6/12/2007	6/20/2007	7/5/2007@IW-2D 7/6/2007@IW-2S
Amount	1457.38 L (385 gal)	1381.68 L (365 gal)	997.90 kg (2200 lb)	1249.19 L (330 gal)
Nutrient Mix	246.05 L (65 gal)	283.91 L (75 gal)	283.91 L (75 gal)	227.13 L (60 gal)
Rate	2.68 lpm (4.7 gpm)	15.14 lpm (4 gpm)	9.46 lpm (2.5 gpm)	
Recirculation Time	2 weeks	5 month	15 days	81 days
Supplement EVO	YES	YES	YES	YES
Date	10/12/2007	10/11/2007	10/15/2007	10/18/2007
Amount	946.35 L (250 gal)	946.35 L (250 gal)	946.35 L (250 gal)	832.79 L (220 gal)
Injection Period	6 h	6 h	6 h	5 h
Nutrient Mix	227.13 L (60 gal)	227.13 L (60 gal)	227.13 L (60 gal)	227.13 L (60 gal)
Injection Period	2 h	2 h	2 h	2 h
Sampled Well	HSE-11	HSE-14	HSE-6	EW-3
VOC Analysis Performed and Methods	cVOC, EPA 8260 TOC, EPA 9060	cVOC, EPA 8260 TOC, EPA 9060	cVOC, EPA 8260 TOC, EPA 9060	cVOC, EPA 8260 TOC, EPA 9060
# of VOC Samples	32	32	32	27
MMM Analysis Performed and Methods	Total Biomass Taxomonic ID with PLFA SRBs, CENSUS(qPCR) DHC, CENSUS (qPCR) Functional Genes CENSUS (qPCR)	Total Biomass, PLFA Taxomonic ID with PLFA SRBs, CENSUS(qPCR) DHC, CENSUS (qPCR) Functional Genes CENSUS (qPCR)	Total Biomass, PLFA Taxomonic ID with PLFA SRBs, CENSUS(qPCR) DHC, CENSUS (qPCR) Functional Genes CENSUS (qPCR)	Total Biomass, PLFA Taxomonic ID with PLFA SRBs, CENSUS(qPCR) DHC, CENSUS (qPCR) Functional Genes CENSUS (qPCR)
# of MMM Samples	6	6	6	6
Analytical Result	Figure 5	Figure 6	Figure 7	Figures 8–11

Source Zone Test Cell (IW-2S, IW-2D/EW-3: EVO and Hydrogen Gas Infusion): The source zone test cell was slightly different than the hotspot test cell in that injection wells within the shallow aquifer and within the source zone were split into two wells: a collocated shallow and deep injection well. This change was implemented because the deeper sediments above the aquitard were lower in permeability compared with the upper portion of the shallow aquifer, and these deeper sediments were significantly more contaminated. Therefore, without separating the injection well into two separate and isolated screen sections, amendments and nutrients injected into a single well would have been disproportionately administered to areas of higher permeability which did not coincide with the highest levels of contamination.

In this source zone test cell, electron donor addition occurred in two separate periods. The recirculation of the first period was three months, and the second period was one month. For the first period, infusion of hydrogen gas was initiated with recirculation prior to amendment injection. After 20 days of hydrogen gas infusion, the EVO/nutrient mix was injected in a fashion as described earlier. Approximately 1249.2 L (330 gallons) of EVO solution and 227.1 L (60 gallons) of nutrient mix were injected as a solution through IW-2S and IW-2D (the shallow and deep injection wells). The second period started with a supplemental dose of nutrient mixture and was followed by injection of EVO. Each of IW-2S and IW-2D received approximately 832.8 L (220 gallons) of EVO injected over a 5.0 h period and 227.1 L (60 gallons) of nutrient mix injected over a 2.0 h period. Recirculation followed the direct injection of EVO and nutrient solution.

Dairy Whey Test Cell (HSI-8/HSE-6): A portion of the dairy whey amendment was metered into the recirculation flow; however, due to a malfunction of the chemical metering equipment the recirculation flow was temporarily stopped, and most of the dairy whey amendment was directly injected into injection well HSI-8. Approximately 997.9 kg (2200 pounds) of whey along with 283.9 L (75 gallons) of concentrate nutrient solution was blended with groundwater extracted from HSE-6, and injected into HSI-8. The solution (whey/nutrient/groundwater) were injected over an approximate 2.0 h period at an approximate rate of 9.5 L per minute (2.5 gallons per minute). Recirculation was reinitiated immediately following the direct injection of the amendments and nutrients. Recirculation was stopped 15 days after the direct injection. About three months after ceasing recirculation, the test cell received a supplemental dose of EVO and nutrient solution. This supplemental dose added to HSI-8 consisted of approximately 946.4 L (250 gallons) of EVO solution added over a 6.0 h period and 227.1 L (60 gallons) of nutrient solution over a 2.0 h period. Recirculation, which began immediately after the EVO injection, operated continuously for a month.

Ethyl Lactate Test Cell (HSI-14/HSE-11): Approximately 1457.4 L (385 gallons) of EL and 246.1 L (65 gallons) of nutrient mix were injected simultaneously as a solution at an approximate feed rate of 2.7 L per hour (4.7 gallons per hour) over 4 days into the recirculation flow between HSE-11 and HSI-14. During and after the injection, recirculation was continued for total of two weeks. About 4 months after ceasing recirculation, a supplemental injection of nutrient mixture and EVO was added to the test cell. The supplemental injection at this test cell consisted of approximately 946.4 L (250 gallons) of EVO solution over a 6.0 h period and 227.1 L (60 gallons) of nutrient mix injected over a 2.0 h period. Recirculation, which began immediately after the supplemental EVO injection, lasted for a month.

Biorecirculation Sampling

Biorecirculation sampling occurred from the beginning of the recirculation period and 1 month after stopping recirculation flow. Table 1 summarizes biorecirculation sampling analysis and test methods. The primary sampling during this phase was from the test cell extraction wells at the manifold; however, in the source zone dual completed monitoring wells located 0.9 m (3 feet) from the extraction well and screened in the lower and upper portions of the shallow aquifer were also sampled. These wells were designated SMW-3S and SMW-3D, where the 'S' and 'D' designated the shallow and deeper zones of the water-table aquifer.

2.3.3. Field Test Verification Sampling

Verification sampling took place about two weeks after the last biorecirculation sampling. Similar to the pre-pumping baseline sampling, the verification testing encompassed collecting and analyzing groundwater from a much larger sample set compared to the sampling performed during the active pilot testing. The results from the verification sampling provided an end-of-test snapshot of aquifer conditions in the test-cell treatment zones, including surrounding, and even outlying aquifer locations. Sufficient groundwater was collected at each well to adequately allow for field screening and laboratory full suite analyses.

2.3.4. Microbiological Sampling

In addition to the routine geochemical and contaminant sampling and analyses, samples were collected during the baseline recirculation testing, electron donor addition, and verification sampling phases of the pilot test for PFLA and Census qPCR analysis. Sampling of the microbiome was achieved by flowing the groundwater through a sterile filter (Bio-flo® samplers supplied by Microbial Insights) to collect the suspended microbes onto the filter. The total volume of groundwater passed through the filter was recorded and the filter was sealed and put on ice for delivery to the laboratory for analysis. The residues captured on the filter, including the microorganisms filtered from the groundwater, were backflushed in buffered sterile water solution for concentration via centrifuge. Extracts for MMM analyses were generated from these concentrated filtrate residues. Three samples were submitted to establish a baseline of the microbial consortia present in the hotspot and source zone areas of the shallow aquifer. A total of six groundwater samples were collected during the biorecirculation testing and 1 was collected during verification sampling period for the MMM analyses submitted to Microbial Insights. All samples were subjected to select MMMs to determine the presence of chlorinated hydrocarbon-degrading microorganisms and genes producing reductive dechlorination enzymes, the basic phylogeny of the microbiome, and their population density. Unlike traditional laboratory-based microcosm studies, these analyses were performed on fresh groundwater samples, without culturing or growth on selective medium.

3. Results

3.1. Tracer Test Results

As discussed above in Section 2.2.1, remedial design modeling suggested a 9.5 lpm (2.5 gpm) recirculation flow rate to achieve an approximate 26-day travel time between injector–extractor wells. The dense, high-resolution geologic characterization achieved during the installation of the wells and the tracer test results, however, indicated that the first arrival and median travel times were much shorter than 26 days, at recirculation flow rates between 15.1 and 20.4 lpm (4.0 and 5.4 gpm). The difference between design and operational flows and travel times was due to heterogeneities in the shallow zone geology, which consisted of a general fining of the materials from the water table to the aquitard. This resulted in higher permeable zones in the upper portions of the aquifer and channels or braids of coarse materials interbedded and intermittent throughout. At the EL, EVO, and source zone test cells, between approximately 1892.7 and 2271.3 L (500 and 600 gallons) of bromide tracer stock solution was injected into the recirculation flow stream at rates ranging between approximately 15.1 and 18.9 L per hour (4 and 5 gallons per hour, respectively). The flows were adjusted upward from the design flow rates until the pressure responses recorded at the extraction and injection wells were approximately equal to the pressure responses at these wells predicted in the numerical flow model used for the remedial design. Average bromide stock solution concentrations ranged between approximately 1250 mg/L and 3000 mg/L. Care was taken to keep the injected bromide concentration below 100 mg/L to reduce its antiseptic properties. Tracer testing was not conducted in the dairy whey test cell.

Breakthrough and peak detections of the bromide tracer at the extraction wells varied from test cell to test cell. At the EL and EVO test cells, breakthrough was detected at the extraction wells 18 h after injection started. The peak detection times and concentrations varied between these test cells. In the EL test cell the peak detection occurred after 6.9 days of injection, and in the EVO test cell the peak detection occurred 5.4 days post injection. The peak concentrations detected at the EL and EVO test cells were approximately 50 mg/L and 80 mg/L bromide, respectively. These peak concentrations were between approximately 50% and 70% of the injected bromide concentration and indicated variability in hydrodynamic dispersion, likely a measure in the degree of heterogeneity in different areas of the water-table aquifer.

In the source zone test cell, a total of 2400.0 L (634 gallons) of tracer stock solution, at an average bromide concentration of 1250 mg/L, was injected. The average feed rate of the tracer stock solution was 20.4 L per hour (5.4 gallons per hour). Breakthrough of the bromide tracer was detected 96 h later in EW-3, with the peak detection occurring after 48 days of injection and recirculation. Compared with the other test cells, these results confirm the low permeability zones present at the base of the shallow aquifer and the need for segregating the source zone injection wells.

The USEPA semi-analytical ground-water flow simulation program RESSQC, a computational module of Wellhead Protection Analysis (WHPA), was used to delineate the time-related capture zone around the extraction wells and tracer fronts around injection wells. The volume of the capture zone at time of maximum breakthrough was modeled for each test cell. Ultimately, the flow model calibrated to the results of the tracer tests was used to assess full-scale operational modes, where injected amendments could be mixed with greater contact and long travel paths could be developed to extend travel times and improve contact with bio-amendments within the hotspot treatment area.

3.2. Dissolved Chlorinated Hydrocarbon Analysis Results

The Appendix A contains heat maps of bio-amendment distribution in the subsurface shown as TOC, as well as the distribution of PCE and its ERD byproducts (TCE, DCE, VC, and ethene) prior to and during pilot testing and the first portion of full-scale remediation operations. The sections below discuss the dissolved chlorinated volatile organic compounds (cVOC) results of each test cell.

3.2.1. Ethyl Lactate Test Cell cVOC Results

The proportions of PCE, TCE, DCE isomers, VC, and their sum (total cVOC) at HSE-11 are summarized in Figure 5. A comparison of VOC analysis results prior to and following the addition of EL showed that the dissolved-phase concentration of total cVOCs decreased from a baseline of 9.88 micromole (μM) to a low of 6.59 μM. Following the supplemental EVO addition, the concentration increased to 14.87 μM before decreasing to 9.39 μM on 8 January 2008. PCE concentrations steadily dropped from a baseline average of 1410 μg/L to non-detectable (<1.0 μg/L) on 19 November 2007, but increased to 2.7 μg/L on 8 January 2008. TCE increased from a baseline average of 307 μg/L to a high of 790 μg/L on 31 July 2007 before steadily decreasing, and was non-detectable (<1.0 μg/L) after 19 November 2007. Cis-DCE increased from a baseline average of 90 μg/L to a high of 1800 μg/L on 19 November 2007 and decreased to 900 μg/L as of 8 January 2008. Trans DCE increased from being non-detectable (<5.0 μg/L) to a concentration of 2.4 μg/L on 8 January 2008. Vinyl chloride was non-detectable (<1.0 μg/L) until the supplemental addition of EVO, after which a peak concentration of 32 μg/L was reached on 19 November 2007 and decreased to 4.1 μg/L by 8 January 2008. It should be noted that an anticipated result of ERD is increased dissolved-phase concentrations in the initial phases of remediation due to the increased solubility of PCE-daughter products (e.g., VC is approximately 18 times more soluble in water than PCE), the desorption of adsorbed-phase contaminants from the addition of amendments, and the increased flow induced by recirculation.

3.2.2. EVO Test Cell cVOC Results

Fractions of and total cVOC concentrations at HSE-14 are summarized in Figure 6. A comparison of cVOC analyses prior to and following the addition of EVO shows that total cVOCs decreased from 7.57 μM to a low of 5.34 μM. Following the supplemental EVO addition, concentrations increased to 10.77 μM, and then decreased to 7.95 μM on 8 January 2008. PCE steadily dropped from a baseline average of 1550 μg/L to non-detectable (<1.0 μg/L) on 26 November 2007, and then increased to 1.2 μg/L by 8 January 2008. TCE increased from the baseline of 45 μg/L to a high of 670 μg/L by 31 July 2007 before steadily decreasing to non-detectable (<1.0 μg/L) by 10 December 2007. Cis-DCE increased from a baseline of <5.0 μg/L to a high of 1200 μg/L on 19 November 2007 and decreased to

750 µg/L by 8 January 2008. Trans-DCE increased from non-detectable (<5.0 µg/L) to a concentration of 2.2 µg/L on 8 January 2008. Vinyl chloride was non-detectable (<2.0 µg/L) until 10 September 2007, after which a peak concentration of 19 µg/L was detected on 26 November 2007, but decreased to 11 µg/L on 8 January 2008.

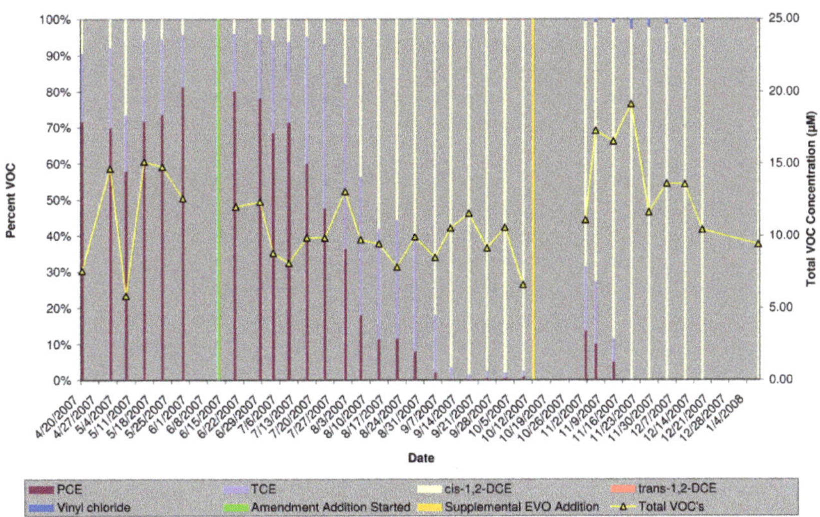

Figure 5. Comprehensive test cell data summary: Ethyl lactate/EVO, HSE-11/HSI-14, HSE-11% VOC, and total VOC concentration.

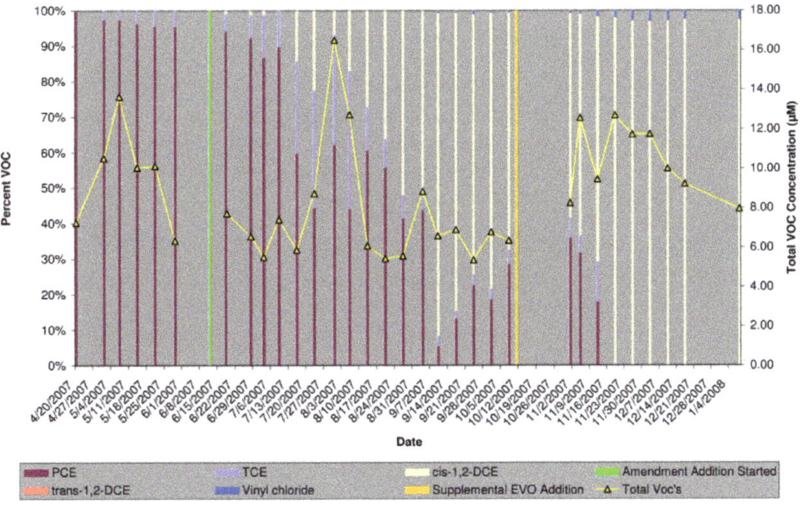

Figure 6. Comprehensive test cell data summary: EVO, HSE-14/HSI-19, HSE-14% VOC, and total VOC concentration.

3.2.3. Dairy Whey Test Cell cVOC Results

Proportions of and total cVOC concentrations at HSE-6 are summarized in Figure 7. A comparison of cVOC analyses prior to and following the addition of whey showed that the average concentration of total cVOCs decreased from 12.89 µM to a low of 8.09 µM. Following the supplemental EVO addition, the cVOCs increased to 16.85 µM with a low of 12.12 μM, and then increased to 14.60 μM on 8 January 2008. PCE concentrations steadily dropped from a baseline average of 1957 μg/L to non-detectable (<1.0 μg/L) by 19 November 2007. TCE increased from a baseline average of 393 μg/L to a high of 900 μg/L on 13 August 2007, and then steadily decreased to non-detectable (<1.0 μg/L) by 19 November 2007. Cis-DCE increased from 87 μg/L to a high of 2000 μg/L on 19 November 2007 and decreased to 1400 μg/L by 8 January 2008. Trans-DCE increased from being non-detectable (<1.0 μg/L) to 4.1 μg/L by 8 January 2008. Vinyl chloride was non-detectable (<1.0 μg/L) until the supplemental addition of EVO, after which a peak concentration of 7.1 μg/L was reached on 10 December 2007, decreasing to 6.6 μg/L on 8 January 2008.

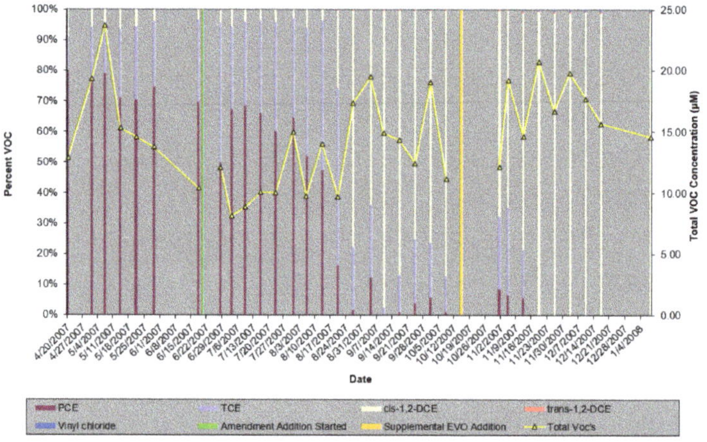

Figure 7. Comprehensive test cell data summary: Whey/EVO, HSE-6/HSI-8, HSE-6% VOC, and total VOC concentration.

3.2.4. Source Area Test Cell

The subsections below provide details on the shallow and deep injection well and source-area monitoring well cVOC sampling and analyses results. Percent cVOC and total cVOC concentrations at EW-3, IW-3D, and SMW-3S and SMW-3D are summarized in Figures 8–11, respectively.

IW-2S/EW-3 cVOC Results

A comparison of cVOC analyses prior to and following the addition of EVO shows that total cVOCs increased from a baseline of 52.73 μM to a peak of 106.03 μM detected on 4 September 2007, followed by a decrease to 43.60 μM by 8 January 2008. PCE concentrations steadily dropped from 8250 μg/L to 9.8 μg/L on 8 January 2008. TCE increased from a baseline average of 48.3 μg/L to a high of 3600 μg/L on 23 July 2007, and then steadily decreased to non-detectable (<2.0 μg/L) after 26 November 2007. Cis-DCE increased from 261 μg/L to a high of 9900 μg/L on 04 September 2007, and then decreased to 4100 μg/L by 8 January 2008. Trans DCE increased from non-detectable (<5.0 μg/L) to a peak of 150 μg/L detected on 21 November 2007, and concentrations fell to 48 μg/L as of 8 January 2008. Vinyl chloride was non-detectable (<5.0 μg/L) until 31 July 2007. A peak was detected at 110 μg/L on 12 November 2007, and concentrations decreased to 46 μg/L by 8 January 2008.

IW-2D/EW-3 cVOC Results

A comparison of cVOC analyses prior to and following the addition of EVO shows that total cVOCs decreased in SMW-3D from 383 μM to 191.5 μM by 8 January 2008. PCE concentrations steadily dropped from 63,500 μg/L to 29,000 μg/L by 8 January 2008. TCE concentrations, which were non-detectable (<20 µg/L) at baseline, increased to 94 µg/L prior to the supplemental EVO addition. Eventually, TCE reached a concentration of 800 µg/L by 8 January 2008. Cis- DCE was not detected from baseline until 30 October 2007. Levels increased to 1000 µg/L on 9 January 2008. Trans-DCE was first detected on 10 December 2007 at a concentration of 9.7 µg/L and reached 16 µg/L on 8 January 2008. Vinyl chloride is and has been non-detectable (<2.0 µg/L) since of 8 January 2008.

An analysis of samples collected from IW-3D showed that total VOCs decreased from a baseline 60.6 µM to 57.9 µM prior to the supplemental EVO injection, and then decreased to 17.1 µM on 09 January 2008. PCE concentrations steadily dropped from a baseline average of 9550 µg/L to being non-detectable (<2.0 µg/L) on 9 January 2008. TCE decreased from a baseline average of 45 µg/L to non-detectable (<2.0 µg/L) on 5 November 2007. Cis-DCE increased from 270 µg/L to a high of 9700 µg/L detected on 8 August 2007, then decreased to 1600 µg/L by 09 January 2008. Trans DCE concentrations increased from being non-detectable (<20 µg/L) to a peak of 160 µg/L detected on 27 October 2007, and then decreased to 18 µg/L on 9 January 2008. Vinyl chloride levels were non-detectable (<20 µg/L) until 8 August 2007, reached a peak of 320 µg/L on 31 October 2007, and then decreased to 24 µg/L on 9 January 2008.

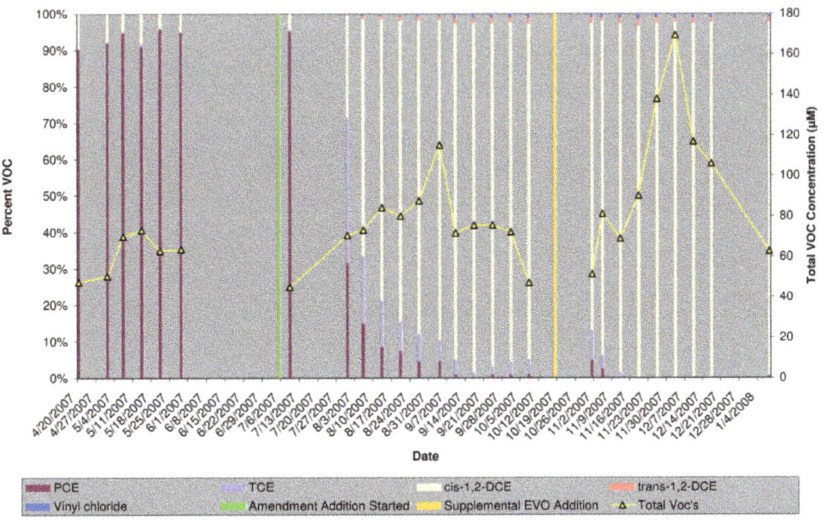

Figure 8. Comprehensive test cell data summary: EVO + hydrogen gas, EW-3% VOC, and total VOC concentration.

3.3. Microbiological Analysis Results

MMM analysis results from each test cell are discussed below.

Dairy Whey Test Cell (HSE-6/HSI-8): An analysis of HSE-6 groundwater samples showed a 1.5X order-of-magnitude (OM) increase in total population and a 4.5XOM increase in sulfate-reducing bacteria (SRBs) following the dairy whey addition. A comparison of PLFA analysis results prior to and following whey addition also showed an OM increase in total biomass. The bacteria *Dehalococcoides* (DHC), an important genus known to completely reduce cVOCs to ethane [16–19], increased 6X from baseline to 11 July 2007 (4.6 cells/mL vs. 27.3 cells/mL). The DHC population returned to baseline levels by the conclusion of the pilot test. The gene for the vinyl chloride reduction enzyme (vinyl chloride reductase (VC-RDase) and the gene for the TCE reduction enzyme (TCE-RDase) were not detected in baseline samples but were detected in samples collected after the dairy whey addition.

EL Test Cell (HSE-11/HSI-14): Analysis HSE-11 groundwater samples showed a 1.5X OM increase in total population and a 6X OM increase in SRBs following the addition of EL.

DHC population levels remained relatively unchanged in the EL test cell at less than ten cells per milliliter of groundwater. Additionally, an increased presence DHC was observed and VC-RDase was detectable after the addition of EL. A comparison of PLFA results prior to and following EL addition showed an order-of-magnitude increase in total biomass and an approximate 20% increase in *Firmicutes* (fermenting bacteria).

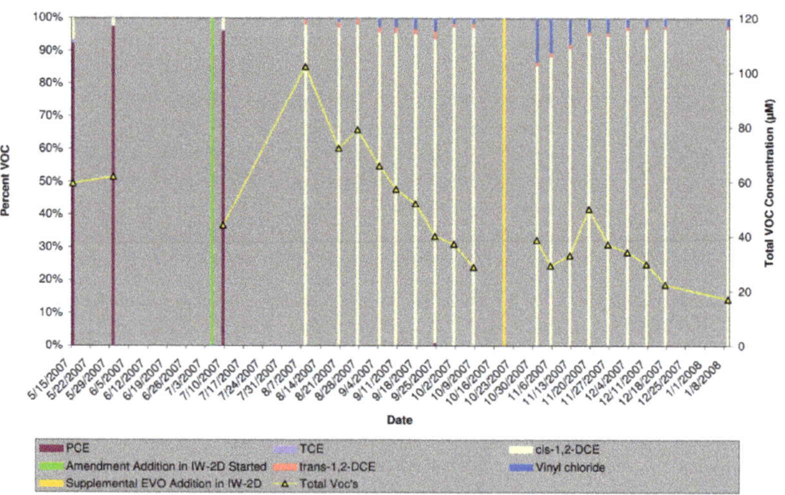

Figure 9. Comprehensive test cell data summary: EVO + hydrogen gas, IW-3D/IW-2D, IW-3D % VOC, and total VOC concentration.

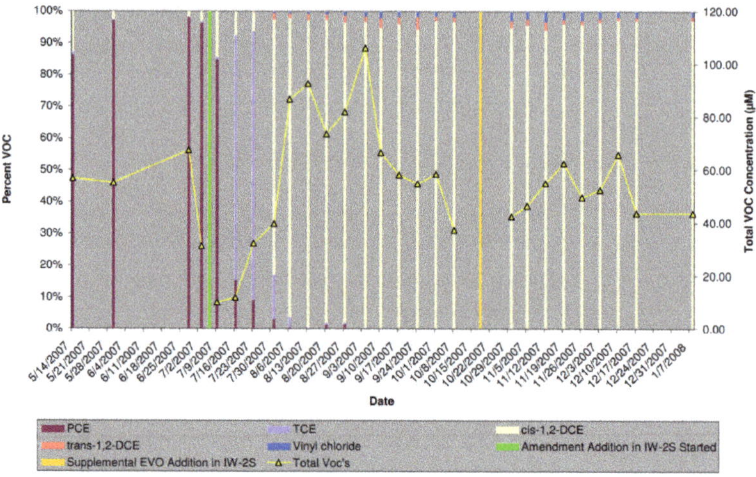

Figure 10. Comprehensive test cell data summary: EVO + hydrogen gas, SMW-3S% VOC, and total VOC concentration.

EVO Test Cell (HSE-14/HSI-19): Analysis of DNA extracted from HSE-14 groundwater samples showed a 1.5XOM increase in total population and a 6.5XOM increase in SRBs following the addition of EVO. The DHC population increased from non-detectable to approximately 100 cells per milliliter at the approximate midpoint of the test duration. By the end of the test the DHC population level dropped but remained detectable. Additionally, VC-RDase, which was not detected prior to EVO addition, was detected as part of the biorecirculation sampling. Comparison PLFA analysis results prior to and following EVO addition showed a 2XOM increase in total biomass, and an approximate 15% increase in *Firmicutes*.

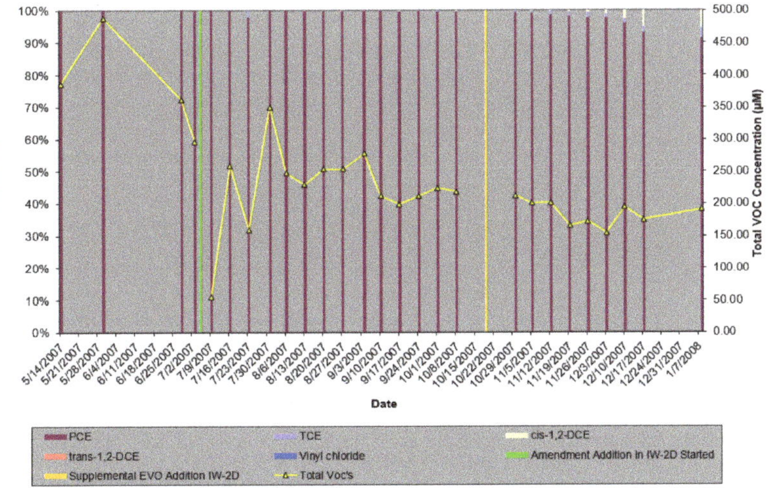

Figure 11. Comprehensive test cell data summary: EVO + hydrogen gas, SMW-3D% VOC, and total VOC concentration.

Source Zone EVO + Hydrogen Gas Infusion Test Cell (EW-3/IW-2S and IW-2D): Analysis of microbial DNA extracted from EW-3 groundwater samples showed a 2XOM in total population and a 4.5XOM increase in SRBs. The most encouraging aspects of the microbiological analysis were the changes in DHC. During the last three months of sampling (November 2007 through January 2008), DHC levels were observed to steadily increase from a background concentration of between 0.164 cells/milliliter and 9.98 cells/milliliter to 235 cells/milliliter. Comparison of PFLA results prior to and following EVO and hydrogen gas addition showed a 2.5XOM increase in total biomass, and an approximate 3-fold increase in Firmicutes.

Summary: The microbial consortia in the shallow aquifer was shown to contain the microbes (e.g., DHC) and the genes necessary for complete reductive dechlorination. Furthermore, EVO increased and supported respiration of cVOCs better when compared with the dairy whey and EL. MMM results tracked changes in the microbial consortium correlated to reductive chlorination. Select groundwater samples from the site were subjected to shotgun metagenomic analysis. Results from baseline to five months after the start of full-scale remedial action have already been published [20]. The MMM results in the present study are supported by the metagenomic analyses. Results incorporating over five years of genomic data, including 16S amplicons, are now available (manuscript in preparation).

4. Discussion

The selection of the electron donor may be the most important parameter for complete and rapid reductive dichlorination of chlorinated solvent contamination in groundwater. Previous studies have indicated that molecular hydrogen (H_2) plays a prominent role in the reductive dechlorination [13,21]. Complex donors (polylactates, vegetable oil, and dairy whey) supply the H2 as they are utilized by the subsurface microbial consortia.

Data from laboratory and field demonstrations of enhanced reductive dechlorination suggest that the rate and quantity of H_2 made available to the microbial population must be carefully controlled to limit the competition for the H_2 from microbial groups other than the dechlorinators [21,22]. These other competing microbial groups include methanogens and sulfate reducers. If too much H_2 is available, the diversity of the consortia can shift and those organisms that are more efficient at utilizing the H_2 than the dechlorinators can become prominent, resulting in inefficient utilization of the bio-amendments [21,23].

Trends in the contaminant concentration changes observed during the pilot testing were assessed using the U.S. EPA freeware package ProUCL 5.0 [24], which provides tools needed to statistically analyze environmental data. Specifically, the ProUCL 5.0 statistical test module was used to determine the confidence of trends in data. Specifically, the nonparametric trend test known as the Mann–Kendall (M-K) test was used with a 95% level of significance (LOC) for assessing the significance of trends in time-series datasets [25]. These trend analyses and other pertinent observations are discussed for each test cell in the subsections below.

4.1. Ethyl Lactate Test Cell (HSI-14/HSE-11)

A steady rate of reductive dechlorination of PCE to DCE was observed during the first half of the test. Trend analysis of the datasets support this conclusion in that the PCE and TCE concentrations over the first half of the test showed evidence of significant decline but no significant change during the second half. Vinyl chloride was not detected in any of the samples collected during the first half of the test. TOC showed evidence of a steady and significant decline, indicating that it was being utilized by the aquifer-borne microorganisms. TOC breakthrough and peak detection time matched well with the tracer test, suggesting that EL has a low adsorption to soil mass and high mobility in the aquifer.

EVO was added at the test duration midpoint. Immediately following total cVOC concentrations increased, likely a result of its desorption from the soil into the EVO. VC was detected in the test cell extraction well for the first time one month following the EVO addition.

4.2. EVO Test Cell (HSI-19/HSE-14)

Total cVOC levels peaked at approximately six weeks following the initial addition of EVO, likely again due to its desorption induced by the EVO. Following the peak, total cVOCs levels steadily dropped over the next month of the test at which point it remained relatively unchanged. M-K trend analyses of these portions of the dataset showed that during the month following the peak in PCE concentration there was evidence of a significant decline, followed by no significant changes in rate. VC was detected following the initial dose of EVO, indicating that complete ERD was occurring. TOC breakthrough and peak detection times did not match well with the conservative tracer test run in this test cell suggesting that EVO has high retention in the aquifer matrix and lower mobility in the groundwater.

Following the second dose of EVO, total dissolved chlorinated ethene concentrations again increased, including an initial peak in PCE and TCE. However, vinyl chloride also increased in the test cell extraction well, peaking about six weeks following the second EVO dose. Within approximately five weeks following EVO addition, PCE and TCE were below detection limits and total chlorinated ethene concentrations dropped from their peak. In addition, DCE and vinyl chloride concentrations dropped along with TOC levels. The results and findings stemming from the EVO test cell are similar to the successful approaches and studies contained in the protocol for in situ bioremediation of chlorinated solvents using edible oil [26].

Using the TOC levels detected at HSE-14 following its peak detections, substrate utilization rates were calculated throughout the test. Employing the Lineweaver–Burke Double Reciprocal Method, the inverses of these rates were plotted against the corresponding inverse of the midpoint substate concentrations to determine the half-saturation constant (K_m) and the maximum utilization rate (V_{max}) for this test cell. The Monod kinetics constants estimated were 153 mg/L and 8.7 mg/L per day for the K_m and V_{max}, respectively. These results were used to refine the electron donor dosing amounts and system recirculation periods as well as to define TOC monitoring targets for full scale operations.

4.3. Dairy Whey Test Cell (HSI-8/HSE-6)

A steady rate of reductive dechlorination of PCE to DCE was observed during the first four months of the test, after which levels remained relatively unchanged. VC was not detected in any of the samples collected between baseline and the midpoint of the test. Similar to the trend analyses performed at the EL and EVO test cells, the declines in PCE and TCE concentrations were statistically significant. Additionally, TOC was observed to steadily increase over the first seven weeks of the test, after which it steadily dropped. TOC breakthrough and peak detection times did not match well with the conservative tracer test run in this test cell suggesting that like EVO, dairy whey is retained in the aquifer matrix and its mobility is limited; however, it appeared to have less retention and greater mobility compared with EVO. The results observed in the dairy whey test cell were similar to those observed by others [27].

As a result of the EVO addition at the midpoint of the test, total dissolved chlorinated ethene concentrations increased, and vinyl chloride was detected in the test cell. Within approximately five weeks following this midpoint EVO addition, PCE and TCE levels were below detection limits and total cVOC levels, including DCE, dropped.

4.4. Source Zone Test Cell (EVO + Hydrogen Gas Infusion)

Results from samples collected at four wells (the extraction and injection wells EW-3 and IW-3D and two monitoring wells located in-between the injection and extraction wells, SMW-3S and SMW-3D) were used to assess the expected performance of full-scale enhanced bioremediation at the source zone. The extraction well used for recirculation at the source zone was EW-3. Approximately 0.9 m (3 feet) from EW-3 and in line with nested injection well IW-3S/D is the nested monitoring well SMW-3S/D. The shallow monitoring well SMW-3S was screened in the higher permeability material in the upper portion of the shallow aquifer at between approximately 1.5 m and 5.5 m (5 feet and 18 feet) bgs. The deep monitoring well SMW-3D was screened in the deeper low-permeability unit between approximately 6.1 m and 7.6 m (20 feet and 25 feet) bgs. Injection well IW-3D, which is located approximately 3.1 m (10 feet) southeast of the nested injection well IW-2S/D and 5.2 m (17 feet) north-northeast of extraction well EW-3, was screened in the low-permeability unit (similar to SMW-3D) and was used as an additional monitoring point. Monitoring at each of these sites is discussed below.

EW-3 Test Cell Extraction Well: After their initial increase, cVOC levels declined over the month before the midpoint EVO addition. During this period a steady rate of reductive dechlorination of PCE to DCE was observed, including the production of vinyl chloride. A trend analysis of PCE and TCE time-series datasets from the EW-3 samples showed evidence of significant decline during this month-long period. The trends in TOC levels at EW-3 matched approximately what was observed at the EVO Test Cell extraction well HSE-14.

SMW-3S Shallow Monitoring Well: The trends observed at SMW-3S closely resemble those observed from analysis of results from samples collected from EW-3. An M-K trend analysis of the time-series dataset showed evidence of significant decline in PCE and TCE as a result of enhanced reductive dichlorination.

IW-3D Test Cell Injection Well: Similar to the EW-3 observations, total cVOC levels steadily declined up to the second dose of EVO after they peaked following the initial injection of EVO. During this month-long period a steady and rapid rate of reductive dechlorination of PCE to DCE, including the production of vinyl chloride, was observed. The M-K trend analyses of the time-series datasets of the total cVOC, PCE, and TCE concentrations showed evidence of a statistically significant decline. Prior to the second dose of EVO, vinyl chloride production peaked and began to drop. The trend in VC levels matched the peak and drop in TOC levels. As a result of the second dose of EVO, total dissolved chlorinated VOC concentrations slightly increased before steadily decreasing. The concentrations of DCE and VC tracked the rise and fall of TOC concentration. By the end of test, PCE and TCE were not detectable.

SMW-3D Deep Monitoring Well: Unlike the results from the shallow monitoring well SMW-3S and the test cell extraction well EW-3, the M-K trend analysis of the time-series datasets showed no evidence of a significant change in total cVOC concentrations or its components (PCE/TCE/DCE) at SMW-3D prior to the administration of the second dose of EVO. The majority (>99%) of the cVOC mass remained as PCE and the total concentration remained relatively unchanged compared with baseline sampling results. After the additional dose of EVO, changes in the distribution of chlorinated ethenes began to appear and TOC levels began to increase slightly. At the end of the test the majority (>91%) of the cVOC mass at SMW-3D remained as PCE. VC was not detected, and TOC levels, although above background levels, were low at approximately 5 mg/L.

Summary: Recent studies have shown that DNAPL source longevity may be reduced by as much as an order of magnitude (e.g., from hundreds of years to decades) by ERD [28,29]. The pilot test results suggest that similar treatment success could be achieved in the source area by ERD. Additionally, moving EVO into the lower portion of the shallow aquifer is difficult due to its low permeability from the greater fractions of silt and clay in the aquifer matrix. However, since IW-3D is constructed like SMW-3D, it can be said that once EVO substrate is distributed in the tighter, deeper portion of the shallow aquifer source zone, complete ERD should in this zone will occur. Careful and frequent addition of substrate to this region of the shallow aquifer will be key in successfully remediating the source zone.

5. Conclusions

Based on the findings through the field pilot test, EVO along with a yeast-extract based nutrient mixture was selected as the bio-amendment of choice for treatment of the affected aquifer. TOC and cVOC monitoring can be used to track progression and assess the need for additional substrate dosing. Electron acceptor and general water quality analyses can be performed on a less frequent (e.g., yearly) basis. Using the existing infrastructure and because of the limited costs, hydrogen gas infusion should continue at the source area during full scale treatment. As the source zone area treatment progresses, recirculation of the treated source zone groundwater with the other portions of the system may be beneficial due to the diversity of the source area consortia as it relates to halogenated hydrocarbon degraders. A periodic (yearly) check for those organisms that respire cVOCs completely (e.g., *Dehalococcoides*) will assist in determining potentially recalcitrant areas of the ERD remediation system.

Author Contributions: Conceptualization, P.G., A.B. and R.A.R.; methodology, P.G., A.B., R.A.R. and J.M.; software, P.G.; validation, P.G., A.B., R.A.R. and J.M.; formal analysis, P.G. and A.B.; investigation, P.G. and A.B.; resources, P.G., A.B., R.A.R. and J.M.; data curation, P.G. and A.B.; writing—original draft preparation, P.G. and A.B.; writing—review and editing, R.A.R. and J.M.; visualization, P.G. and A.B.; supervision, R.A.R. and J.M.; project administration, P.G. and A.B. All authors have read and agreed to the published version of the manuscript.

Funding: The project was funded by US EPA Superfund Program: North Railroad Avenue Plume (USEPA ID: NMD986670156): https://cumulis.epa.gov/supercpad/cursites/csitinfo.cfm?id=0604299.

Institutional Review Board Statement: Not applicable.

Informed Consent Statement: Not applicable.

Data Availability Statement: Data presented in this paper can be obtained from the New Mexico Environment Department by public records request (https://www.env.nm.gov/public-record-request/). Requestor should cite the North Railroad Avenue Plume project, data presented contained in the Field Test Plan Results Report, North Railroad Avenue Plume Superfund Site dated 25 February 2008.

Conflicts of Interest: The authors declare no conflict of interest.

Appendix A

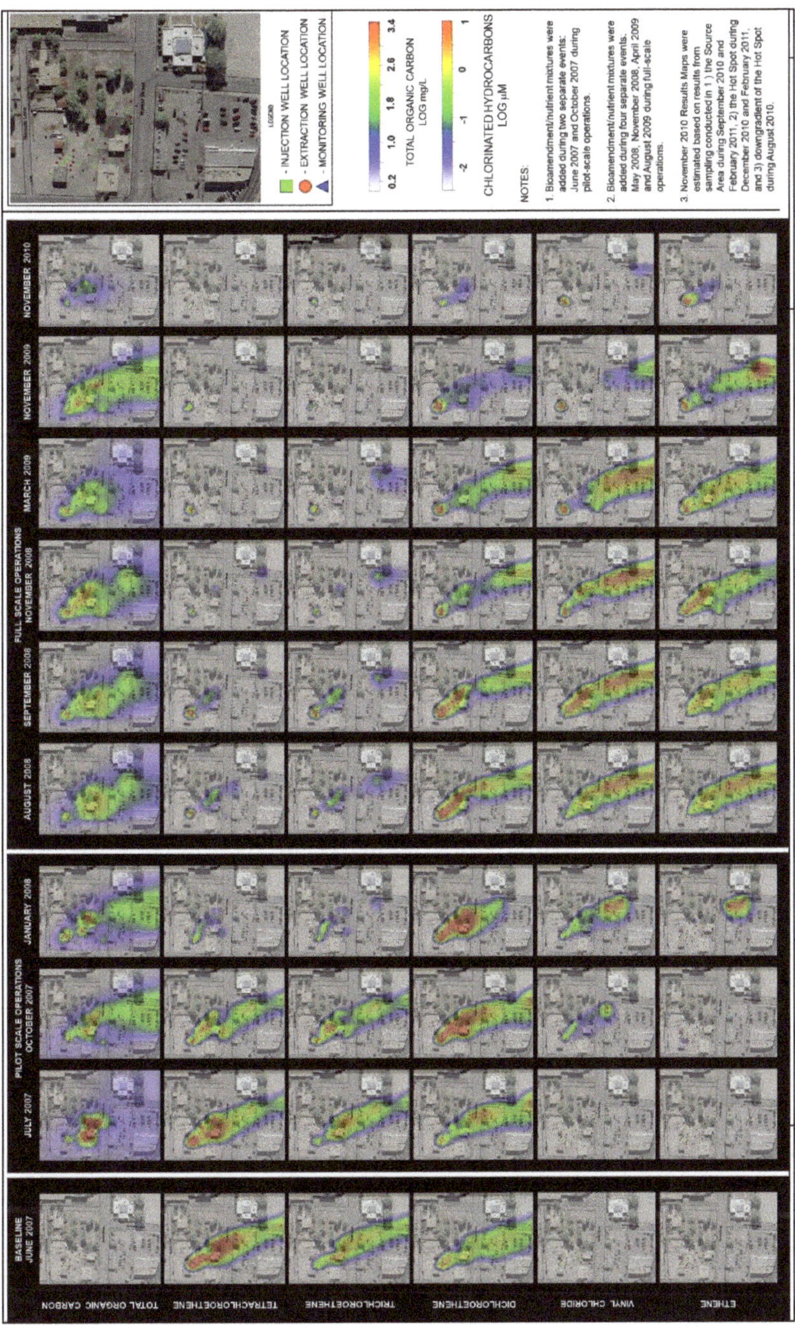

Figure A1. Heat map of bio-amendment addition (TOC) and dissolved chlorinated hydrocarbon during and following the pilot test.

References

1. Duke Engineering & Services. *Remedial Investigation Report, North Railroad Avenue Plume Superfund Site*; NPL #NMD986670156; New Mexico Environmental Department Superfund Oversight Section Archive: Española, NM, USA, 2001.
2. Duke Engineering & Services. *Feasibility Study Report, North Railroad Avenue Plume Superfund Site*; NPL #NMD986670156; New Mexico Environmental Department Superfund Oversight Section Archive: Española, NM, USA, 2001.
3. INTERA Incorporated. *Final Design Report, North Railroad Avenue Plume*; New Mexico Environmental Department Superfund Oversight Section Archive: Española, NM, USA, 2002.
4. AMEC Earth & Environmental. *Final Revised Field Test Plan Hot-Spot Bioremediation System North Railroad Avenue Plume Superfund Site*; NPL# NMED986670156; New Mexico Environmental Department Superfund Oversight Section Archive: Española, NM, USA, 2006.
5. John, C. Fry, Direct Method and Biomass Estimation. In *Method in Microbiology*; Elsevier Science: Amsterdam, The Netherlands, 1990; Volume 22, pp. 41–85. Available online: https://www.sciencedirect.com/science/article/abs/pii/S0580951708702393#aep-abstract-id7 (accessed on 8 October 2021).
6. Peacock, A.D.; White, D.C. *Hydrocarbon and Lipid Microbiology Protocols*; Microbial Biomass and Community Composition Analysis Using Phospholipid Fatty Acids; Springer: Berlin, Germany, 2016; pp. 65–76. Available online: https://link.springer.com/protocol/10.1007%2F8623_2016_213 (accessed on 8 October 2021).
7. Muyzer, G.; de Waal, E.C.; Uitterlinden, A.G. Profiling of complex microbial populations by denaturing gradient gel electrophoresis analysis of polymerase chain reaction-amplified genes coding for 16S rRNA. *Appl. Environ. Microbiol.* **1993**, *59*, 695–700. [CrossRef] [PubMed]
8. Löffler, F.E.; Sun, Q.; Li, J.; Tiedje, J.M. 16S rRNA gene-based detection of tetrachloroethene-dechlorinating Desulfuromonas and Dehalococcoides species. *Appl. Environ. Microbiol.* **2000**, *66*, 1369–1374. [CrossRef] [PubMed]
9. Jugder, B.E.; Ertan, H.; Bohl, S.; Lee, M.; Marquis, C.P.; Manefield, M. Marquis and Michael Manefield Organohalide Respiring Bacteria and Reductive Dehalogenases: Key Tools in Organohalide Bioremediation. *Front. Microbiol.* **2016**, *7*, 249. [CrossRef] [PubMed]
10. Chiang, S.Y.D.; Mora, R.; Diguiseppi, W.H.; Davis, G.; Sublette, K.; Gedalanga, P.; Mahendra, S. Characterizing the intrinsic bioremediation potential of 1,4-dioxane and trichloroethene using innovative environmental diagnostic tools. *J. Environ. Monit.* **2012**, *14*, 2317–2326. [CrossRef] [PubMed]
11. Anantharaman, K.; Brown, C.T.; Hug, L.A.; Sharon, I.; Castelle, C.J.; Probst, A.J.; Thomas, B.C.; Singh, A.; Wilkins, M.J.; Karaoz, U.; et al. Banfield Thousands of microbial genomes shed light on interconnected biogeochemical processes in an aquifer system. *Nat. Commun.* **2016**, *7*, 13219. [CrossRef] [PubMed]
12. Zinder, S.H.; Gossett, J.M. Reductive Dechlorination of Tetrachloroethene by a High Rate Anaerobic Microbial Consortium. *Environ. Health Perspect.* **1995**, *103* (Suppl. 5), 5–7. [PubMed]
13. Persons Corporations, Principles and Practices of Enhanced Anaerobic Bioremediation of Chlorinated Solvents Prepared for Air Force Center for Civil Engineer Center, Naval Facilities Engineering Command and Environmental Security Technology Certification Program. 2004. Available online: https://frtr.gov/matrix/documents/Enhanced-In-Situ-Reductive-Dechlorinated-for-Groundwater/2004-Principles-and-Practices-of-Enhanced-Anaerobic-Bioremediation-of-Chlorinated-Solvents.pdf (accessed on 19 October 2021).
14. United States Environmental Protection Agency. *Method 8260 (SW-846): Volatile Organic Compounds by Gas Chromatography/Mass Spectrometry (GC/MS), Revision 3*; United States Environmental Protection Agency: Washington, DC, USA, 2006.
15. United States Environmental Protection Agency. *Method 9060A (SW-846): Total Organic Carbon by Carbonaceous Analyzer, Revision 2*; United States Environmental Protection Agency: Washington, DC, USA, 2004.
16. Scheutz, C.; Durant, N.D.; Dennis, P.; Hansen, M.H.; Jørgensen, T.; Jakobsen, R.; Cox, E.E.; Bjerg, P.L. Concurrent ethene generation and growth of Dehalococcoides containing vinyl chloride reductive dehalogenase genes during an enhanced reductive dechlorination field demonstration. *Environ. Sci. Technol.* **2008**, *42*, 9302–9309. [CrossRef] [PubMed]
17. van der Zaan, B.; Hannes, F.; Hoekstra, N.; Rijnaarts, H.; de Vos, W.M.; Smidt, H.; Gerritse, J. Correlation of Dehalococcoides 16S rRNA and chloroethene-reductive dehalogenase genes with geochemical conditions in chloroethene-contaminated groundwater. *Appl. Environ. Microbiol.* **2010**, *76*, 843–850. [CrossRef] [PubMed]
18. Clark, K.; Taggart, D.M.; Baldwin, B.R.; Ritalahti, K.M.; Murdoch, R.W.; Hatt, J.K.; Löffler, F.E. Normalized Quantitative PCR Measurements as Predictors for Ethene Formation at Sites Impacted with Chlorinated Ethenes. *Environ. Sci. Technol.* **2018**, *52*, 13410–13420. [CrossRef] [PubMed]
19. Åkesson, S.; Sparrenbom, C.J.; Paul, C.J.; Jansson, R.; Holmstrand, H. Characterizing natural degradation of tetrachloroethene (PCE) using a multidisciplinary approach. *Ambio* **2021**, *55*, 1074–1088. [CrossRef] [PubMed]
20. Reiss, R.A.; Guerra, P.; Makhnin, O. Metagenome phylogenetic profiling of microbial community evolution in a tetrachloroethene-contaminated aquifer responding to enhanced reductive dechlorination protocols. *Stand. Genom. Sci.* **2016**, *11*, 88. [CrossRef] [PubMed]
21. Yang, Y.; McCarty, P.L. Competition for Hydrogen within a Chlorinated Solvent Dehalogenating Anaerobic Mixed Culture. *Environ. Sci. Technol.* **1998**, *32*, 3591–3597. [CrossRef]
22. Koenigsberg, S.S.; Sandefur, C.A. The Use of Hydrogen Release Compound for the Accelerated Bioremediation of Anaerobically Degradable Contaminants: The Advent of the Time-Release Electron Donors. *Remediation* **1999**, *10*, 31–53. [CrossRef]

23. Fennell, D.E.; Gossett, J.M.; Zinder, S.H. Comparison of Butyric Acid, Ethanol, Lactic Acid, and Propionic Acid as Hydrogen Donors for the Reductive Dechlorination of Tetrachloroethene. *Environ. Sci. Technol.* **1997**, *31*, 918–926. [CrossRef]
24. United States Environmental Protection Agency. *ProUCL Version 5.0.00 Technical Guide, Statistical Software for Environmental Data Sets with and without Nondetect Observations*; United States Environmental Protection Agency: Washington, DC, USA, 2013.
25. Hollander, M.; Wolfe, D.A. *Nonparametric Statistical Methods*, 2nd ed.; John Wiley & Sons: New York, NY, USA, 1999.
26. Solutions IES, Inc.; Terra Systems, Inc. (TSI); Parsons Infrastructure & Technology Group, Inc. Final Protocol for In Situ Bioremediation of Chlorinated Solvents Using Edible Oil for United State Air Force Center for Engineering and the Environmental. 2007. Available online: https://clu-in.org/download/remed/Final-Edible-Oil-Protocol-October-2007.pdf (accessed on 19 October 2021).
27. Tarnawski, S.-E.; Rossi, P.; Brennerova, M.V.; Stavelova, M.; Holliger, C. Validationofan Integrative Methodology to Assess and Monitor Reductive Dechlorination of Chlorinated Ethenes in Contaminated Aquifers. *Front. Environ. Sci.* **2016**, *4*, 7. [CrossRef]
28. Christ, J.A.; Ramsburg, C.A.; Abriola, L.M.; Pennell, K.D.; Löffler, F.E. Coupling Aggressive Mass Removal with Microbial Reductive Dechlorination for Remediation of DNAPL Source Zones: A Review and Assessment. *Environ. Health Perspect.* **2005**, *113*, 465–477. [CrossRef] [PubMed]
29. Durant, N.; Smith, L.; Condit, W. Technical Memorandum: Design Considerations for Enhanced Reductive Dechlorination Prepared for Naval Facilities Engineering Command. 2015. Available online: https://www.navfac.navy.mil/content/dam/navfac/Specialty%20Centers/Engineering%20and%20Expeditionary%20Warfare%20Center/Environmental/Restoration/er_pdfs/d/navfacexwc-ev-tm-1501-erd-design-201503f.pdf (accessed on 19 October 2021).

 *applied sciences*

Article

Application of Combined In Situ Chemical Reduction and Enhanced Bioremediation to Accelerate TCE Treatment in Groundwater

Min-Hsin Liu [1,*], Chung-Ming Hsiao [1], Chih-En Lin [2] and Jim Leu [2]

[1] Department of Environmental Engineering and Management, Chaoyang University of Technology, 168, Jifeng E. Rd., Wufeng District, Taichung 413310, Taiwan; xiao10225088@gmail.com
[2] Clean EnviroEngineering Technology Co., Ltd., Xitun District, Taichung 407608, Taiwan; scorpiobear@cleanenv.com.tw (C.-E.L.); rjleu@yahoo.com (J.L.)
* Correspondence: jliu@cyut.edu.tw; Tel.: +886-910599767

Abstract: Groundwater at trichloroethylene (TCE)-contaminated sites lacks electron donors, which prolongs TCE's natural attenuation process and delays treatment. Although adding electron donors, such as emulsified oil, accelerates TCE degradation, it also causes the accumulation of hazardous metabolites such as dichloroethylene (DCE) and vinyl chloride (VC). This study combined in situ chemical reduction using organo-iron compounds with enhanced in situ bioremediation using emulsified oil to accelerate TCE removal and minimize the accumulation of DCE and VC in groundwater. A self-made soybean oil emulsion (SOE) was used as the electron donor and was added to liquid ferrous lactate (FL), the chemical reductant. The combined in situ chemical reduction and enhanced in situ bioremediation achieved favorable results in a laboratory microcosm test and in an in situ biological field pilot test. Both tests revealed that SOE+FL accelerated TCE degradation and minimized the accumulation of DCE and VC to a greater extent than SOE alone after 160 days of observation. When FL was added in the microcosm test, the pH value decreased from 6.0 to 5.5; however, during the in situ biological pilot test, the on-site groundwater pH value did not exhibit obvious changes. Given the geology of the in situ pilot test site, the SOE+FL solution that was injected underground continued to be released for at least 90 days, suggesting that the solution's radius of influence was at least 5 m.

Keywords: ferrous lactate; in situ chemical reduction; bioremediation; trichloroethylene (TCE); green and sustainable remediation (GSR)

Citation: Liu, M.-H.; Hsiao, C.-M.; Lin, C.-E.; Leu, J. Application of Combined In Situ Chemical Reduction and Enhanced Bioremediation to Accelerate TCE Treatment in Groundwater. *Appl. Sci.* **2021**, *11*, 8374. https://doi.org/10.3390/app11188374

Academic Editor: Elida Nora Ferri

Received: 31 July 2021
Accepted: 7 September 2021
Published: 9 September 2021

Publisher's Note: MDPI stays neutral with regard to jurisdictional claims in published maps and institutional affiliations.

1. Introduction

The remediation of sites contaminated by dense nonaqueous phase liquid (DNAPL) is extremely difficult, which is why the development of economic and effective remediation technologies for DNAPL-contaminated sites is crucial. Trichloroethylene (TCE), a common DNAPL, is used in textile processing, refrigeration, vapor degreasing, metal washing, dry wash facilities, lubricants, and adhesives [1]. TCE is a common pollutant in sites with contaminated groundwater. The International Agency for Research (IARC) on Cancer listed TCE as "carcinogenic to humans" [2,3]. Thus, TCE is one of the most common and hazardous pollutants.

Microbes can convert chlorinated pollutants into hazard-free final products through dechlorination under an anaerobic state. However, the lack of electron donors in the environment often prolongs the time required for microbes to degrade chlorinated pollutants. Adding a commercial emulsified vegetable oil (CEVO) as an electron donor to accelerate reductive dechlorination of contaminated groundwater is a frequently adopted in situ bioremediation technology [4–6]. Direct injection of edible oil into the contamination plume does not yield favorable remediation results because edible oil has poor transmissibility;

therefore, large amounts of groundwater must be drawn and replaced continuously to enable the oil to diffuse into soil pores [7]. Emulsified edible oil (emulsified vegetable oil) has fine and evenly distributed droplets, and can thus more easily diffuse into the pores of different types of soil [8–10]. Emulsified edible oil releases carbon sources and fatty acid, which can stimulate anaerobic reductive dechlorination. In addition, microbes can be applied to facilitate the complete removal of chlorine to form nontoxic chlorinated ethenes (CEs) [11–13]. Previous studies have injected emulsified oil into contaminated aquifers to form a bioreactor system. Through the slow release of a CEVO, microbial activity in the aquifer is stimulated over a long period of time to achieve complete dichlorination [14] and remove heavy metal and organic pollutants in groundwater [15,16].

Although CEVOs are favorable hydrogen-releasing substrates [17], in situ bioremediation using CEVOs to stimulate anaerobic reductive dechlorination of contaminated groundwater has a few defects. Biodegradation of a CEVO generates organic acid, which decreases the environment's pH value and causes soil acidification problems. In anaerobic environments, sulfates may form hydrogen sulfide and generate odors. Moreover, when vinyl chloride (VC), the metabolite of reductive dechlorination of tetrachloroethylene (PCE) and TCE, accumulates to a certain level, microbial activity may decline [18] or the growth of microbes that can degrade VC may be partially inhibited [19]. Subsequently, the VC degradation rate decreases and prolongs the time required for bioremediation of the contaminated site. Adding zerovalent iron can effectively overcome these defects. Zerovalent iron has three main mechanisms in water: (1) in anaerobic environments, it releases electrons from its surface to facilitate the reductive dechlorination of chlorinated organic compounds (RCl); (2) Fe^{2+} is oxidized into Fe^{3+} and releases electrons to facilitate the reductive dechlorination of RCl; and (3) zerovalent iron is oxidized in aqueous solutions to reduce water to H_2 and OH^-, and H_2 can be used as the electron donor to facilitate catalyzed hydrogenolysis with RCl, thereby achieving the goal of dechlorination [20]. Additionally, as a zerovalent iron solution added into sulfide can generate sulfides and iron oxides, many scholars have added zerovalent iron to biosludge from husbandry wastewater containing high concentrations of sulfides, and found that the generation of hydrogen sulfide was significantly controlled [21,22]. This indicates that the addition of zerovalent iron could resolve the odorous water problem caused by hydrogen sulfide. Furthermore, a study by Herrero et al. showed the benefits of zerovalent iron on bioremediation of PCE [23]. The study conducted remediation tests in four groups: (a) natural attenuation, (b) addition of lactic acid, (c) addition of zerovalent iron, and (d) addition of lactic acid integrated with zerovalent iron. The results revealed that, in the group with lactic acid added, when the microbes degraded PCE to *cis*-1,2-dichloroethene (DCE), the concentration of *cis*-1,2-DCE could not be reduced effectively and the generation of VC was not observed. In the zerovalent iron group, decreases in PCE and TCE concentrations were observed; however, the *cis*-1,2-DCE concentration continued to increase and could not be effectively reduced. In the group with lactic acid integrated with zerovalent iron, PCE was gradually degraded into VC and the concentration of VC gradually decreased. Therefore, the addition of zerovalent iron also helps overcome the bottleneck of in situ bioremediation in degrading compounds into VC through reductive dechlorination.

Although the addition of zerovalent iron can promote the degradation of VC derivatives and solve many in situ bioremediation-derived problems, such as acidification and odors, zerovalent iron is a solid that may, after injection, be synthesized with environmental substances and precipitate, thereby limiting the distance of transmission. Su et al. used pneumatic injection and direct injection methods and observed the distance of transmission of emulsified zerovalent iron after injection [24]. The results revealed that the transmission distance through pneumatic injection was approximately 2.1 m and the maximal transmission distance through direct injection was 0.89 m. The majority of the nano-iron contained in the chemical was converted into magnetite after 2.5 years and the particle size varied from 35–140 nm to 0.1–1 μm. The results suggested that zerovalent iron reacts in underground environments over time and forms coordination complexes, which have enlarged particle sizes and might affect the distance of transmission. Ferrous lactate (FL) can be dissolved in water to form aqueous solutions for injection, which is convenient for in situ operation and may have a longer distance of transmission than solid zerovalent iron.

This study evaluated the feasibility of using two types of self-made emulsions combined with FL in the bioremediation of TCE-contaminated sites. First, a microcosm test was conducted to evaluate the differences between the two types of self-made emulsions and commercial emulsion and to evaluate if remediation effectiveness increased after the addition of FL. Then, in situ biological field tests were conducted to understand the steps required to implement bioremediation with self-made emulsion combined with FL. The results can serve as a reference for in situ remediation.

2. Materials and Methods

2.1. Materials and Analytical Methods

The first emulsion used in this study was soybean oil emulsion (SOE), which was mainly comprised of soybean oil (69.5%), sodium lactate (4%), sodium bicarbonate (1%), simple green (10.5%), sucrose fatty acid ester (4.5%), and water (10.5%). The emulsion was blended using a 10,000 rpm homogenizer for 30 min. The second emulsion was fatty acid ester (FAE), which was mainly composed of oleic acid (10.4%), glycerol (5%), ethyl lactate (5.9%), buffer solution (1%), and water (77.7%). The emulsion was blended using a 10,000 rpm homogenizer for 5 min. The FL selected was a commercial product from Henan Jindan Lactic Acid Technology Co., Ltd., Dancheng County, Henan, China. During the experiment, in situ groundwater sampling and water quality analysis were performed in accordance with the standard analysis methods of Taiwan. Volatile organic compounds in water were detected with a gas chromatograph/mass spectrometer; total organic carbon in water was detected with a method of peroxide pyrosulfate thermal oxidation/infrared spectrometry; and nitrate, nitrite, and sulfate were detected by ion exchange chromatography. The analytical methods used in the article follow National Institute of Environmental Analysis, Taiwan EPA [25–28].

2.2. Microcosm Test

The microcosm test consisted of five test sample groups. The five test sample groups were blank, self-made SOE, FAE, CEVO, and self-made SOE+FL (1%). Each of the groups used 250 mL serum vials to prepare six vials of isolated microcosms. When the predetermined time for analysis was up, the vials were opened one by one for analysis. Except for the blank group, which did not have any emulsion added, each serum vial consisted of 60 g of soil (sifted through a #20 mesh net), 15 mL of emulsion, a small amount of nutrients, and 2 mg/L of TCE. Subsequently, groundwater samples from the in situ field test location were added to the vials until no headspace was left; then, the vials were sealed. The vial bodies were wrapped with aluminum foil to minimize light irradiation. The total duration of the microcosm test was 90 days. Multiple samples were prepared for each of the five test groups using the jar test method. On days 0, 10, 25, 45, 60, and 90, a serum vial from each group was opened for pH, oxidation-reduction potential (ORP), dissolved oxygen (DO), electrical conductivity (EC), total organic carbon (TOC), and volatile organic compounds (VOCs) analyses. On days 0, 10, 45, and 90, additional nitrate, nitrite, and sulfate analyses were performed. On days 0 and 90, the blank group and the test group with the greatest effectiveness to date were selected for real-time polymerase chain reaction analysis [29] to evaluate the variation in the amount of representative strains of *Dehalococcoides* (DHC) and *Dichloroeliminans* (DCA1), which are microbes that can facilitate reductive dechlorination.

2.3. In Situ Biological Field Test

The in situ field test was conducted at a metal processing plant in central Taiwan. The plant had used TCE for surface processing without proper contamination control measures, resulting in TCE contamination at the site. The site's geological structure consisted of 0–1.5 m of sandy silt, 1.5–2 m of silty sand with sporadic clay, 2–4 m of silty clay with sandy clay, 4–5.5 m of silty medium-to-fine sand, 5.5–6 m of silty clay, and 6–10 m of silty medium-to-fine sand with sporadic thin layers of silt. The interbedded geologic structure was frequently observed at the site and the groundwater level was from 4 m to 5 m below the ground surface. The groundwater flow direction was mainly in the northeast to southwest direction. The hydraulic conductivity of the site was approximately 2.8×10^{-5} to 1.6×10^{-6} m/s. The injection well and monitoring well distribution is presented in Figure 1. At the upstream spot 1 m away from the injection well (IW), an observation well (OW1) was established. OW2 and OW3 were established at downstream spots 3 m and 5 m away from the IW, respectively. In addition, a contaminated monitoring well was selected as the control well (CW) within the contamination range of the site. The CW was approximately 30 m away from the in situ biological field test site and was free from the effects of the in situ chemical injection.

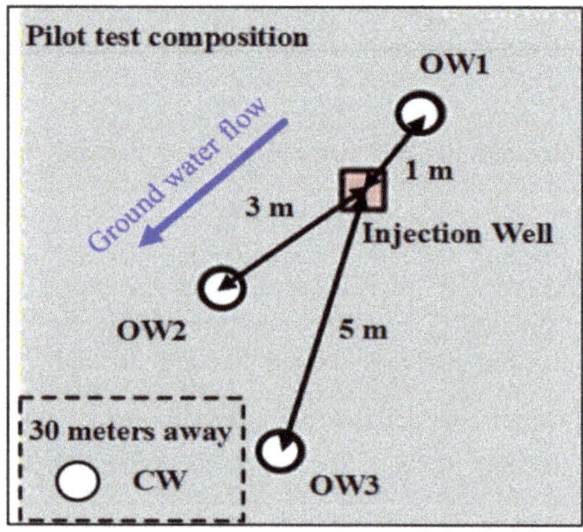

Figure 1. In situ biological field test well distribution.

Given that the geologic structures were silty fine sand and interbedded silt and clay, the double packer injection method was adopted. The double packer injection well structure is depicted in Figure 2b. The outer layer of the Manchette tube was Bentonite cement slurry. On the Manchette tube, injection holes were designed at 50 cm intervals. During injection of the chemicals at the target depth, the double packer system was extended to the predetermined injection depth and filled with water in the water bag at both ends to ensure chemical exposure at each contamination depth. During injection, the injection pressure was controlled at 2–4 bar to avoid cracks on soil pores and the formation of short-circuiting flows. During the testing period, chemicals were injected once every 2 months. The amount injected was adjusted according to the TOC detection results during the testing period. During the entire testing period, water quality characteristics including TOC, VOCs, nitrate, nitrite, ammonia nitrogen, sulfate, sulfides, total iron, and iron (II) were periodically monitored.

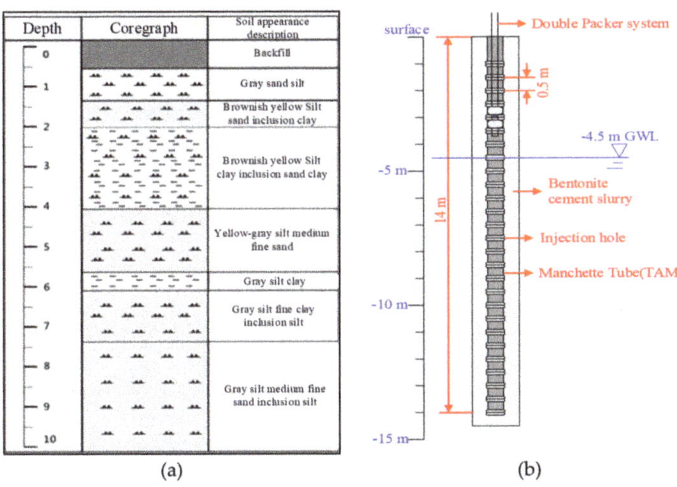

Figure 2. In situ geological structure and layered formation of the injection well: (**a**) in situ biological field geologic structure; (**b**) double packer injection well structure.

3. Results and Discussion

3.1. Microcosm Testing Results

The effects of the addition of different slow-release substrates on environmental parameters are depicted in Figure 3. The testing results revealed that the use of emulsion products may accumulate organic acid, thus decreasing the pH in groundwater. The effect was particularly substantial with the addition of SOE+FL. The microcosm was an enclosed nonflowing space, which further increased the likelihood of acid accumulation. Regardless, the pH value in each of the groups was higher than 5.5, falling into a range suitable for microbial growth. The initial ORP of each group was approximately 200 mV. After the addition of different chemicals, the ORPs were reduced to 0 mV or lower. The greatest decrease was observed in the SOE+FL group. Because the in situ groundwater originally contained approximately 70 mg/L TOC, the blank group (BK) also exhibited a decreasing trend in ORP. However, the ORP of the BK group remained higher than 0 mV for the duration of the test, which was different from that of the other groups with emulsions added. The initial DO of each group was 6.48 mg/L. The results revealed that the CEVO and SOE+FL test groups showed lower DO levels compared with the other groups. Moreover, the groups with emulsions added (CEVO, SOE, and SOE+FL) had higher EC values, with the SOE+FL group having the highest EC. The higher EC in these groups was assumed to be caused by the electron donors provided by FL. The overall results indicated that the addition of FL accelerated ORP and DO reduction; however, the addition of FL also caused lower pH values.

The water quality variation of each group during the testing period is depicted in Figure 4. The initial TOC of the groundwater used in this test was 76.8 mg/L. After the addition of various slow-release substrates, the TOC concentration in each of the non-BK test groups was higher than 3000 mg/L. The TOC concentrations following the addition of CEVO and self-made SOE were similar (approximately 3500 mg/L). The addition of FL further increased the TOC concentration to 4936 mg/L, suggesting that carbon sources in FL increased the electron donor content in emulsions. Additionally, on day 45 of the test, the chemical consumption rate of the SOE+FL group was faster than the consumption of the pure SOE or CEVO groups. It is possible that adding FL accelerated the environmental reduction, thereby increasing the chemical consumption rate. The nitrate concentration in each testing group increased in the initial period of the testing, possibly because nutrients were added in all groups. The nitrogen, phosphorous, vitamin B12, and other substances included in the nutrients reacted with residual DO in the vials, causing increased concentrations of nitrate. The sulfate content in each group exhibited an overall decreasing trend, suggesting that the later stages of the test entered an environment for sulfate reduction.

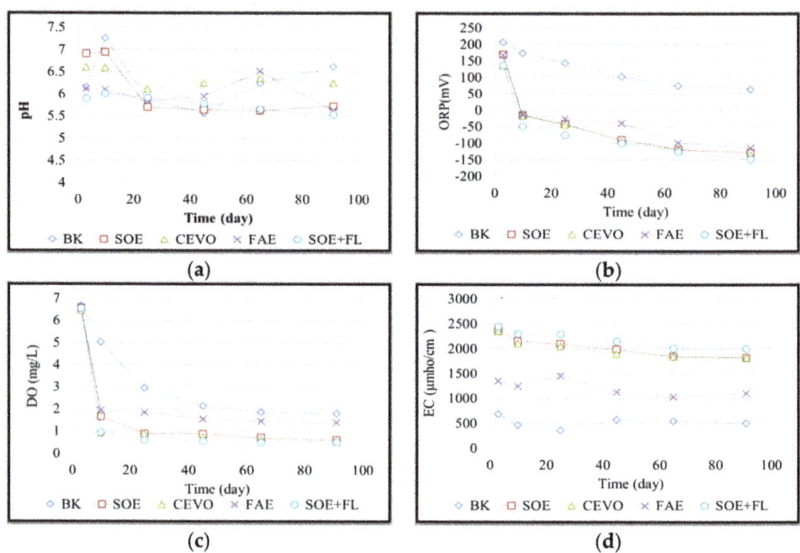

Figure 3. Environmental parameter variation in the microcosm tests: (**a**) pH; (**b**) ORP; (**c**) DO; and (**d**) EC.

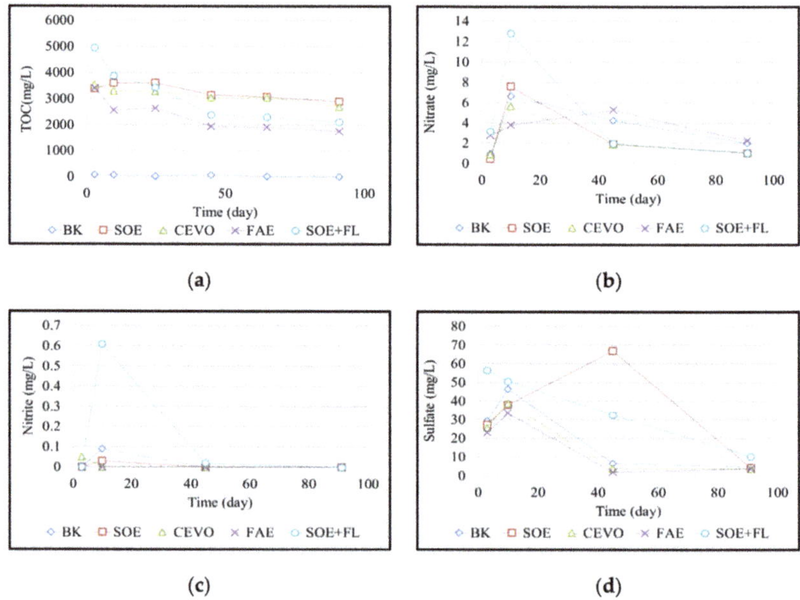

Figure 4. Water quality parameter variation in the microcosm tests: (**a**) TOC; (**b**) nitrate; (**c**) nitrite; and (**d**) sulfate.

The TCE removal rates and derivative concentration variations in each group of the microcosm test are depicted in Figure 5. The BK group exhibited a TCE removal rate of approximately 22% on day 45 because the groundwater used contained TOC. The TCE removal rate of the self-made SOE, CEVO, FAE, and SOE+FL groups was approximately 71%, 64%, 52%, and 94%, respectively. The results indicated that adding FL accelerated environmental reduction and increased biodegradation efficiency. The TCE removal rates of the self-made SOE and CEVO were similar, suggesting that the self-made SOE was as stable as the commercial product. Moreover, the self-made SOE reached 95% and 99% TCE removal on days 65 and 90, respectively, which were higher than the 87% and 82% TCE removal rates, respectively, of the CEVO. According to the CEVO degradation tendency and the approximately 2500 mg/L TOC in the CEVO group on day 90, the CEVO group might have been able to reach 99% removal of TCE if the testing period had been extended. The FAE group reached approximately 55% TCE removal without any further increase in removal because of its low oil content. In summary, under the same oil content and fixed TCE concentration, only the SOE+FL and SOE test groups could reach an eventual TCE removal rate of 99%. The addition of FL accelerated TCE degradation. The groups with emulsions alone required a longer time for degradation to gradually reach a reducing condition in the environment. Adding FL shortened the time required to achieve the reducing condition, so microbes could start anaerobic reductive dechlorination earlier. Moreover, Figure 5b reveals that the SOE+FL group generated the highest concentration of cis-1,2-DCE, suggesting that the SOE+FL group had the greatest reductive dechlorination efficiency. Therefore, the subsequent in situ biological field test was conducted using SOE+FL. On day 90, the amount of representative strains of DHC and DCA1 in the SOE+FL group was 5.64×10^3 and 6.17×10^3 gene copies/L, respectively.

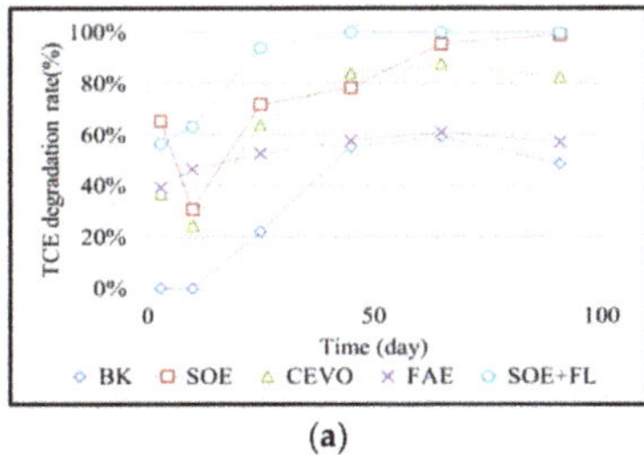

(a)

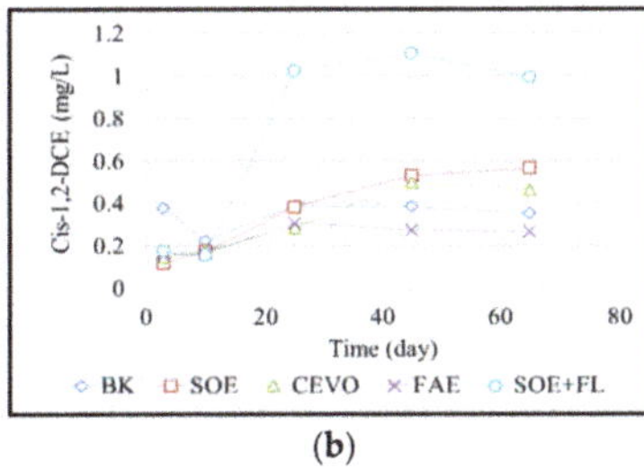

(b)

Figure 5. Effects of different emulsions on chlorinated pollutants in the microcosm tests: (**a**) TCE and (**b**) *cis*-1,2-DCE.

3.2. In Situ Biological Field Test Results

3.2.1. TOC Concentration Variation

On days 0, 36, and 130 of the in situ biological field test, 40 L SOE + 5 kg FL, 230 L SOE + 20 kg FL, and 130 L SOE + 10 kg FL were injected, respectively. The FL was adjusted to 2% concentration with in situ groundwater before being blended with SOE. After each chemical injection, a groundwater volume of three times the chemical volume was injected to propel the chemicals. Layered injection was adopted. Because the TCE contamination depth of the site was 4.5 to 12 m, a double packer system was used to inject the chemicals to 5, 6, 7, 10, 11, and 12 m beneath the surface. The TOC concentration variations in each monitoring well after the chemical injections are presented in Figure 6. The results revealed that, because of the special structure of the Machette tube in the double packer injection system, the IW retained no groundwater or chemicals. Therefore, no water quality data were obtained in the IW. After the first injection, the TOC concentration of each monitoring well exhibited no apparent increase, probably because the first injection calculation underestimated the hydraulic conductivity, causing an insufficient amount of chemicals to be injected. After adjustment, the second injection was conducted on day 36 and TOC concentrations increased in each monitoring well. The CW TOC concentration remained at 1–3 mg/L, suggesting that the study site lacked electron donors, which prevented reductive dechlorination through natural attenuation of TCE. Because OW1 was only 1 m away from the IW, a TOC increase was observed soon after each injection. The TOC concentration in OW1 after the third injection (393 mg/L) was higher than the highest TOC concentration (243 mg/L) observed after the second injection. These results indicated that SOE+FL was effectively adsorbed into soil pores and continued to release electron donors to reach effective reductive dechlorination. In addition, the TOC concentration of 94 mg/L observed at OW1 on day 90 after the SOE+FL injection demonstrated that SOE+FL had at least 90 days of slow-release effect. Because OW2 and OW3 were farther from the IW, their TOC concentrations began to increase on day 90, with the increasing slope greater at OW2 than at OW3. The maximum TOC concentration reached at OW2 and OW3 was 282 mg/L at OW2. On day 160, the TOC concentration at OW2 and OW3 was 144 mg/L. Because the TOC concentration of the IW upstream area was maintained at 200 mg/L or higher, a decrease in OW3's TOC concentration was not observed. These overall results revealed that the SOE+FL transmission reached at least 5 m when a double packer system under 2 bar pressure of injection was used.

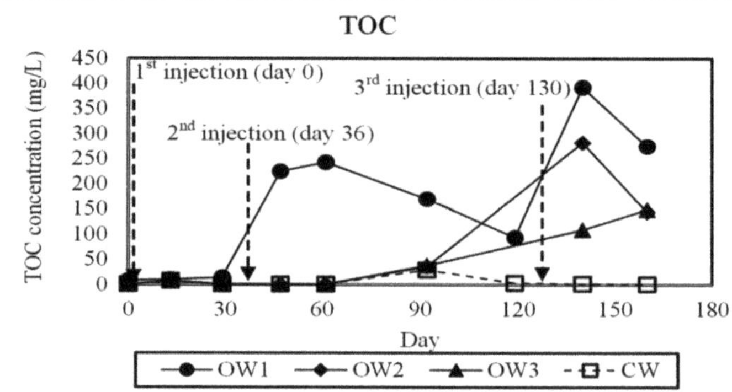

Figure 6. TOC concentration variations before and after injection for each monitoring well in the in situ field testing.

3.2.2. In Situ Water Quality Parameter Variation

Because OW1 was closer to the IW, OW2 and CW were selected for comparing underground environments and water quality parameters before and after injection. The results are presented in Table 1. OW2 had family vegetable gardens in its peripheral areas; thus, the initial nitrate concentration in the groundwater was higher than that at CW. Because the groundwater at the in situ field testing site was flowing, the pH value of OW2 after injection did not decrease at a level comparable to the microcosm test results. The TOC results, as stated previously, indicated that chemicals were transmitted to OW2 only after day 90. Thus, only an ORP decrease was observed before day 90. Prior to day 90, the nitrate and nitrite concentrations did not decrease, and the iron (II) concentration did not increase. On day 140, the nitrate concentration of OW2 decreased substantially, the iron (II) concentration increased to 3.6 mg/L, and sulfide was observed at a concentration of 0.22 mg/L. The results from the CW indicated that the groundwater was originally in an anaerobic state; however, little difference in water quality was observed between day 0 and day 140 because of the lack of electron donors.

Table 1. Environmental water quality parameter comparison of OW2 and CW in the in situ field testing.

Well	Sampling Day	pH	DO (mg/L)	ORP (mV)	Nitrate (mg/L)	Nitrite (mg/L)	Fe (II)(mg/L)	Sulfate (mg/L)	Sulfide (mg/L)
OW2	0	6.41	1.94	328	7.92	ND	0.09	45.7	ND
	29	6.42	2.32	316	8.99	0.002	0.06	43.7	ND
	61	6.41	0.52	72	9.49	ND	0.5	39.8	ND
	119	6.44	1.83	85	8.62	0.005	0.18	44.1	<0.04
	140	6.26	1.11	-2	0.84	ND	3.6	50.3	0.22
CW	0	6.78	1.6	-5	0.02	ND	0.08	122	ND
	29	6.67	1.73	23	0.03	ND	0.35	128	ND
	61	6.57	0.65	-8	0.02	ND	1.05	130	0.02
	92	6.26	0.07	103	0.05	N.D.	0.02	122	0.03
	119	6.45	1.38	202	0.08	0.003	0.35	127	<0.04
	140	6.16	1.8	64	0.04	ND	1.98	127	ND

3.2.3. TCE and Derivatives' Concentration Variations

The TCE concentrations monitored after injection during the in situ biological field test are presented in Figure 7. Following the first injection, because the injection volume was insufficient, the TCE concentrations in OW1, OW2, and OW3 did not decrease in the first 30 days. After the second injection, the TCE concentrations in OW1, OW2, and OW3 decreased. As the CW lacked electron donors, the TCE concentration at CW remained at 0.8–1.0 mg/L, suggesting a low natural attenuation rate of TCE at the site. The distance between OW1 and the IW was the shortest, so TOC concentrations at OW1 were higher than those at OW2 and OW3. TCE concentrations at OW1 decreased from 0.225 mg/L to 0.033 mg/L on day 60, reaching a TCE removal rate of approximately 85%. The results of the microcosm test revealed that SOE+FL accelerated the TCE degradation rate, corresponding to the substantial decrease in TCE concentrations. On day 160 of the in situ field test, TCE concentrations at OW1 were lower than 0.01 mg/L, which is lower than the TCE monitoring standard of groundwater-related regulations. Prior to day 90, TOC concentrations at OW2 and OW3 were already substantially low, so further decreases in concentration were not observed. After day 90, TCE concentrations at OW2 decreased observably. On day 160, TCE concentrations at OW2 decreased from 0.164 mg/L to 0.0685 mg/L, reaching a TCE removal rate of approximately 58%. At OW3, TOC concentrations increased slowly and, as a result, TCE concentrations at OW3 did not decrease in a notable manner. On day 140 (OW3 was not analyzed on day 160), the TCE concentration of OW3 decreased from 0.636 mg/L to 0.048 mg/L, satisfying the groundwater regulation standards of TCE, Taiwan Groundwater Pollution Control Standards, as shown on Table 2 [30]; however, the removal rate was a mere 25%. A possible cause of the low removal rate can be attributed to the long distance between OW3 and the IW, which required a long transmission time for the chemicals to reach OW3, thus shortening the time allotted for stabilizing the groundwater environment. A longer testing time was required for observable effectiveness to be achieved.

Table 2. Taiwan Groundwater Control Standards.

Control Item	Control Standards (mg/L)	
	Class I Groundwater	Class II Groundwater
TCE	0.005	0.05
cis-1,2-DCE	0.07	0.7
trans-1,2-DCE	0.1	1.0
VC	0.002	0.02

Note: Class I is for groundwater in drinking water resource protection area; Class II is for groundwater outside the Class I area.

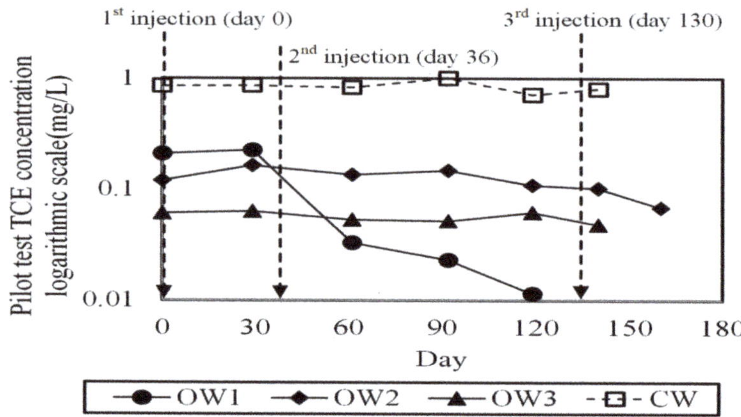

Figure 7. TCE concentration variation of each monitoring well in the in situ field testing.

Anaerobic reductive dechlorination of TCE derives chlorinated byproducts. Scientific literature has indicated that microbial degradation of soybean oil may generate organic acid and hydrogen atoms, which replace the chlorine atoms. During the reductive dechlorination process, intermediates such as *cis*-1,2-DCE, VC, and the final product ethene are produced [11,31]. The *cis*-1,2-DCE and VC concentrations of each monitoring well during the in situ biological field tests are presented in Figures 8 and 9, respectively. The results revealed that, although substantial TCE degradation was observed to begin on day 60 at OW1, no apparent accumulation of *cis*-1,2-DCE and VC concentrations was observed. On day 160, the *cis*-1,2-DCE concentration of OW1 increased from 0.012 mg/L to 0.124 mg/L, exhibiting no considerable accumulation. VC concentrations at OW1 were lower than the detection limit (<0.01 mg/L). At OW2, *cis*-1,2-DCE concentrations varied similarly to the TCE concentration variations observed in the well. The concentrations clearly began to decrease on day 90. The *cis*-1,2-DCE concentration at OW2 decreased from 2.63 mg/L to 0.742 mg/L on day 160, reaching a removal rate of 72%. The VC concentration at OW2 increased from below the detection limit (<0.01 mg/L) to 0.112 mg/L. The OW1 monitoring results revealed that, although the TCE concentration decreased considerably, the metabolites did not accumulate in a notable manner. DCE concentration even decreased rapidly after day 90. These results suggested that adding FL in emulsions accelerated metabolite degradation and decreased metabolite accumulation. However, as the distance from the IW increased, the FL concentration became insufficient and caused the accumulation of metabolites such as VC, as observed in OW2. Sheu [32] compared PCE and TCE degradation effectiveness in microcosm tests in 2015 using two substrates—a general emulsified oil substrate (EOS) and a long-lasting emulsified colloidal substrate (LECS), a combination of EOS and nanoscale zero-valent iron (nZVI). The test results indicated that the VC concentration was about 0.5 mg/L 130 days after the EOS application. However, VC concentration met Taiwan Groundwater Control Standards 130 days after the LECS application. It was verified that nZVI of LECS did not result in the cumulation of metabolic byproducts compared with EOS. Ferrous lactate used in the test had similar effectiveness as in Sheu's tests.

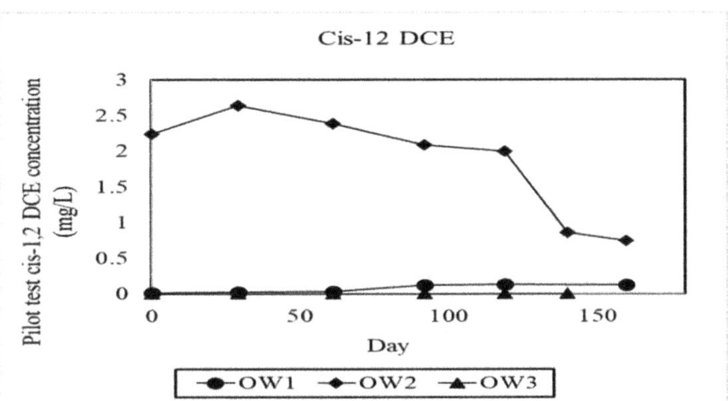

Figure 8. *cis*-1,2-DCE concentration variation of each monitoring well in the in situ field testing.

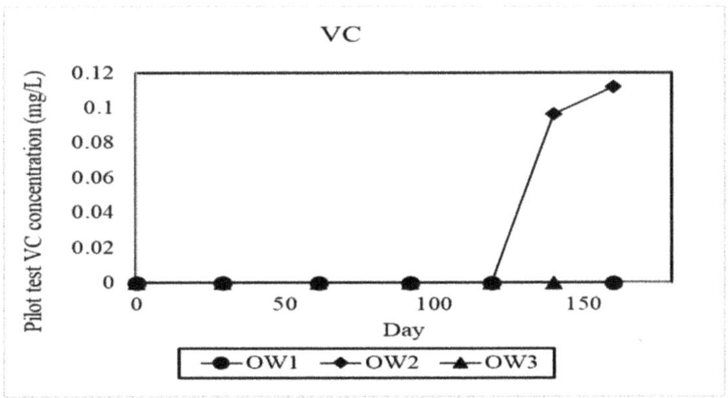

Figure 9. VC concentration variation of each monitoring well in the in situ field testing.

4. Summary

This study evaluated the effect of SOE+FL on anaerobic reductive dechlorination of TCE. The following conclusions were drawn: (1) the results of laboratory microcosm test and pilot test revealed that SOE+FL accelerated TCE degradation and minimized the accumulation of DCE and VC more than SOE alone; (2) a double packer injection well, constructed using a Manchette tube, can provide surgical injection of chemicals at contaminated intervals; (3) the in situ bioremediation pilot test results revealed that the SOE+FL solution transported in the underground environment continued to be released for at least 90 days, with a radius of influence of at least 5 m.

Author Contributions: Experiment design and execution, M.-H.L. and C.-M.H.; manuscript writing/review, C.-E.L., M.-H.L., and J.L.; analysis, M.-H.L. and J.L.; project administration, M.-H.L. All authors have read and agreed to the published version of the manuscript.

Funding: This study was funded by the research project supported by the Taiwan EPA (grant number 109GA0006002009). The views or opinions expressed in this article are those of the writers and should not be construed as opinions of the Taiwan EPA. Mention of trade names, vendor names, or commercial products does not constitute endorsement or recommendation by Taiwan EPA.

Institutional Review Board Statement: Not applicable.

Informed Consent Statement: Informed consent was obtained from all subjects involved in the study.

Data Availability Statement: The data presented in this study are available on request from the corresponding author.

Conflicts of Interest: The authors declare no conflict of interest.

References

1. Morrison, R.D. Application of forensic techniques for age dating and source identification in environmental litigation. *Environ. Forensics* **2000**, *1*, 131–153. [CrossRef]
2. International Agency for Research on Cancer. *World Cancer Report 2014*; Stewart, B.W., Wild, C.P., Eds.; IARC: Lyon Cedex 08, France, 2014; p. 152. ISBN 978-92-832-0443-5.
3. Purdue, M.P.; Stewart, P.A.; Friesen, M.C.; Colt, J.S.; Locke, S.J.; Hein, M.J.; Waters, M.A.; Graubard, B.I.; Davis, F.; Ruterbusch, J.; et al. Occupational exposure to chlorinated solvents and kidney cancer: A case-control study. *Occup. Environ. Med.* **2017**, *74*, 268–274. [CrossRef] [PubMed]
4. Yang, Y.; McCarty, P.L. Comparison between donor substrates for biologically enhanced tetrachloroethene DNAPL dissolution. *Environ. Sci. Technol.* **2002**, *36*, 3400–3404. [CrossRef]
5. Hunter, W.J. Bioremediation of chlorate or perchlorate contaminated water using permeable barriers containing vegetable oil. *Curr. Microbiol.* **2002**, *45*, 287–292. [CrossRef] [PubMed]
6. Hunter, W.J. Injection of innocuous oils to create reactive barriers for bioremediation: Laboratory studies. *J. Contam. Hydrol.* **2005**, *80*, 31–48. [CrossRef]
7. Hunter, W.J.; Shaner, D.L. Nitrogen limited biobarriers remove atrazine from contaminated water: Laboratory studies. *J. Contam. Hydrol.* **2009**, *103*, 29–37. [CrossRef] [PubMed]
8. Lieberman, M.T.; Borden, R.C.; Zawtocki, C.; May, I.; Casey, C. Long term TCE source area remediation using short term emulsified oil substrate (EOS®) recirculation. In Proceedings of the 21st Annual International Conference on Soils, Sediments, and Water, University of Massachusetts, Amherst, MA, USA, 17 October 2005; pp. 17–20.
9. Borden, R.C. Effective distribution of emulsified edible oil for enhanced anaerobic bioremediation. *J. Contam. Hydrol.* **2007**, *94*, 1–12. [CrossRef] [PubMed]
10. Marks, C. *Comparison of EHC®, EOS®, and Solid Potassium Permanganate Pilot Studies for Reducing Residual TCE Contaminant Mass*; URS Corporation: San Francisco, CA, USA, 2012.
11. Aulenta, F.; Fuoco, M.; Canosa, A.; Papini, M.P.; Majone, M. Use of poly-β-hydroxy-butyrate as a slow-release electron donor for the microbial reductive dechlorination of TCE. *Water Sci. Technol.* **2008**, *57*, 921–925. [CrossRef]
12. Fuller, M.E.; Schaefer, C.E.; Steffan, R.J. Evaluation of a peat moss plus soybean oil (PMSO) technology for reducing explosive residue transport to groundwater at military training ranges under field conditions. *Chemosphere* **2009**, *77*, 1076–1083. [CrossRef]
13. Hunter, W.J.; Shaner, D.L. Studies on removing sulfachloropyridazine from groundwater with microbial bioreactors. *Curr. Microbiol.* **2011**, *62*, 1560–1564. [CrossRef]
14. Chambers, J.; Wilkinson, P.B.; Wealthall, G.P.; Loke, M.H.; Dearden, R.A.; Wilson, R.; White, D.; Ogilvy, R.D. Hydrogeophysical imaging of deposit heterogeneity and groundwater chemistry changes during DNAPL source zone bioremediation. *J. Contam. Hydrol.* **2010**, *118*, 43–61. [CrossRef]
15. Boni, M.R.; Sbaffoni, S. The potential of compost-based biobarriers for Cr (VI) removal from contaminated groundwater: Column test. *J. Hazard. Mater.* **2009**, *166*, 1087–1095. [CrossRef] [PubMed]
16. Hunter, W.J.; Shaner, D.L. Biological remediation of groundwater containing both nitrate and atrazine. *Curr. Microbiol.* **2010**, *60*, 42–46. [CrossRef] [PubMed]
17. Leombruni, A.; Mueller, M.; Seech, A.; Leigh, D. Field application of a reagent for in site chemical reduction and enhanced reductive dichlorination treatment of an aquifer contaminated with tetrachloroethylene (PCE), trichloroethylene, 1,1-dichloroethylene, dichloropropane and 1,1,2,2-tetrachloroethane (R-130). *Environ. Eng. Manag. J.* **2020**, *19*, 1791–1796.
18. Philips, J.; Haest, P.J.; Springael, D.; Smolders, E. Inhibition of geobacter dechlorinators at elevated trichloroethene concentrations is explained by a reduced activity rather than by an enhanced cell decay. *Environ. Sci. Technol.* **2013**, *47*, 1510–1517.
19. Sleep, B.E.; Seepersad, D.J.; Mo, K.; Heidorn, C.M.; Hrapovic, L.; Morrill, P.L.; McMaster, M.L.; Hood, E.D.; LeBron, C.; Lollar, B.S.; et al. Biological enhancement of tetrachloroethene dissolution and associated microbial community changes. *Environ. Sci. Technol.* **2006**, *40*, 3623–3633. [CrossRef]
20. Matheson, L.J.; Tratnyek, P.G. Reductive dehalogenation of chlorinated methanes by iron metal. *J. Environ. Sci. Technol.* **1994**, *28*, 2045–2053. [CrossRef]
21. Su, C.; Puls, R.W.; Krug, T.A.; Watling, M.T.; O'Hara, S.K.; Quinn, J.W.; Ruiz, N.E. A two and half-year-performance evaluation of a field test on treatment of source zone tetrachloroethene and its chlorinated daughter products using emulsified zero valent iron nanoparticles. *Water Res.* **2012**, *46*, 5071–5084. [CrossRef]
22. Li, X.Q.; Brown, D.G.; Zhang, W.X. Stabilization of biosolids with nanoscale zero-valent iron (nZVI). *J. Nanopart. Res.* **2007**, *9*, 233–243. [CrossRef]
23. Herrero, J.; Puigserver, D.; Nijenhuis, I.; Kuntze, K.; Carmona, J.M. Combined use of ISCR and biostimulation techniques in incomplete processes of reductive dehalogenation of chlorinated solvents. *Sci. Total Environ.* **2019**, *648*, 819–829. [CrossRef]

24. Su, C.; Puls, R.W.; Krug, T.A.; Watling, M.T.; O'Hara, S.K.; Quinn, J.W.; Ruiz, N.E. Travel distance and transformation of injected emulsified zerovalent iron nanoparticles in the subsurface during two and half years. *Water Res.* **2013**, *47*, 4095–4106. [CrossRef] [PubMed]
25. National Institute of Environmental Analysis (NIEA); Taiwan EPA. T, NIEA W785.57B. 2020. Available online: https://www.epa.gov.tw/niea/3617C6A64E37A35A (accessed on 28 July 2021).
26. National Institute of Environmental Analysis (NIEA); Taiwan EPA. Determination of Total Organic Carbon in Water-Peroxypyrosulfate Heating Oxidation/Infrared Measurement Method, NIEA W532.52C. 2009. Available online: https://www.epa.gov.tw/niea/99FCF9EFD3CB163C (accessed on 28 July 2021).
27. National Institute of Environmental Analysis (NIEA); Taiwan EPA. Determination of Nitrate Nitrogen and Nitrite Nitrogen in Water-Cadmium Reduction Flow Analysis Method, NIEA W436.52C. 2015. Available online: https://www.epa.gov.tw/niea/C8FB816F3E6542A2 (accessed on 28 July 2021).
28. National Institute of Environmental Analysis (NIEA); Taiwan EPA. Determination of Inorganic Anions in Drinking Water by Ion Chromatography, NIEA W415.54B. 2018. Available online: https://www.epa.gov.tw/niea/2C82301AFEA0B53A (accessed on 28 July 2021).
29. Ritalahti, K.M.; Amos, B.K.; Sung, Y.; Wu, Q.; Koenigsberg, S.S.; Löffler, F.E. Quantitative PCR targeting 16S rRNA and reductive dehalogenase genes simultaneously monitors multiple *Dehalococcoides* strains. *Appl. Environ. Microbiol.* **2006**, *72*, 2765–2774. [CrossRef] [PubMed]
30. Taiwan Environmental Protection Administration. Groundwater Pollution Control Standards. In *Soil and Groundwater Pollution Remediation Act*; Taiwan Environmental Protection Administration: Taipei, Taiwan, 2000.
31. Tarnawski, S.E.; Rossi, P.; Brennerova, M.V.; Stavelova, M.; Holliger, C. Validation of an integrative methodology to assess and monitor reductive dechlorination of chlorinated ethenes in contaminated aquifers. *Front. Environ. Sci.* **2016**, *4*, 1–16. [CrossRef]
32. Sheu, Y.T. Application of a Long-Lasting Colloidal Substrate with pH and Hydrogen Sulfide Control Capabilities for TCE-Contaminated Groundwater Remediation. Ph.D. Dissertation, Institute of Environmental Engineering, National Sun Yat-Sen University, Kaohsiung, Taiwan, 2015.

applied
sciences

Article

Insights into the Restoration of Tributyltin Contaminated Environments Using Marine Bacteria from Portuguese Fishing Ports

Hugo R. Monteiro [1,*], Ariana B. Moutinho [1], Maria J. Campos [1], Ana C. Esteves [2] and Marco F. L. Lemos [1,*]

[1] MARE–Marine and Environmental Sciences Centre, ESTM, Polytechnic of Leiria, 2520-641 Peniche, Portugal; ariana.moutinho@ipleiria.pt (A.B.M.); mcampos@ipleiria.pt (M.J.C.)

[2] CESAM–Centre for Environmental and Marine Studies, Department of Biology, University of Aveiro, 3810-193 Aveiro, Portugal; acesteves@ua.pt

* Correspondence: hugo.monteiro@ipleiria.pt (H.R.M.); marco.lemos@ipleiria.pt (M.F.L.L.); Tel.: +351-262-783-607 (H.R.M. & M.F.L.L.); Fax: +351-262-783-088 (H.R.M. & M.F.L.L.)

Abstract: Tributyltin (TBT) is an organotin chemical mainly used as biocide in marine antifouling paints. Despite the restrictions and prohibitions on its use, TBT is still an environmental problem due to its extensive application and subsequent release into the environment, being regarded as one of the most toxic chemicals released into the marine ecosystems. Microorganisms inhabiting impacted sites are crucial for their restoration since they have developed mechanisms to tolerate and break down pollutants. Nonetheless, transformation products resulting from the degradation process may still be toxic or, sometimes, even more toxic than the parent compound. The determination of the parent and degradation products by analytical methods, although necessary, may not be ecologically relevant since no information is provided regarding their ecotoxicity. In this study, marine bacteria collected from seven Portuguese fishing ports were isolated and grown in the presence of TBT. Bacteria that exhibited higher growth were used to bioremediate TBT-contaminated waters. The potential of these bacteria as bioremediation agents was evaluated through ecotoxicological assays using the sea snail *Gibbula umbilicalis* as model organism. Data suggested that some TBT-tolerant bacteria, such as *Pseudomonas putida*, can reduce the toxicity of TBT contaminated environments. This work contributes to the knowledge of TBT-degrading bacteria.

Keywords: bioremediation; ecotoxicology; *Gibbula umbilicalis*; marine bacteria; tributyltin; TBT-tolerance

Citation: Monteiro, H.R.; Moutinho, A.B.; Campos, M.J.; Esteves, A.C.; Lemos, M.F.L. Insights into the Restoration of Tributyltin Contaminated Environments Using Marine Bacteria from Portuguese Fishing Ports. *Appl. Sci.* **2021**, *11*, 6411. https://doi.org/10.3390/app11146411

Academic Editor: Elida Nora Ferri

Received: 17 June 2021
Accepted: 9 July 2021
Published: 12 July 2021

Publisher's Note: MDPI stays neutral with regard to jurisdictional claims in published maps and institutional affiliations.

1. Introduction

Tributyltin (TBT) is an organotin chemical that acts as a biocide in antifouling systems. Since the 1960s up until its ban in the late 2000s, TBT was widely employed in marine paints, successfully preventing the attachment of crustaceans, mollusks, algae and slime on boat hulls and various immersed structures and equipment [1–3]. The massive use of TBT led to the contamination of the environment, mainly marine and freshwater ecosystems, being ports, harbors, and marinas the most affected and historically contaminated sites due to high boat traffic [4–7]. Research undertaken since the 1970s unveiled the haunting toxicity of TBT to non-target organisms [8–12].

As a result, the European Union (EU) restricted its use on ships smaller than 25 m with the directive 89/677/CEE [13] and in 2003 with the regulation 782/2003/EC, banned the use of paints containing organotins in EU ships [14]. In 2008, the United Nations' International Maritime Organization called for a global prohibition of organotin-based paints on ship hulls [6,8,15]. Despite restrictions, TBT is still present on a global scale, in water and sediments at higher levels than those reported to cause detrimental effects on living organisms. This evidence suggests a recent illegal use of TBT-based antifouling paints [16–18]. Once released from an antifouling coating, TBT can be adsorbed to suspended particles in the water, biota, and sediment, and persist in the environment for long periods because of its low degradation rates [16,18,19].

Tributyltin is a known endocrine disruptor compound that causes several adverse effects in different organisms, from invertebrates to mammals [20–22]. A well-documented adverse effect is imposex: the development of male sexual characteristics in female gastropods [16]. Imposex has been reported in sea snails at concentrations as low as $1\,ng\,L^{-1}$ [23], significantly impacting sea snails' population dynamics. Despite the decreasing levels of imposex in the Portuguese Coast since its ban in 2008, a complete recovery of imposex on marine gastropods has not yet been achieved [16]. For this reason, TBT is still recognized as an important environmental problem on a global scale and often considered one of the most toxic substances released into the marine environment [19].

Microorganisms from historically impacted sites can develop a tolerance to pollutants and, in some cases, degrade them, which can be a crucial factor in the recovery of contaminated environments [24]. Despite TBT toxicity, tolerance in some bacteria has been reported [25]. Under favorable conditions, TBT may be biodegraded via sequential dealkylation processes to form dibutyltin (DBT), monobutyltin (MBT) and eventually inorganic tin, decreasing its toxicity in the process [8,26]. Another proposed mechanism of bacterial TBT degradation is the use of TBT as a carbon source [25–27]. Despite this increasing number of studies, the use of naturally occurring microorganisms for remediation of TBT is still far from a full-scale application as information on the actual mechanisms involved in the degradation process is still scarce [25,28].

Moreover, the breakdown of TBT must be carefully followed as these by-products may still exert some toxicity to marine organisms. Hence, besides plain analytical reasoning, ecotoxicological testing on bioremediated media is needed to determine the actual biological damage of these degradation products.

Gibbula umbilicalis is a sea snail with a very wide distribution across the rocky shores of the European Atlantic coast. This species is very abundant and easy to collect and maintain in the laboratory, features that make it a good model in ecotoxicology [29,30] and suitable model for monitoring TBT pollution.

In the present study, TBT-tolerant bacteria collected from Portuguese maritime ports were isolated and characterized, with the main goal of determining the potential of these isolates to bioremediate TBT into less toxic compounds. Additionally, to evaluate bioremediation efficiency, *G. umbilicalis* sea snails were exposed to potential bioremediated media, using survival as an endpoint.

2. Materials and Methods

2.1. Study Area and Sample Collection

Seawater samples were collected from seven Portuguese ports with high maritime traffic: Póvoa de Varzim (41.376120, −8.766945); Leixões (41.195238, −8.684177); Aveiro (40.645899, −8.727098); Figueira da Foz (40.146848, −8.849176); Peniche (39.355422, −9.375479); Setúbal (38.521228, −8.887277); and Sines (37.950219, −8.864599). All samples were collected during low tide, 50 cm above sediment level. Physicochemical parameters (temperature, pH, salinity, conductivity, and dissolved oxygen) were measured using a multiparameter water meter (Hanna HI 9828; Hanna Instruments, Villafranca Padovana, Italy). At each sampling location, water samples were collected for bacteria isolation and for chemical analyses (butyltins and metallic elements). Samples were processed within 8 h of collection.

2.2. Chemical Analyses

Chemical analyses of butyltins (tributyltin, dibutyltin and monobutyltin) were made at the Instituto Superior Técnico (Lisbon, Portugal). Samples were derivatized with tetraethylborate and analyzed by GC-MS (RTX-5MS column (30 mm × 0.25 mm × 0.25 μm)) using tributyltin chloride, dibutyltin chloride and monobutyltin chloride as standards. Metals analyses were conducted at the University of Aveiro (Aveiro, Portugal) for the following metallic elements: silver (Ag), cadmium (Cd), cobalt (Co), chromium (Cr), copper (Cu), iron (Fe), manganese (Mn), nickel (Ni), lead (Pb), strontium (Sr), zinc (Zn), tin (Sn) and mercury (Hg). Samples were acidified and analyzed by ICP-MS using an internal standard (ISO 17294), except for mercury which was determined using gold as a stabilizing agent, as described by Allibone et al. [31]. The limits of detection were as follows: 50 μg L^{-1} for Fe, 20 μg L^{-1} for Zn, 10 μg L^{-1} for Ni and Ag, 5 μg L^{-1} for Cr, Cu and Sn, 2 μg L^{-1} for Mn, Co and Pb, 1 μg L^{-1} for Cd, and 0.5 μg L^{-1} for Hg.

2.3. Bacteria Isolation and Selection

2.3.1. Bacteria Isolation

Tributyltin (tributyltin chloride, 96%; Sigma-Aldrich, Madrid, Spain) 1.23 M stock solution was prepared in ethanol. Bacteria were grown in Tryptic Soy Agar (TSA; Sharlau, Barcelona, Spain) with 1.5% (w/v) NaCl and contaminated with 0.1, 1, and 3 mM TBT, adjusting the total volume of ethanol in all media to 1% (v/v). Control groups were prepared with no TBT and 1% ethanol.

Water samples were vacuum filtered through a 0.45 μm pore cellulose membrane filter (Sartorius Stedim Biotech, Göttingen, Germany), placed on TSA/TBT media (three replicates per concentration for each sampling site) and incubated at 30 °C. After 48 h, colony-forming units (CFU) were counted and 3 mM TBT-tolerant bacteria from one plate per location were isolated and purified by streak plate method.

2.3.2. REP-PCR Genomic Fingerprinting

Genomic DNA extraction was made with 250 μL of a fresh culture in Tryptic Soy Broth (TSB, Sharlau, Barcelona, Spain). Cells were centrifuged for 5 min at 15,680$\times$ g and resuspended in 50 μL Tris-EDTA (TE) buffer. Cells were lysed with 5 μL of Lysozyme (10 mg/mL) and incubated for 1 h at 37 °C, followed by a 10 min incubation at 65 °C with 50 μL of Lysis Solution (Genomic DNA purification kit, Fermentas, Baden-Württemberg, Germany). Next, 100 μL of chloroform was added to the lysate and spun 5 min at 15,680$\times$ g. The aqueous phase was collected and mixed with 100 μL of isopropanol, followed by centrifugation at 15,680$\times$ g for 10 min. The resulting pellet was washed with 100 μL of 70% ethanol, centrifuged (5 min, 15,680$\times$ g), dried and finally resuspended in 50 μL of TE buffer and stored at -20 °C. Prior to PCR reaction, an agarose gel electrophoresis was performed to verify DNA quality and extraction efficiency.

The REP-PCR reaction was performed with the following primer sequences: REP1R (5' III ICG ICG ICA TCI GGC 3') and REP2I (5' ICG ICT TAT CIG GCC TAC 3') [32]. A reaction mixture was prepared as follows: 11.15 μL of water, 5 μL of 5x buffer, 3 μL of MgCl$_2$ (25 mM), 1.5 μL of dNTP solution (2 mM of each dNTP), 1.25 μL of DMSO, 1 μL of each primer (10 mM) and 0.1 μL of *Taq* polymerase. One μL of DNA was used as a template. PCR amplification was performed using a MyCycler Thermal Cycler (Bio-Rad, Hercules, CA, USA) with an initial denaturation step at 95 °C for 5 min followed by 30 cycles of denaturation at 94 °C for 1 min, annealing at 40 °C for 1 min and extension at 65 °C for 8 min. The final extension was carried at 65 °C for 16 min. PCR products were evaluated by ethidium bromide-stained agarose gel electrophoresis and characterized by Dice/UPGMA clustering [33]. Image acquisition was carried using a GelDoc (Bio-Rad, Hercules, CA, USA) imaging system and banding patterns were analyzed with GelCompar II 3.0 (Applied Maths, Kortrijk, Belgium).

2.3.3. Bacteria Selection

Selected bacterial phylotypes were grown in TSB until reaching an optical density (O.D.) of 0.600 at 570 nm. Afterwards, the isolates were grown in TSB with 0, 1, 3 and 6 mM of TBT and incubated at 160 rpm and 30 °C. Absorbance was measured in a microplate reader (Labsystems Multiskan EX, Helsinki, Finland) every 30 min until the stationary phase. The faster-growing bacteria in high concentrations of TBT were selected for bioremediation assay.

2.4. Bioremediation Assay

Two isolates (S13 and F3) were selected (Supplementary data S1, Figure S2). The experimental setup consisted of glass vessels containing sterile seawater with 500 μg L^{-1} of TBT. Each bacterial inoculum (0.1% TSB with O.D. = 0.600) was added to the contaminated media. In parallel, vessels containing sterile seawater and each inoculum were prepared to evaluate possible toxic by-products of bacteria growth. In addition, a negative control consisting of 0.1% TSB in sterile seawater was carried out. Three replicates were made per treatment and kept away from light for 48 h at 20 ± 1 °C, with continuous aeration. For the ecotoxicological testing, contents were filtered by pore size 0.45 μm cellulose membranes and stored in amber glass bottles at 4 °C until used (<8 h).

2.5. Gibbula umbilicalis Acute Toxicity Tests
2.5.1. Preliminary TBT Ecotoxicity Testing

The acute toxicity of TBT was determined using the marine gastropod *G. umbilicalis* (da Costa, 1778) as model organism. Sea snails were collected during low tide in Peniche (central Portuguese coast), acclimated in glass tanks with clean natural seawater (34 PSU) at 20 ± 1 °C for two weeks with a 16 h:8 h light:dark cycle. The snails were fed *ad libitum* with the seaweed *Ulva lactuca* (Linnaeus, 1753). Snails with similar shell size (10 ± 2 mm) were selected and starved 24 h prior to the tests. Sea snails were exposed in 60 mL glass vessels filled with 0.01, 0.05, 0.2, 0.5, 1, 2, 5, 10, 25, 50, and 100 μg L^{-1} of TBT (nominal concentrations) and covered with tulle netting. Eight replicates consisting of one individual per vessel were used per concentration. After 48 and 96 h, mortality was checked by mechanical stimulation of the operculum and the median lethal concentration (LC$_{50}$) determined.

2.5.2. TBT Remediation Ecotoxicity Testing

A similar setup was used to assess the ability of bacteria collected from marine ports to bioremediate TBT-contaminated waters. An acute test was performed exposing the snails to increasing dilutions of each treated media: 20, 12.2, 7.6, 4.6, 2.8, 1.8, 1, 0.7, and 0.4% of the initial media (500 μg L^{-1} or 100%). In parallel, the following control groups were added to the experiment: (1) a contaminated (untreated) media with TBT using the same dilution range; (2) sterile seawater with each bacterial inoculum and (3) sterile seawater with 0.1% TSB (negative control), aerated for 48 h. Concentrations were chosen based on preliminary range-finding tests and literature search of environmentally relevant TBT concentrations. After 48, 72, and 96 h of exposure, mortality was recorded and the LC$_{50}$ estimated.

2.6. Bacterial Identification

Selected isolates were identified using a MALDI-TOF MS at the Hospital San Pedro de Alcántara (Cáceres, Spain). A characteristic spectrum was generated, specific for each species and compared to a reference database. A similarity over 2.0 indicated an identification at the species level [34].

2.7. Statistical Analysis

One-way analysis of variance (ANOVA) was performed to analyze the number of colony-forming units amongst treatments, followed by Dunnett's post hoc test to determine significant differences between experimental treatments and controls. Data were checked for homoscedasticity and residual normality and when needed, data were log-transformed to correct for non-normality. Median lethal concentrations (LC$_{50}$'s) were determined using four-parameter logistic curves and compared using the extra sum-of-squares F test. All statistical analyses were performed using GraphPad Prism 9 (GraphPad Software Inc., San Diego, CA, USA) and the significance level (*p*-value) was set at 0.05.

3. Results

3.1. Chemical Analyses

Physicochemical parameters and chemical analyses of metallic elements are presented in supplementary data (Supplementary data S1, Tables S1 and S2). The concentration of butyltins (TBT, DBT, and MBT) was <50 ng L^{-1} and inorganic tin concentration was <5 µg L^{-1} on all sampling locations. Leixões was the port with the highest level of metal contamination, with Pb, Mn, and Zn detected at 2.2, 17.7, and 36 µg L^{-1}, respectively.

3.2. Colony-Forming Units and Percentage of Tolerant Bacteria

The concentration of CFUs decreased with increasing TBT concentration, as did the relative abundance (%) of tolerant bacteria when compared to the control. At 0.1 mM TBT (Figure 1), Póvoa de Varzim sample presented the highest percentage of tolerant bacteria (45.5 ± 6.8%) while Setúbal had the lowest (2.6 ± 0.7%). At 1 mM and 3 mM TBT, Peniche port water exhibited the highest percentage of tolerants (8.2 ± 0.5% and 7.7 ± 1.8%, respectively) as well as the highest absolute abundance of CFUs (410 ± 26 CFU mL^{-1} and 383 ± 91 CFU mL^{-1}, respectively). Aveiro (0.3 ± 0.1%) and Setúbal (0.1 ± 0.03%) exhibited the lowest percentage of tolerant bacteria to 1 mM and 3 mM, respectively. Aveiro also presented the lowest abundance of cultivable isolates even in the absence of TBT, while Peniche presented the highest. Significant differences between TBT-contaminated media and control group were verified for all sampling sites. In total, 157 isolates tolerant to 3 mM TBT were purified by streaking technique and used in subsequent assays.

3.3. Bacterial Selection for Bioremediation

Following a cluster analysis of REP-PCR fingerprints, a total of 111 isolates from 157 genetic profiles were regarded as different phylotypes (cut off set at 94% similarity; Supplementary data S1, Figure S3). Similarity ranged from 37.5% to 100%, with the sample from Figueira da Foz showing the highest variability and Peniche the lowest. To further select the most promising bioremediation candidates, bacteria were grown in high concentrations of TBT (from 1 to 6 mM).

Two strains were selected according to their growth in the presence of 6 mM TBT (Supplementary data S1, Figure S2). Bacteria selected for bioremediation assays were S13 and F3. Strain S13 was identified as the gram-negative bacteria *Pseudomonas putida* and strain F3 was identified as the gram-negative *Serratia marcescens*.

3.4. Acute Toxicity Tests

The 48 h and 96 h LC_{50}'s (95% Confidence Interval) of TBT for *G. umbilicalis* were estimated as 61.96 µg L^{-1} (60.62–63.33) and 15.73 µg L^{-1} (13.84–17.88), respectively (Supplementary data S1, Figure S1). Regarding the exposure to potentially bioremediated media, LC_{50} values are presented as a percentage of the initial TBT solution (500 µg L^{-1}; Table 1). After 48 h of exposure, it was only possible to estimate the LC_{50} for the TBT-untreated media, as 17.02% (16.20–17.94) since no mortality was recorded in the remaining treatments. For the remediation treatments, the LC_{50} was estimated to be higher than 20% (the highest concentration tested). After 72 h, the LC_{50} for the TBT-untreated media was estimated as 8.40% (8.19–8.62), with higher LC_{50} values estimated for the remediation treatments, suggesting a decrease in toxicity of treated media. For the treatment with bacteria S13, the LC_{50} was still estimated to be higher than 20%. A similar outcome was observed after 96 h of exposure, with the lowest LC_{50} estimated for the TBT-untreated media (5.91% (8.89–5.93)) and the highest for the treatment with bacteria S13 (Table 1).

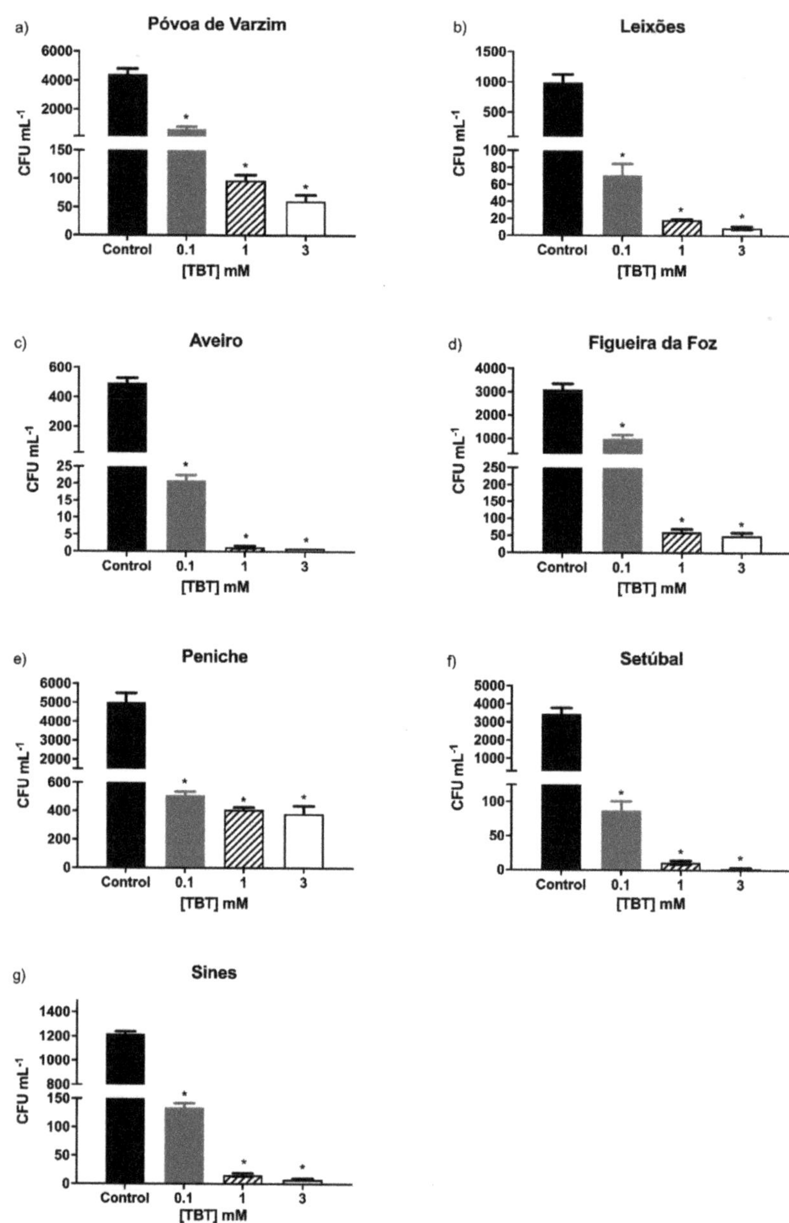

Figure 1. Relationship between concentration of tributyltin (0, 0.1, 1, and 3 mM) and CFUs for all sampling sites (**a**) Póvoa de Varzim; (**b**) Leixões; (**c**) Aveiro; (**d**) Figueira da Foz; (**e**) Peniche; (**f**) Setúbal; (**g**) Sines. Results presented as mean + SEM. Asterisk indicates significant differences between treatments and control, $p < 0.05$.

Table 1. Lethal concentration values estimated for *Gibbula umbilicalis* exposed to untreated and bioremediated media. Strain S13: *Pseudomonas putida*; Strain F3: *Serratia marcescens*. Asterisk (*) indicates significant differences in LC_{50} between bioremediated treatments and untreated media. n.d.-not determined.

Exposure Period	Treatment	LC_{50} (%)	95% Confidence Interval
	TBT	17.02	16.20–17.94
48 h	S13 + TBT	>20	n.d.
	F3 + TBT	>20	n.d.
	TBT	8.40	8.19–8.62
72 h	S13 + TBT	>20	n.d.
	F3 + TBT	19.90 *	13.89–45.32
	TBT	5.91	5.89–5.93
96 h	S13 + TBT	19.97 *	19.63–20.34
	F3 + TBT	10.21 *	8.04–13.45

As mentioned, glass vessels with seawater (without TBT) inoculated with selected bacteria were maintained in the same conditions as described for the other treatments. Sea snails were exposed to increasing dilutions of these media. At 20% of the initial solution, no mortality was observed during the 96 h exposure period.

4. Discussion

Naturally occurring marine bacteria from Portuguese ports were grown and isolated in high concentrations of TBT and their decontamination potential was evaluated. The results suggest that TBT is highly selective and can negatively affect bacterial growth. A decrease in the CFUs abundance with increasing TBT concentrations was observed for all sampled sites. At 0.1 mM, the bacterial growth was affected, with a decrease of CFUs over 50% compared to the control, suggesting this low concentration is highly toxic.

In Peniche, the percentage of tolerant bacteria was constant with increasing TBT concentrations. Peniche was also the location with the highest percentage of tolerant bacteria at 3 mM (7.7% compared to control). These results may be explained by the initial pool of bacteria, which was the highest from all sampling sites. Butyltin's levels in collected waters were below the limit of quantification (<50 ng L^{-1}) in all sampling sites. Marinas and ports are hotspots of TBT contamination due to boat traffic and shipyard activities [35,36]. Thus, it is expected that these sampling sites would be contaminated with TBT and that smaller concentrations of this highly toxic chemical might be present, particularly in sediments, where it persists [8,37]. Previous studies reported levels of butyltins in Portuguese waters. In 1999/2000, samples collected along Portuguese rivers and coastal waters revealed butyltins levels ranging from 23 to 79 ng L^{-1} [35]. In 1999, on Tagus estuary, TBT levels ranged from 1.1 to 21.1 ng L^{-1} [38]. Although those studies date back to the pre-ban period, a more recent study revealed that in 2014 TBT was still present in Portuguese coastal waters, since quantifiable amounts were detected in dog-whelks (*Gastropoda*) tissue samples [16].

Additionally, some studies refer the occurrence of TBT-tolerant bacteria to be related to the occurrence of other contaminants such as metals, making them cross-resistant to TBT [39]. Seaports are traditionally contaminated sites, making these heavily impacted sites more selective for bacteria that develop mechanisms to adapt to such conditions.

The REP-PCR method was used to characterize the 157 isolates tolerant to 3 mM of TBT. One hundred and eleven different patterns were identified. Peniche port samples had the lowest level of diversity, with just 5 distinctive patterns identified using REP-PCR, sharing a similarity of more than 79%, suggesting that the isolates from this location were genetically close to each other. This low diversity and higher initial pool of microorganisms (5000 ± 513 CFU mL^{-1}), may explain the high percentage of tolerant CFUs observed for this location.

To select the most suitable strains for bioremediation assays, bacteria were exposed to 6 mM TBT and two different strains were selected based on their growth in this condition. Although TBT inhibits the growth of these isolates, there were no apparent differences in growth between 1 mM, 3 mM, and 6 mM. Cruz et al. [40] reported that gradually transferring bacteria to higher TBT concentrations increases their tolerance, suggesting a "memory response" mechanism, which could justify why some bacteria grew in such high concentrations since they were cultured and maintained in TBT-contaminated media.

Despite the importance of determining the levels of a toxicant and its by-products in the ecosystems, this does not provide information on the contaminant's toxicity and its degradation products. Such data can be obtained through ecotoxicological assays. In the present study, ecotoxicological testing with *G. umbilicalis* proved to be a solid approach to determine toxic effects of TBT. Still, the use of sub-lethal, and thus more sensitive endpoints [41] should also be considered in future studies to better estimate the harmful impacts of low levels of butyltins on gastropods. *Pseudomonas putida* (S13) showed the highest biodegradation potential of TBT-contaminated waters, as evidenced by the observed decrease in toxicity mediated by this bacterium. *Pseudomonas putida* was reported as a TBT-degrading bacteria that uses TBT as a carbon source [27]. The cultivation of *P. putida* in TBT-contaminated media suggests that the breaking down of TBT by this bacterium generates degradation products with a lower toxicity than the parent compound. The ecotoxicological approach used in this study, allows to show the relevant effect on important species. Nonetheless, follow up studies should be complemented with chemical analyses. An increase in LC_{50} was also observed for the treatment with *Serratia marcescens* (F3), suggesting that remediation of the media had occurred, as the medium became less toxic. *Serratia marcescens* has been regarded as a potential organotin-resistant bacteria [26] and here its tolerance is confirmed. To our knowledge, this is the first study concerning the potential of *S. marcescens* to bioremediate TBT-contaminated waters. Nonetheless, despite the decrease in toxicity of the treated media with this bacterium after 96 h of exposure, the media was still toxic to *G. umbilicalis*. This hints the possibility that some TBT may be still present in the media and a longer remediation period (>48 h) may be needed for a more effective TBT degradation. Additionally, toxic by-products may be generated, which should be monitored in future studies, considering that toxic degradation products may hamper the in situ application of TBT-degrading bacteria.

As demonstrated in this study, naturally present bacteria from Portuguese ports can tolerate TBT at high levels and eventually transform it into less toxic compounds, contributing towards ecosystem restoration. The results derived from the ecotoxicological testing demonstrated that *P. putida* and *S. marcescens* have the potential to step up TBT removal from contaminated marine environments and thus representing candidate remediation agents for application in bioremediation approaches. Since heavy metal and antibiotic resistance is frequent in TBT-resistant bacteria, exploring TBT-tolerant bacteria applications might be vital for the restoration of polluted environments in general. Knowledge of their genetics and physiology is key for their future application as natural decontamination facilitators.

Overall, the present study contributed to the growing knowledge of TBT-degrading bacteria. The decrease in toxicity observed strongly suggest that *P. putida* can transform TBT into less toxic products and that ecotoxicological testing can be a valuable tool in bioremediation studies as a complement of chemical analyses. Future research should focus on determining mechanisms underlying TBT resistance and biodegradation. Despite the decreasing trend observed over the past decades, TBT levels in water and biota are still matters of concern [42]. Existing prohibitions have not translated into a complete elimination of TBT and its by-products from the environment, particularly considering the persistent nature of these chemicals, and therefore there is still a paramount need to develop tools to transform and safely remove this chemical from the environment.

Supplementary Materials: The following are available online at https://www.mdpi.com/article/10.3390/app11146411/s1, Table S1: Physicochemical properties of near-sediment waters from sampling sites at Portuguese ports, Table S2: Metal quantification in near-sediment waters from sampling sites at Portuguese ports, Figure S1: Dose-response curves of *Gibbula umbilicalis* exposed to TBT for 48 and 96 h, Figure S2: Growth curves of bacteria in the presence of TBT (1, 3 and 6 mM), Figure S3: Dendrogram showing genetic similarity of isolated bacteria determined by analysis of REP-PCR fingerprint patterns using Dice similarity coefficient and UPGMA cluster methods.

Author Contributions: Conceptualization, H.R.M., A.C.E. and M.F.L.L.; Investigation, H.R.M. and M.J.C.; Resources, A.C.E. and M.F.L.L.; Supervision, A.C.E. and M.F.L.L.; Writing—original draft, H.R.M. and A.B.M.; Writing—review and editing, H.R.M., A.B.M., A.C.E. and M.F.L.L. All authors have read and agreed to the published version of the manuscript.

Funding: This study had the support of Fundação para a Ciência e a Tecnologia (FCT) through the Strategic Project UID/MAR/04292/2020 granted to MARE and UIDP/50017/2020+UIDB/50017/2020 granted to CESAM, through national funds and the co-funding by the FEDER, within the PT 2020 Partnership Agreement and Compete 2020.

Conflicts of Interest: The authors declare no conflict of interest.

References

1. Turner, A. Marine pollution from antifouling paint particles. *Mar. Pollut. Bull.* **2010**, *60*, 159–171. [CrossRef]
2. Omae, I. Organotin antifouling paints and their alternatives. *Appl. Organomet. Chem.* **2003**, *17*, 81–105. [CrossRef]
3. Li, Z.H.; Li, P. Effects of the tributyltin on the blood parameters, immune responses and thyroid hormone system in zebrafish. *Environ. Pollut.* **2021**, *268*, 115707. [CrossRef] [PubMed]
4. Bandara, K.R.V.; Chinthaka, S.D.M.; Yasawardene, S.G.; Manage, P.M. Modified, optimized method of determination of Tributyltin (TBT) contamination in coastal water, sediment and biota in Sri Lanka. *Mar. Pollut. Bull.* **2021**, *166*, 112202. [CrossRef] [PubMed]
5. Castro, Í.B.; Iannacone, J.; Santos, S.; Fillmann, G. TBT is still a matter of concern in Peru. *Chemosphere* **2018**, *205*, 253–259. [CrossRef] [PubMed]
6. International Maritime Organization. Focus on IMO Anti-Fouling Systems. London. 2002, pp. 1–31. Available online: https://www.imo.org/en/OurWork/Environment/Pages/Anti-fouling.aspx (accessed on 1 June 2021).
7. Undap, S.L.; Nirmala, K.; Miki, S.; Inoue, S.; Qiu, X.; Honda, M.; Shimasaki, Y.; Oshima, Y. High tributyltin contamination in sediments from ports in Indonesia and Northern Kyushu, Japan. *J. Fac. Agric. Kyushu Univ.* **2013**, *58*, 131–135. [CrossRef]
8. Antizar-Ladislao, B. Environmental levels, toxicity and human exposure to tributyltin (TBT)-contaminated marine environment. A review. *Environ. Int.* **2008**, *34*, 292–308. [CrossRef]
9. Choi, M.; An, Y.R.; Park, K.J.; Lee, I.S.; Hwang, D.W.; Kim, J.; Moon, H.B. Accumulation of butyltin compounds in finless porpoises (*Neophocaena asiaeorientalis*) from Korean coast: Tracking the effectiveness of TBT regulation over time. *Mar. Pollut. Bull.* **2013**, *66*, 78–83. [CrossRef] [PubMed]
10. De Castro, Í.B.; Perina, F.C.; Fillmann, G. Organotin contamination in South American coastal areas. *Environ. Monit. Assess.* **2012**, *184*, 1781–1799. [CrossRef]
11. Kannan, K.; Guruge, K.S.; Thomas, N.J.; Tanabe, S.; Giesy, J.P. Butyltin residues in southern sea otters (*Enhydra lutris nereis*) found dead along California coastal waters. *Environ. Sci. Technol.* **1998**, *32*, 1169–1175. [CrossRef]
12. Ünlü, S. Marine pollution from ships in the Turkish straits system. In *The Sea of Marmara: Marine Biodiversity, Fisheries, Conservation and Governance*; Özsoy, E., Çağatay, M.N., Balkıs, N., Balkıs, N., Öztürk, B., Eds.; Turkish Marine Research Foundation (TUDAV): Istanbul, Turkey, 2016; pp. 755–767. ISBN 9789758825349.
13. European Union. Council Directive of 21 December 1989 amending for for the eighth time Directive 76/769/EEC on the approximation of the laws, regulations and administrative provisions of the Member States relating to restrictions on the marketing and use of certain dange. *Off. J. Eur. Communities* **1989**, *L398*, 19–23. Available online: http://data.europa.eu/eli/dir/1989/677/oj (accessed on 1 June 2021).
14. European Union. Regulation (EC) No 782/2003 of the European Parliament and of the Council of 14 April 2003 on the prohibition of organotin compounds on ships. *Off. J. Eur. Union* **2003**, 1–11. Available online: http://data.europa.eu/eli/reg/2003/782/oj (accessed on 1 June 2021).
15. Wang, X.; Fang, C.; Hong, H.; Wang, W.X. Gender differences in TBT accumulation and transformation in *Thais clavigera* after aqueous and dietary exposure. *Aquat. Toxicol.* **2010**, *99*, 413–422. [CrossRef]
16. Laranjeiro, F.; Sánchez-Marín, P.; Oliveira, I.B.; Galante-Oliveira, S.; Barroso, C. Fifteen years of imposex and tributyltin pollution monitoring along the Portuguese coast. *Environ. Pollut.* **2018**, *232*, 411–421. [CrossRef] [PubMed]
17. Abreu, F.E.L.; Lima da Silva, J.N.; Castro, Í.B.; Fillmann, G. Are antifouling residues a matter of concern in the largest South American port? *J. Hazard. Mater.* **2020**, *398*, 122937. [CrossRef] [PubMed]
18. Turk, M.; Ivanić, M.; Dautović, J.; Bačić, N.; Mikac, N. Simultaneous analysis of butyltins and total tin in sediments as a tool for the assessment of tributyltin behaviour, long-term persistence and historical contamination in the coastal environment. *Chemosphere* **2020**, *258*, 127307. [CrossRef]

19. Okoro, H.K.; Fatoki, O.S.; Adekola, F.A.; Ximba, B.J.; Snyman, R.G. Organotin Compounds. *Encycl. Toxicol. Third Ed.* **2014**, *3*, 720–725. [CrossRef]
20. Guo, S.; Qian, L.; Shi, H.; Barry, T.; Cao, Q.; Liu, J. Effects of tributyltin (TBT) on *Xenopus tropicalis* embryos at environmentally relevant concentrations. *Chemosphere* **2010**, *79*, 529–533. [CrossRef]
21. Zou, E. Aquatic invertebrate endocrine disruption. In *Encyclopedia of Animal Behavior*, 2nd ed.; Choe, J.C., Ed.; Academic Press: Oxford, UK, 2019; pp. 470–482. ISBN 9780128132524.
22. Blaber, S.J.M. The occurrence of a penis-like outgrowth behind the right tentacle in spent females of *Nucella lapillus* (L.). *J. Molluscan Stud.* **1970**, *39*, 231–233. [CrossRef]
23. Gooding, M.P.; Wilson, V.S.; Folmar, L.C.; Marcovich, D.T.; LeBlanc, G.A. The biocide tributylin reduces the accumulation of testosterone as fatty acid esters in the mud snail (*Ilyanassa obsoleta*). *Environ. Health Perspect.* **2003**, *111*, 426–430. [CrossRef]
24. Kawai, S.; Kurokawa, Y.; Harino, H.; Fukushima, M. Degradation of tributyltin by a bacterial strain isolated from polluted river water. *Environ. Pollut.* **1998**, *102*, 259–263. [CrossRef]
25. Cruz, A.; Anselmo, A.M.; Suzuki, S.; Mendo, S. Tributyltin (TBT): A review on microbial resistance and degradation. *Crit. Rev. Environ. Sci. Technol.* **2015**, *45*, 970–1006. [CrossRef]
26. Dubey, S.K.; Roy, U. Biodegradation of tributyltins (organotins) by marine bacteria. *Appl. Organomet. Chem.* **2003**, *17*, 3–8. [CrossRef]
27. Sampath, R.; Venkatakrishnan, H.; Ravichandran, V.; Chaudhury, R.R. Biochemistry of TBT-degrading marine pseudomonads isolated from Indian coastal waters. *Water. Air. Soil Pollut.* **2012**, *223*, 99–106. [CrossRef]
28. Yáñez, J.; Riffo, P.; Santander, P.; Mansilla, H.D.; Mondaca, M.A.; Campos, V.; Amarasiriwardena, D. Biodegradation of tributyltin (TBT) by extremophile bacteria from atacama desert and speciation of tin by-products. *Bull. Environ. Contam. Toxicol.* **2015**, *95*, 126–130. [CrossRef] [PubMed]
29. Cabecinhas, A.S.; Novais, S.C.; Santos, S.C.; Rodrigues, A.C.M.; Pestana, J.L.T.; Soares, A.M.V.M.; Lemos, M.F.L. Sensitivity of the sea snail *Gibbula umbilicalis* to mercury exposure-Linking endpoints from different biological organization levels. *Chemosphere* **2015**, *119*, 490–497. [CrossRef]
30. Giménez, V.; Nunes, B. Effects of commonly used therapeutic drugs, paracetamol, and acetylsalicylic acid, on key physiological traits of the sea snail *Gibbula umbilicalis*. *Environ. Sci. Pollut. Res.* **2019**, *26*, 21858–21870. [CrossRef]
31. Allibone, J.; Fatemian, E.; Walker, P.J. Determination of mercury in potable water by ICP-MS using gold as a stabilising agent. *J. Anal. At. Spectrom.* **1999**, *14*, 235–239. [CrossRef]
32. Adiguzel, A.; Ozkan, H.; Baris, O.; Inan, K.; Gulluce, M.; Sahin, F. Identification and characterization of thermophilic bacteria isolated from hot springs in Turkey. *J. Microbiol. Methods* **2009**, *79*, 321–328. [CrossRef]
33. Tacão, M.; Alves, A.; Saavedra, M.J.; Correia, A. BOX-PCR is an adequate tool for typing *Aeromonas* spp. Antonie van Leeuwenhoek. *Int. J. Gen. Mol. Microbiol.* **2005**, *88*, 173–179. [CrossRef]
34. Wieser, A.; Schneider, L.; Jung, J.; Schubert, S. MALDI-TOF MS in microbiological diagnostics—identification of microorganisms and beyond (mini review). *Appl. Microbiol. Biotechnol.* **2012**, *93*, 965–974. [CrossRef]
35. Díez, S.; Lacorte, S.; Viana, P.; Barceló, D.; Bayona, J.M. Survey of organotin compounds in rivers and coastal environments in Portugal 1999–2000. *Environ. Pollut.* **2005**, *136*, 525–536. [CrossRef]
36. Dafforn, K.A.; Lewis, J.A.; Johnston, E.L. Antifouling strategies: History and regulation, ecological impacts and mitigation. *Mar. Pollut. Bull.* **2011**, *62*, 453–465. [CrossRef] [PubMed]
37. Xiao, X.; Sheng, G.D.; Qiu, Y. Improved understanding of tributyltin sorption on natural and biochar-amended sediments. *Environ. Toxicol. Chem.* **2011**, *30*, 2682–2687. [CrossRef] [PubMed]
38. De Bettencourt, A.M.M.; Andreae, M.O.; Cais, Y.; Gomes, M.L.; Schebek, L.; Vilas Boas, L.F.; Rapsomanikis, S. Organotin in the Tagus estuary. *Aquat. Ecol.* **1999**, *33*, 271–280. [CrossRef]
39. Suehiro, F.; Mochizuki, H.; Nakamura, S.; Iwata, H.; Kobayashi, T.; Tanabe, S.; Fujimori, Y.; Nishimura, F.; Tuyen, B.C.; Tana, T.S.; et al. Occurrence of tributyltin (TBT)-resistant bacteria is not related to TBT pollution in Mekong River and coastal sediment: With a hypothesis of selective pressure from suspended solid. *Chemosphere* **2007**, *68*, 1459–1464. [CrossRef]
40. Cruz, A.; Caetano, T.; Suzuki, S.; Mendo, S. *Aeromonas veronii*, a tributyltin (TBT)-degrading bacterium isolated from an estuarine environment, Ria de Aveiro in Portugal. *Mar. Environ. Res.* **2007**, *64*, 639–650. [CrossRef]
41. Trenfield, M.A.; van Dam, J.W.; Harford, A.J.; Parry, D.; Streten, C.; Gibb, K.; van Dam, R.A. A chronic toxicity test for the tropical marine snail *Nassarius dorsatus* to assess the toxicity of copper, aluminium, gallium, and molybdenum. *Environ. Toxicol Chem.* **2016**, *35*, 1788–1795. [CrossRef]
42. Finnegan, C.; Ryan, D.; Enright, A.M.; Garcia-Cabellos, G. A review of strategies for the detection and remediation of organotin pollution. *Crit. Rev. Environ. Sci. Technol.* **2018**, *48*, 77–118. [CrossRef]

applied sciences

Article

Influencing Multi-Walled Carbon Nanotubes for the Removal of Ismate Violet 2R Dye from Wastewater: Isotherm, Kinetics, and Thermodynamic Studies

Khamael M. Abualnaja [1], Ahmed E. Alprol [2,*], Mohamed Ashour [2] and Abdallah Tageldein Mansour [3,4,*]

[1] Department of Chemistry, College of Science, Taif University, P.O. Box 11099, Taif 21944, Saudi Arabia; k.ala@tu.edu.sa
[2] National Institute of Oceanography and Fisheries (NIOF), Cairo 11516, Egypt; microalgae_egypt@yahoo.com
[3] Animal and Fish Production Department, College of Agricultural and Food Sciences, King Faisal University, P.O. Box 420, Al-Ahsa 31982, Saudi Arabia
[4] Fish and Animal Production Department, Faculty of Agriculture (Saba Basha), Alexandria University, Alexandria 21531, Egypt
* Correspondence: ah831992@gmail.com (A.E.A.); amansour@kfu.edu.sa (A.T.M.)

Abstract: In this study, a multi-walled carbon nanotube (MWCNT) was synthesized and used as an adsorbent for the removal of Ismate violet 2R dye from contaminated water. The morphology and structure of the synthesized adsorbent were examined via the Brunauer–Emmett–Teller (BET) surface area, X-ray powder diffraction (XRD) analysis, infrared spectroscopy (FT-IR), scanning electron microscopy (SEM), and Raman spectroscopy. The effects of an MWCNT on the removal of IV2R were examined via a batch method using different factors such as pH, agitation time, adsorbent dosage, temperature, and initial dye concentration. The results showed that, at the acidic pH 4, 0.08 g of an MWCNT with 10 mg L^{-1} at 120 min realized the favorable removal of IV2R dye using an MWCNT. Under these operation conditions, the maximum elimination efficiency for real wastewater reached 88.2%. This process benefits from the ability to remove a large amount of dye (approximately 85.9%) in as short as 10 min using 0.005 g of MWCNTs. Moreover, the investigational isotherm data were examined by different models. The equations of error functions were used in the isotherm model to show the most appropriate isotherm model. The highest adsorption capacity for the removal of the dye was 76.92 mg g^{-1} for the MWCNT. Moreover, the regression data indicated that the adsorption kinetics were appropriate with a pseudo-second order and an R^2 of 0.999. The thermodynamic study showed that the removal of IV2R is an endothermic, spontaneous, and chemisorption process. The MWCNT compound appears to be a new, promising adsorbent in water treatment, with 91.71% regeneration after three cycles.

Keywords: adsorption; dye; MWCNT; kinetics; isotherm models; thermodynamic

Citation: Abualnaja, K.M.; Alprol, A.E.; Ashour, M.; Mansour, A.T. Influencing Multi-Walled Carbon Nanotubes for the Removal of Ismate Violet 2R Dye from Wastewater: Isotherm, Kinetics, and Thermodynamic Studies. *Appl. Sci.* **2021**, *11*, 4786. https://doi.org/10.3390/app11114786

Academic Editor: Elida Nora Ferri

Received: 27 April 2021
Accepted: 18 May 2021
Published: 23 May 2021

Publisher's Note: MDPI stays neutral with regard to jurisdictional claims in published maps and institutional affiliations.

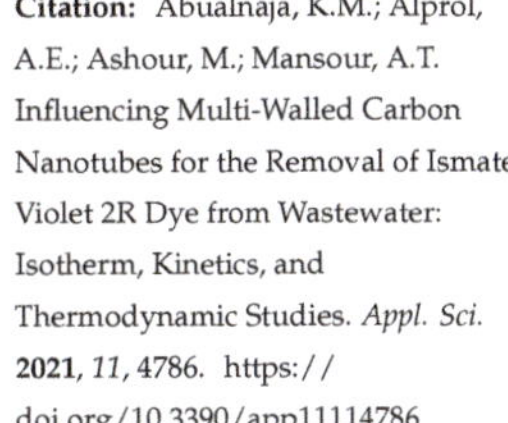

1. Introduction

Aquatic pollution by industrial wastewater is a threat to human and marine lives and is a major environmental problem [1]. Industrial wastewater contains a mixture of poisonous pulp mills, surfactants, oil, organic materials, heavy metals, salts, dyestuff, and industrial discharge, as well as a massive amount of highly colored effluent, causing severe environmental concerns all over the world [2]. The dyes from textile effluents are highly resistant to pH, light, and microbial attack, which results in them remaining in the environment for a longer time [3]. Textile effluents are rich in dyes and chemical compounds, some of which are carcinogenic, non-biodegradable, toxic metals, and pose a major threat to the environment and humankind [4,5]. The majority of dyes pose a potential health hazard to all forms of life, with long-term and accidental over-exposure conditions such as eczema, skin dermatoses, and allergic responses, potentially also affecting the lungs, the liver, the immune system, the vasco-circulatory system, and the reproductive systems of experimental animals and human systems, in addition to having a great effect on the photosynthetic activity in marine biota [6]. Therefore, various treatment methods are applied to remove color effluents and organic pollutants from wastewater, such as ion-exchange, ozonation, chemical reaction, membrane separation, filtration, adsorption, coagulation, electrodialysis, flocculation, photocatalytic degradation, reverse osmosis, and biological degradation, as well as immobilization methods [4,7–10]. These techniques have numerous disadvantages, such as chemical requirements, high energy, and low efficiency, and generally produce large quantities of sludge [11]. Although there have been improvements in several color effluent treatment technologies, the development of an effective, economic, and rapid water method at a trading level is still a defying problem. Previous efforts have concentrated on the adsorption technology for the treatment of color effluents [12,13]. This method can handle fairly high flow rates, generating high-quality wastewater that does not result in the creation of harmful materials, such as free radicals [14]. Moreover, it can minimize or eliminate various kinds of inorganic and organic contaminants, and thus has extensive applicability in the field of contamination control [15]. Furthermore, the adsorption process is successful for contaminant removal from wastewater because of its simple design, high adsorption capacity, ease of operation, insensitivity, and flexibility to toxic pollutants [16]. Over the years, several scientists and researchers all over the world have become interested in nanocompounds. The use of MWCNTs for the elimination of dye pollutant ions has recently been studied [11–15], and is comparable to additional carbon-dependent adsorbents that are utilized for commercial purposes [1].

Comparatively, MWCNTs, as newly developed carbonaceous substances, have received significant interest for application as potential adsorbents in wastewater treatment because of their distinct structure, shorter equilibrium time, great specific surface area, higher sorption capacity, flexible surface chemistry, and lower mass loss in reactivation [17,18]. MWCNTs have been confirmed to be more efficient for environmental contamination control in terms of the elimination of organic materials such as dyes from aquatic effluents, as reported by several researchers such as Machado et al. [19], Mashkoor et al. [20], and Yao et al. [21]. In addition, Shirmardi et al. [22] studied the elimination of malachite dye from aquatic solutions using MWCNTs. The results were found to be q_{max}, 142.82 mg g^{-1} of dye at 80 min, and pH 7. Recently, Alkaim et al. [23] synthesized carbon nanocompounds for the removal of malachite dyes in aqueous solutions. The maximum capacity of the adsorption process was observed to be 187.69 mg g^{-1} at 120 min and pH 10. Machado et al. [24] studied the adsorption of a reactive blue 4 dye from an aqueous mixture by carbon nanotubes. The maximum sorption capacity was observed to be 502.5 mg/g. Moreover, Saber-Samandari et al. [18] studied the adsorption of anionic and cationic dyes from a liquid mixture using gelatin-based magnetic nanocomposite beads, including carboxylic acid functionalized carbon nanotubes. The results showed the percentage removal to be 76.3% of methylene blue. Gang Yu et al. [25] reported that the treatment of organic chemicals compounds by MWCNTs may include more mechanisms, such as hydrophobic interactions, π–π interactions, electrostatic interactions, Lewis acid–base interactions, and hydrogen bonding, which reveals that this requires essential additional investigation.

Ismate violet 2R was selected as a model compound in this research due to its wide application range, which includes coloring paper, and dyeing cottons, wools, and coating for paper stock and medical purposes, in addition to its potential dangerous effects.

Accordingly, the aims of this work were the preparation, functionalization, and purification of multi-walled carbon nanotubes (MWCNTs), as a novel approach, and the application of MWCNTs as an adsorption substance to remove Ismate violet 2R (IV2R) dye from aquatic effluents under various conditions using the batch adsorption method. The main parameters evaluated were the initial dye concentration, pH, contact time, temperature, and adsorbent dose. Moreover, the changes in the physical, chemical, and morphological structure owing to imitation reactions were studied through Raman, XRD,

BET surface, FTIR, and SEM characterizations. Additionally, thermodynamic and kinetics studies were conducted, and the experimental equilibrium was investigated using various sorption isotherm models to assess the adsorption mechanism. In addition, to estimate the regenerative capacity of MWCNTs, the interference of wastewater on the adsorption of IV2R dye was also examined. The novelty of this study can be summarized as the first case, in the oxidized MWCNT literature, of MWCNTs modified with such functionalities and applied to the elimination of IV2R dye. Furthermore, previous studies on dye removal by a single functional group indicated lower elimination than that presented in this work. Moreover, the highest suitable model was achieved by using error functions.

2. Materials and Methods

2.1. Materials

MWCNTs were created using the chemical vapor deposition technique, in which acetylene with cobalt and an iron solution was placed in an inert gas atmosphere connected to a reaction chamber. In this process, nanotubes were made on the substrate through the decay of hydrocarbon at an atmospheric pressure of 600–900 °C. This method can be scaled up to prepare, purify, and functionalize MWCNTs according to Mohammed et al. [26] and Bahgat et al. [27], who reported the synthesis of MWCNTs via the following steps: 2 g of the supporting catalyst was injected into an alumina boat and displayed on a cylindrical quartz tube fitted inside a furnace at 600 °C, and the catalyst was heated in the presence of N_2 gas for 10 min with a rate of 90 mL min^{-1}. The movement of acetylene gas was tolerated by the quartz tube undergoing catalyzation at 90 mL min^{-1} with a flow rate of 40 min after the catalyst was heated. The flow of acetylene ceased after the required period, and the new product was cooled at room temperature.

2.2. Purification and Functionalization of MWCNTs

An enormous surface area prompts a strong tendency to shape agglomerates. Surface functionalization helps in steadying the scattering, since it can repress the reaggregation of nanotubes and can prompt the coupling of polymeric grids with MWCNTs. Covalent functionalization of MWCNTs can be determined by identifying some functional groups on sites of MWCNTs through oxidizing agents, such as strong acids, which causes the formation of hydroxyl or carboxylic groups (–OH, –COOH) on the outside of the nanotubes. Such groups are known as functionalization-type defect groups. The functionalization process was performed using 10 mL of concentrated sulfuric acid and 30 mL of nitric acid in a 250 mL flask loaded with 10 g of the produced MWCNTs and 5 g of phosphorous pentaoxide. The mixture solution was refluxed for 120 min at 350 °C to obtain an MWCNT suspension mixture. Then, the mixture solution was washed with deionized water, followed by drying for 24 h at 50 °C to obtain carboxylate MWCNTs [27].

2.3. Dye Solution Preparation

Stock solutions of IV2R were set by dissolving accurately weighed dye in distilled water at a concentration of 1 L, without additional purification. The characteristics of the dye used in this study are presented in Table 1. The experimental mixtures were achieved by diluting the dye stock mixture in accurate proportions to obtain various initial concentrations.

2.4. Adsorption Experiments

The effects of pH (2–10), contact time (10–180 min), initial dye concentration (10–80 mg L^{-1}), adsorbent dose (0.005–0.08 g), and temperature (25–55 °C) for IV2R removal were examined. The particular conditions are cited in the related plots. The tests were conducted in triplicate by the batch technique to gather equilibrium data. The reaction mixture containing 50 mL of the dye solution and the solution were shaken at room temperature (25 ± 2 °C) at a speed of 110 rpm under fixed conditions in an incubator shaker. The pH value was adjusted by HCl or NaOH before adding the adsorbent. After shaking, the concentration of

IV2R dye was determined in a clear supernatant at any time at 550 nm using a UV–visible spectrophotometer. Concentrations of dye in a solution can be assessed quantitatively, as stated by the Lambert–Beer law, by linear regression equations, achieved through plotting a calibration curve for the concentration ranges of dye.

Table 1. Chemical and physical characteristics of the ISMATE violate 2R dye [28].

Characteristics	Value
Dye name	ISMATE violate 2R
Wavelength (λ max)	550 nm
Mol. wt.	700
Molecular formula	$C_{22}H_{14}N_4O_{11}S_3CuCl$
C.I. name	IV2R
Molecular structure	

The adsorption capacity (q_e) and dye elimination percentage were calculated using Equations (1) and (2), as follows [29]:

$$q_e = \frac{\left(C_i - C_f\right) \times V}{W} \tag{1}$$

$$\text{Percentage removal } (\%) = \frac{\left(C_i - C_f\right)}{C_i} \times 100 \tag{2}$$

where C_i and C_f (mg L^{-1}) are the primary concentration at the initial time and the final concentration of IV2R at a certain period, respectively, while V represents the volume of the dye mixture (L) and W relates to the weight of the dry adsorbent (g).

2.5. Characterization of MWCNTs

The morphologies of the adsorbents were characterized using a scanning electron microscope (JEOL JSM 6360 LA). Moreover, Brunauer–Emmett–Teller (BET) desorption–adsorption tests were conducted on a BELSORP mini-II, BEL Japan, Inc. The specific external area, mean pore diameter, and pore volume of the MWCNTs were calculated by means of N_2 desorption–adsorption measurements at an N_2 solution with an 89.62 kPa saturated vapor pressure and an adsorption temperature of 77 K.

Furthermore, FTIR and Raman analysis spectrophotometry were used to measure the influence of the prepared MWCNTs materials—a Shimadzu FTIR-8400 S (Japan) and a Bruker Senterra Raman spectrometer were utilized, respectively. Additionally, the samples were assessed using a particle size analyzer (Beckman Coulter; Miami, FL, USA) at angles of 11.1° and 90° to determine the distribution of particle sizes. Additionally, an X-ray diffraction apparatus (D8 Advance X-Ray Di-ractometer; Bruker AXS, Wisconsin, USA) was used to investigate the crystal structure of the MCNTs with Panalytical in a scope from 10° to 90° with Kα Cu radiation.

2.6. Adsorption Isotherm Study

2.6.1. Isotherm Experiment

The extent of removal of IV2R from aqueous solutions depends heavily on the initial dye concentration. In order to assess isotherm study, different IV2R concentrations varying from 10 to 100 mg L^{-1} were examined at constant parameters and pH 6 by 0.01 g of MWCNTs adsorbents, at 30 °C, for 3 h, at 150 rpm, and were mixed with 50 mL of dyes solution [30]. The data were fitted and calculated in terms of the following isotherms:

2.6.2. Theoretical Background of Isotherm Models

The Freundlich Model

The ability of the Freundlich model to fit the experimental data, via plotting a curve of log q_e with respect to log C_e, was employed to generate a slope of n and the intercept value of K_f. The Freundlich model can be easily linearized by plotting it in a logarithmic form [31]:

$$\log q_e = \log K_f + 1/n \log C_e \tag{3}$$

Freundlich constants K_f and n were determined from the isotherm equation according to Equation (3).

The Langmuir Model

The Langmuir model is presented by mathematical expression [32] as the following equation:

$$q_e = q_{\max} bC_e/(1 + bC_e) \tag{4}$$

where $q_{\max}$ (mg g^{-1}) is the maximum sorption capacity corresponding to the saturation capacity (representing the total binding sites of a fiber) and b (L mg^{-1}) is a coefficient related to the affinity among the MCNTs and dye ions.

The linear relationship can be achieved from plotting curve $(1/q_e)$ vs. $(1/C_e)$:

$$1/q_e = 1/(bq_{max} C_e) + 1/q_{max} \tag{5}$$

in which b and $q_{\max}$ are determined from the slope and intercept, respectively.

The Henderson and Halsey Isotherm Models

These models are suitable for heteroporous solids and the multilayer sorption technique [33,34]. The Halsey model was calculated using the following equation:

$$\ln q_e = \frac{1}{n}\ln K + \frac{1}{n} \ln C_e \tag{6}$$

where n and K are Halsey constants. Meanwhile, the Henderson model was obtained from the following equation:

$$\ln[-\ln(1 - C_e)] = \ln K + \left(\frac{1}{n}\right) \ln q_e \tag{7}$$

while the Henderson constants are nh and Kh.

The Harkins–Jura Model

This model explains multilayer sorption, in addition to the presence of heterogeneous pore scattering in a sorbent [34]. This model was obtained from the following equation:

$$\ln q_e = \frac{1}{n}\ln K + \frac{1}{n} \ln C_e \tag{8}$$

where the isotherm constants are AHJ and BHJ.

The Smith Model

The Smith model is appropriate for heteroporous solids and multilayer adsorption. This model is usually obtained from the following equation:

$$q_e = W_{bS} - W_S \ln(1 - C_e) \tag{9}$$

where the Smith model parameters are W_{bS} and W_s.

The Tempkin Model

The Tempkin isotherm model assumes that the heat of the sorption of each particle in the layer declines with coverage because of adsorbate–adsorbent interactions; moreover, the sorption process is described through a uniform distribution of the binding energies, up to some higher binding energy [35]. The Tempkin model was calculated from the following equation:

$$q_e = B \ln A + B \ln C_e \tag{10}$$

where A is the constant of equilibrium binding (L g^{-1}) associated with a higher binding energy, b (J mol^{-1}) is a constant corresponding to the heat of sorption, and $B = (\text{RT b}^{-1})$ (J mol^{-1}) is the Tempkin constant and is related to the heat of adsorption and the gas constant (R = 8.314 J mol^{-1} K^{-1}).

2.6.3. Error Functions Test

To determine the greatest and most appropriate model to investigate the equilibrium data, various error functions were studied. The following models were used for the error functions test [36].

Average Percentage Error (ABE)

The ABE model shows a tendency or appropriateness among the predicted and experimental values of the sorption capacity used for plotting model curves (ABE) and can be designed consistent with the following equation:

$$\text{APE}(\%) = \frac{100}{N} \sum_{i=1}^{N} \left| \frac{q_{e,isotherm} - q_{e,calc}}{q_{e,isotherm}} \right|_i \tag{11}$$

where n shows the number of investigational data points.

Nonlinear Chi-Square Test (χ^2)

The nonlinear chi-square test is a statistical approach for the best appropriate treatment system. The approach of the chi-square error, χ^2, model is assumed as the following equation:

$$X^2 = \frac{(q_{e,isotherm} - q_{e,calc})^2}{q_{e,isotherm}} \tag{12}$$

Sum of Absolute Errors (EABS)

An augmentation in errors provides a better fit, leading to a bias toward great concentration data. EABS tests can be assessed using the following equation:

$$\text{EABS} = \sum_{i=1}^{N} \left| q_{e,calc} - q_{e,isotherm} \right|_i \tag{13}$$

2.7. Adsorption Kinetics

Adsorption kinetics studies were conducted in 50 mL conical flasks. A solution of pH 2 with 0.01 g of adsorbent was mixed separately with 50 mL of the IV2R solution of 10 mg L^{-1} concentrations, and the solution was agitated at room temperature under requisite time intervals viz. 10, 20, 30, 60, 120, and 180 min [37]. The clear solutions were analyzed for residual IV2R concentrations in the solution.

2.7.1. Pseudo-First-Order Kinetics Model

The linear form of the generalized pseudo-first-order equation [38] is given by the following equation:

$$d_q/d_t = K_1 (q_e - q_t) \tag{14}$$

where q_e is the amount of dye adsorbed at equilibrium (mg g^{-1}), q_t is the quantity of dyes adsorbed at time t (mg g^{-1}), and K_1 is expressed as a pseudo-first-order rate constant (min^{-1}). The integrating equation was assessed as follows:

$$\text{Log } (q_e/q_e - q_t) = k_1 t/2.303 \tag{15}$$

The pseudo-first-order equation is given by the following formula in a linear equation:

$$\text{Log } (q_e - q_t) = \log q_e - k_1 t/2.303 \tag{16}$$

Plots of $\log(q_e - q_t)$ against (t) should provide a linear relationship from k_1 and q_e, which can be assessed via the slope and intercept, respectively [39].

2.7.2. Pseudo-Second-Order Kinetics Model

The pseudo-second-order equation is expressed as follows [40]:

$$dq_t/d_t = K_2(q_e - q_t)^2 \tag{17}$$

where K_2 indicates the constant of the second-order rate (g mg^{-1} min). The integrated equation is presented as follows:

$$1/(q_e - q_t) = 1/q_e + K_2 \tag{18}$$

Ho et al. [11] obtained a linear form of the pseudo-second-order equation as follows:

$$t/q_t = 1/K_2 q_e^2 + t/q_e \tag{19}$$

Plots of (t/q_t) against (t) would provide a linear relationship, and the values of the q_e and K_2 parameters can be calculated from the slope and intercept, respectively.

2.7.3. The Intraparticle Diffusion Model

The intraparticle diffusion equation [14] is as follows:

$$q_t = K_{dif} t^{1/2} + C \tag{20}$$

where C is the intercept and K_{dif} (mg g^{-1} min$^{-0.5}$) reflects the intraparticle diffusion rate constant, which is estimated through the slope of the regression line.

2.8. Adsorption Thermodynamics

A thermodynamic study is necessary for taking information about whether the nature of sorption techniques is spontaneous or not, and the Gibbs and the value of free energy change, $\Delta G°$, are important principles of non-spontaneity. The changes in the enthalpy $\Delta H°$ and entropy $\Delta S°$ parameters should be considered to calculate the Gibbs free energy for sorption techniques at various temperatures (e.g., 25, 35, 45, and 55 °C) using the following nonlinear forms [41]:

$$K_d = q_e/C_e \tag{21}$$

$$\Delta G° = -RT \ln K_d \tag{22}$$

$$\Delta G° = \Delta H° - T \Delta S°, \text{ or} \tag{23}$$

$$\Delta G° = T (\Delta S°) + \Delta H° \tag{24}$$

The values of $\Delta H°$ and $\Delta S°$ were calculated from the intercept and slope of the plotted curve of T vs. $\Delta G°$ of Equation (23) or (24), in addition to $\Delta G°$, which can be obtained from Equation (22).

2.9. Application on Real Wastewater

A wastewater sample was collected from the El-Emoum drain, which neighbors Lake Maruit, Alexandria, Egypt, and which comprises numerous industrial sewage and agricultural wastes. The wastewater had a pH of 9.5, a TSS of 250 mg L^{-1}, and a TDS of 3240 mg L^{-1}. Owing to the low concentrations of dyes, an amount of IV2R dye was added to obtain 10 mg L^{-1} of this dye. The dye solution was filtered to remove precipitates and suspended matters. The application of MWCNTs to remove the IV2R dye from the wastewater was carried out under optimized conditions (pH 2 for 120 min at 30 °C and 0.08 g of adsorbents in the solution). Deionized water comprising a similar concentration of dyes was prepared to use as a control to estimate the effect of adsorbent on IV2R dye removal. The absorbance was determined using a Schimadzu UV 2100 spectrophotometer. Furthermore, the pH was determined by a digital electrode pH meter, while the total suspended solids (TSS) and total dissolved solids (TDS) were measured by standard methods [42].

3. Results and Discussion

3.1. Characterization of the MWCNTs

3.1.1. FT-IR Analysis

Infrared spectroscopy is a useful qualitative tool for determining the composition of compounds. Figure 1 shows the typical FTIR spectra in the range of 500–4000 cm^{-1} for MWCNTs. In the spectra of the functionalized MWCNT material, the broad peaks at 3510–3780 cm^{-1} are assigned to N-H and O-H stretching of the carboxylic group, respectively. The increased intensity of the OH stretching broadband at ~3510 cm^{-1} suggests a greater proportion of defective CNTs in the sample, which can be easily oxidized in the air [43]. However, the band at 2358 cm^{-1} is recognized as thiol S-H stretching. The sharp band at 1730 cm^{-1} corresponds to C-O stretching vibration of COOH. The FTIR of amine-functionalized CNT shows new bands at 1654 cm^{-1} and corresponds to C=C stretching, C=N due to the reaction with hydroxyl amine, and N-H deformation [44]. The difference in the FTIR spectra of the MWCNTs in the 1000–1270 cm^{-1} region can be attributed to C-N stretching due to carbon atom and terminal amino group bonding, and confirms the presence of amine groups on the MWCNTs [44].

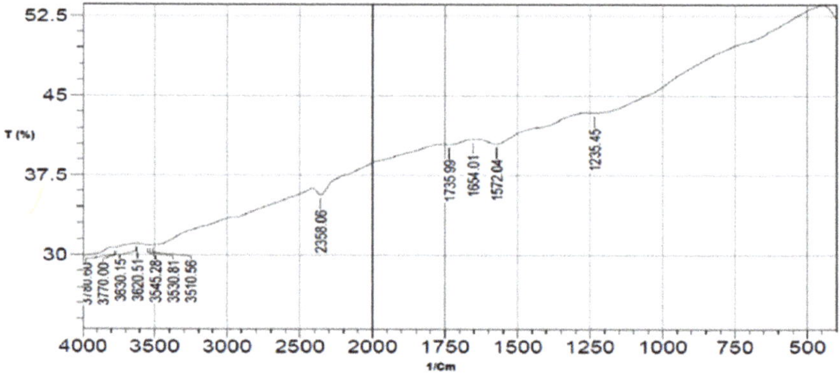

Figure 1. FTIR pattern of MWCNTs.

3.1.2. BET Surface Analysis

The BET and nitrogen adsorption–desorption study investigation method was utilized to quantify the pore diameter and specific surface space of the MWCNTs (Figure 2). A sample of the MWCNTs was optimized at a deposition temperature of 750 °C. The results of the BET surface area examination of the MWCNTs show that the pore diameter was 6.67 nm, and the whole surface area was 181.99 m^2 g^{-1}. The surface area of CNTs is affected by their diameter, agglomeration, surface functionalization, presence of impurities, and so on [45].

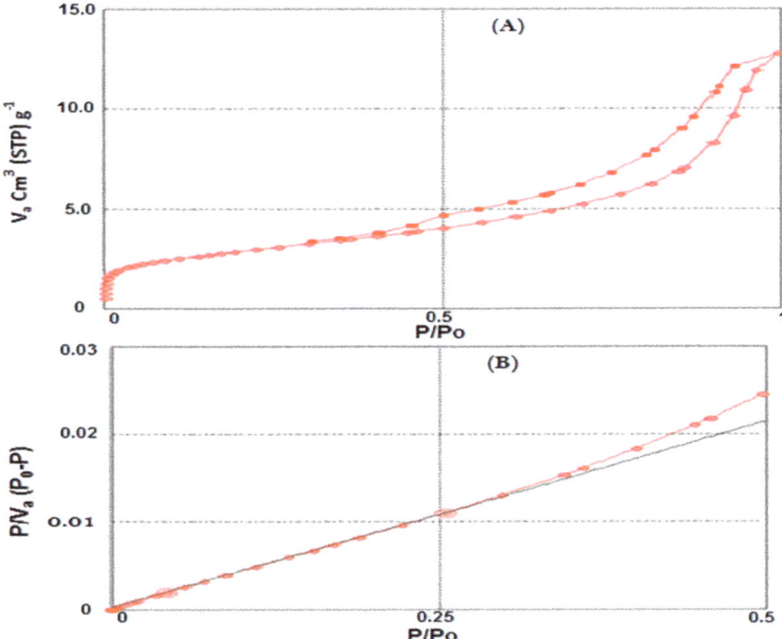

Figure 2. BET specific surface area (**A**) and adsorption–desorption isotherm examination (**B**) of the MCNTs.

3.1.3. SEM Examination

The external morphology of the MWCNTs was examined by scanning electron microscopy. Figure 3A,B displays the SEM images as prepared and purified MWCNTs, respectively. These figures show greatly tangled tubes and the existence of amorphous carbon in the unpurified sample. This observation, after acid treatment, confirms the disappearance of the amorphous texture, and there is no evidence of damage to the tubular structure of the MWCNTs, nor did the multi-walled arrangement vary. The tubular diameter of the MWCNTs ranged between 75 and 100 nm.

3.1.4. XRD Measurement Analysis

X-ray diffraction was applied in order to investigate the changes in the crystalline structure of the MWCNTs. Figure 4 shows the XRD patterns of the MWCNTs. It is clear from the figure that there are strong bands at 20 at 28.08°, 31.4°, 34°, and 44.66° for the oxidized MWCNTs, corresponding to an interplanar space of 3.41 Å and 2.05 Å, respectively. This could be due to the (002) and (100)/(101) planes of the MWCNTs (graphite structure) providing improved crystallinity and a decline in the amorphous structure due to acid treatment. Similar observations have been described for MCNTs prepared via the CVD method [46,47].

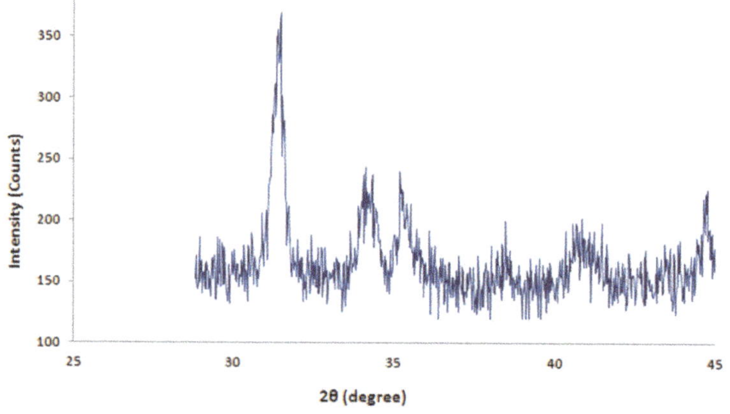

Figure 3. SEM examination at 2 μm and magnification (8000×) for the prepared (**A**) and purified (**B**) MWCNTs.

Figure 4. XRD diffraction of the MWCNTs.

3.1.5. Raman Characterization

A Raman spectrum displays different peaks that undergo alterations with changes in the characteristics of a sample to identify the chemical composition or microstructure constituents of the sample. A Raman spectrum of the MWCNTs is represented in Figure 5. The peak intensity ranging from ~1300 to 1350 cm^{-1} represents disordered carbon (D band), while peaks ranging from ~1580 to 1600 cm^{-1} represent graphitic carbon (G band). The D peak for the MCNTs occurred at ~1325 cm^{-1}, while G peaks were observed at ~1580 cm^{-1}. The ratio of ID/IG was calculated to estimate the variation of CNTs' quality and quantity of different defects. The ratio for the CNTs was 0.83; a smaller ID/IG ratio (<1.0) indicates a higher quality of CNTs [48]. A sample has quite a high number of defects if the intensity of these peaks is proportional [49]. Additionally, the peaks at 10–200 cm^{-1} are assigned to the lattice vibrations in crystals. The absorption peaks at 640, 830, 1080, and 1450 cm^{-1} are attributed to the υ(C-S), υ(C-O-C), υ(C=S), δ(CH$_2$), and δ(CH$_3$) asymmetric, respectively. Furthermore, the absorption peak at 1630 cm^{-1} was caused by –C=N group stretching. The absorption band at 1806 cm^{-1} is assigned to the C=C group. The carbonyl group was found in 2000 cm^{-1}. However, the peaks at 3150 and 2570 cm^{-1} are assigned to υ (O-H) and υ (-S-H), respectively.

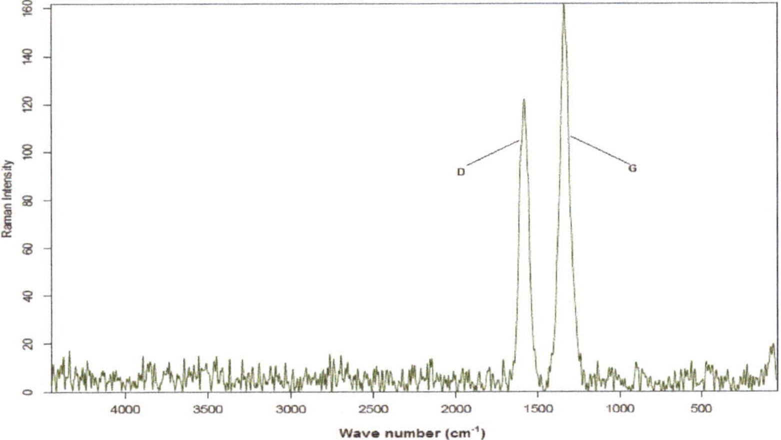

Figure 5. Raman spectrum for the MWCNTs.

3.2. Adsorption Experiments

3.2.1. Influence of pH

The pH of a solution may change the form and chemistry of the target dye ions and the binding sites on the adsorbent. The effect of pH on the removal of IV2R ions by MCNTs was performed in a pH range of 2–10, the results of which are shown in Figure 6. These results show that when the pH of the solution was augmented from 2 to 4, the adsorption of dye increased to reach its maximum for the MCNTs of 92.98% at pH 4; then, it slightly decreased. Therefore, pH 4 was selected as a compromise value. This can be explained by the adsorbent containing several functional groups that are affected by the pH of the solution. Moreover, an acidic solution increases in the percentage of anionic dye removal because of the electrostatic attraction between anionic dye and the positive charge of the surface of the adsorbent [50]. At a higher pH, electrostatic repulsion occurs among dye molecules and the negatively charged surface, therefore reducing the percentage elimination of anionic dyes and the sorption capacity [11]. Additionally, Ghoneim et al. [29] concluded that, at a higher pH, elimination is low compared to the maximum condition. This can be clarified as the binding site not being able to activate under basic conditions.

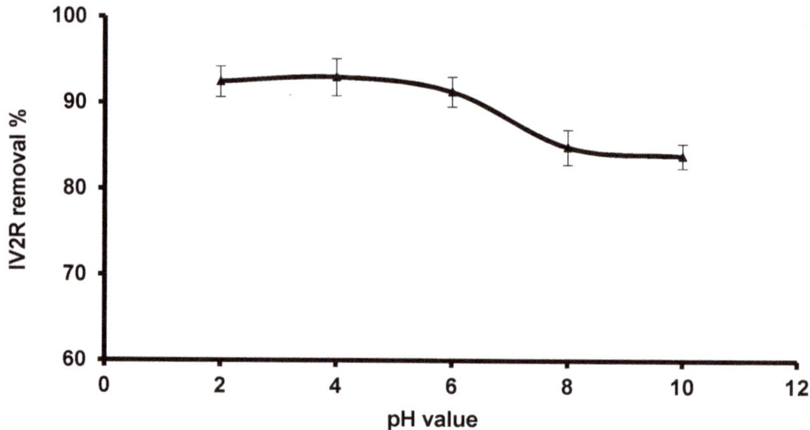

Figure 6. Influence of the pH value on the sorption of IV2R dye.

3.2.2. Influence of Adsorbent Dose

The mass of an adsorbent shows its capacity for a given concentration of dye. The influence of the adsorbent dose was studied at 25 °C by varying the sorbent dose to 0.005, 0.01, 0.2, 0.04, and 0.08 g. For all of these runs, the concentration of the IV2R dye was fixed at 10 mg L^{-1}, and the results are illustrated in Figure 7. An increase in the percentage of removal was observed when the dose was increased from 0.005 to 0.08 g, and the maximum removal was obtained at the adsorbent dose of 0.08 g with a percentage of 98.32% for the IV2R dye in the MWCNTs. The obtained data show that the adsorption of dye increased gradually with an augmentation in the amount of MWCNTs due to the greater availability of dye-binding sites of the surface area at higher concentrations of the adsorbent [51]. Zhao et al. [17] studied the treatment of methyl orange dye by MWCNTs, and it was found that, as the mass of MWCNTs increases, the whole adsorptive capacity of MO increases because of the increase in the number of reaction sites available and surface area to MO.

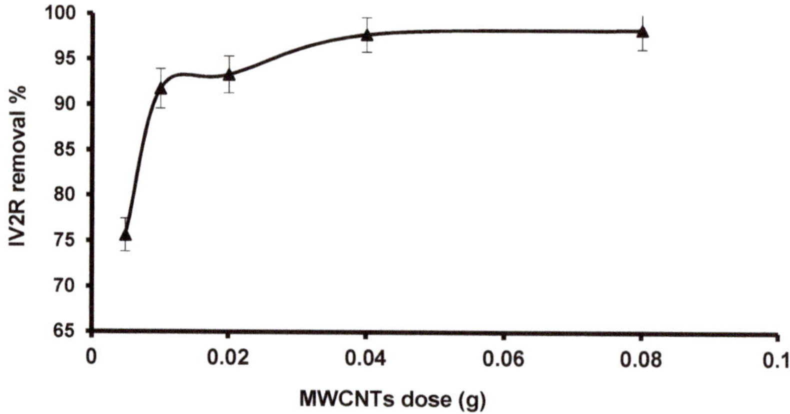

Figure 7. Influence of the MWCNT dose on the removal method.

3.2.3. Influence of Contact Time

Contact time is an essential factor in all transfer phenomena for the adsorption process, and the equilibrium time is important when considering economical water in addition to wastewater applications. The results are presented in Figure 8, indicating that the adsorption of IV2R dye ion removal increased to reach equilibrium, with 91.67% at 120 min.

The adsorption was rapid at the first time of contact, which occurred in the early stage of adsorption after a few minutes; this was due to the fact that most of the binding sites were free and the adsorbent sites were empty [52], which allowed the quick binding of dyes ions on the adsorbent. Adsorption happens rapidly and is usually controlled via the diffusion method from the bulk of the solution on the surface [53]. The increasing contact time increases the adsorption, which is possible because of a greater surface area of the MWCNTs being available at the beginning and exhaustion of the conversion of external adsorption sites, at which the adsorbate (dye particles) is transported from the external to the internal sites of the MWCNTs adsorbent molecules [54]. Hashem [55] recorded that adsorption occurs in the first 40 min of the sorption method, which shows the high diffusion of the dye particles into the outer surface of the adsorbent. In later stages, fewer vacant adsorption sites are converted, which reduces the rate of adsorption, in addition to reaching an equilibrium state.

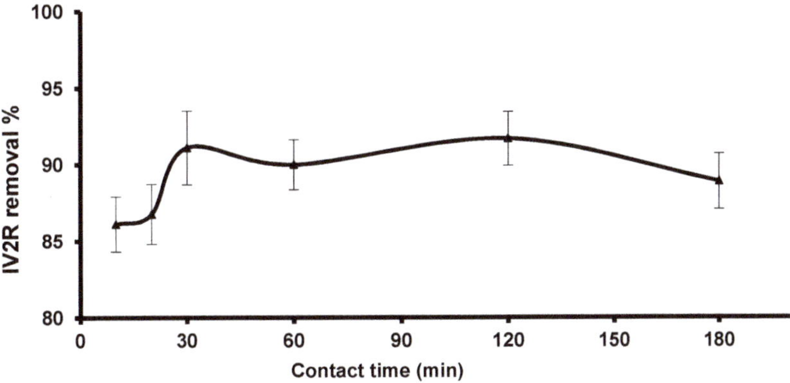

Figure 8. Influence of agitation time on the removal of IV2R.

3.2.4. Influence of the Initial Dye Concentration

The extent of removal of dyes from an aqueous mixture depends heavily on the dye concentration. The influence of the initial concentration of dye ions in a series of 10–70 mg L^{-1} on adsorption was examined, and the results are shown in Figure 9. By using MWCNTs, the maximum uptake was found at the concentration of 10 mg L^{-1} (91%) for IV2R dye. After this, the dye removal rate decreased with the rising concentration of IV2R dye until the end of the experiments, at the same contact time and adsorption temperature. However, the results showed that the amount of dye adsorbed per unit mass of adsorbent increased from 1.2 to 5.5 mg g^{-1} when the concentration of dye increased from 10 to 70 mg L^{-1}. The increase in the adsorption capacity is possibly due to greater interaction among the dye and adsorbent, in addition to an increase in the driving force of the concentration gradient with the increase in the initial concentration [17]. Moreover, the higher quantity of dye removal at higher concentrations is probably because of increased diffusion and decreased resistance to dye uptake [56]. Furthermore, the rapid initial stage of dye removal with this adsorbent is ascribed to the availability of active sites on the adsorbent and to the gradual occupancy of these sites; the sorption converts to be less efficient. Generally, increasing the dye concentration leads to a decrease in the percentage of removal of dye, which may be because of the saturation of sorption sites on the surface of the adsorbent [1,39]. Moreover, the major quantity of dye adsorbed at a high dye concentration could be related to two main effects, namely, the high probability of collision between dye ions with the adsorbent's surface and the high rate of dye ion diffusion onto the adsorbent's surface [57].

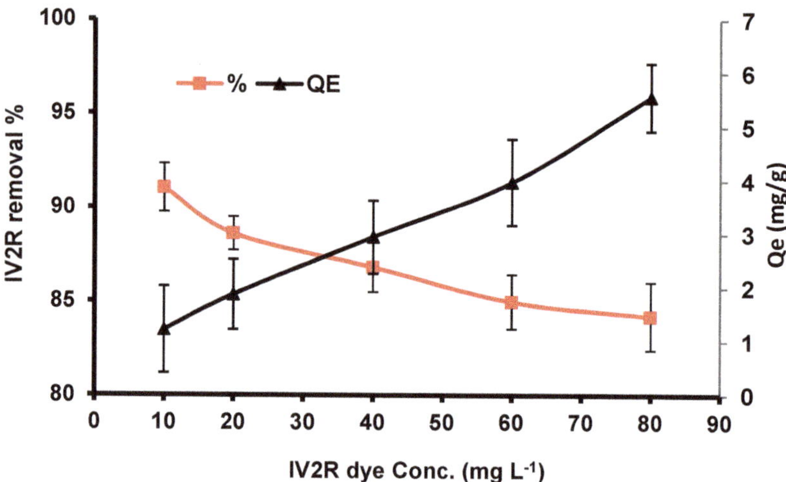

Figure 9. Influence of the initial IV2R dye concentration.

3.2.5. Influence of Temperature

Figure 10 shows the variation of the percentage of removal of IV2R dye ions in the MWCNTs. It can be concluded that the maximum percentage of removal of dye was obtained at 30 °C with 91.15% and reduced with the rise in temperature from 30 to 55 °C for the IV2R dye in the MWCNTs. With a rise in temperature from 25 to 30 °C, the diffusion rate of the adsorbate molecules within the pores, because of the decreasing solution viscosity, also modified the equilibrium capacity of the MWCNTs for a particular adsorbate. A further increase in temperature (above 30 or 35 °C) led to a decrease in the percentage removal. This is mainly due to the decreased surface activity [58,59]. This decrease in the efficiency of the adsorption may be ascribed to several causes, such as deactivation of the adsorbent's surface, a growing tendency for the dyes to escape from the solid stage to the bulk stage, or the destruction of specific active sites on the surface of the adsorbent owing to bond ruptures [16].

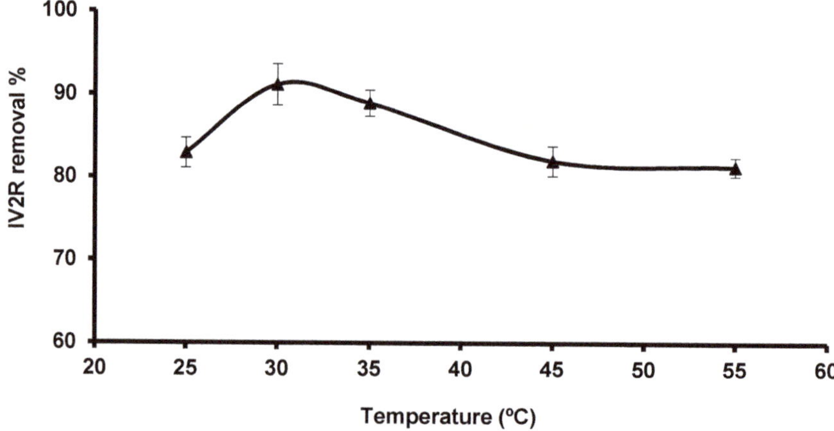

Figure 10. Influence of temperature on the removal of IV2R.

3.3. Isothermal Analysis

3.3.1. Freundlich Isotherm

The obtained results were fit with the experimental data of the Freundlich isotherm model, which were confirmed by a high correlation coefficient of $R^2 = 0.989$ for the MWCNTs, indicating that this model is favorable for the adsorption process (Figure 11). A strong bond occurred between the IV2R dye and the adsorbents, as indicated by the value of $1/n$, which is called the heterogeneity factor, describing the deviation from the linearity of sorption as follows: When $1/n$ is equivalent to 1, the adsorption is linear, and the ammonia particle concentration does not influence the division between the two stages. When $1/n$ is below 1, chemical adsorption occurs; however, when $1/n$ is more than 1, cooperative adsorption occurs, and this adsorption is more favorable physically and includes strong interactions among the particles of the adsorbate [6]. In this work, the value of factor "$1/n$" was less than 1; the data shows chemical sorption method on a surface with this isotherm equation being favorable.

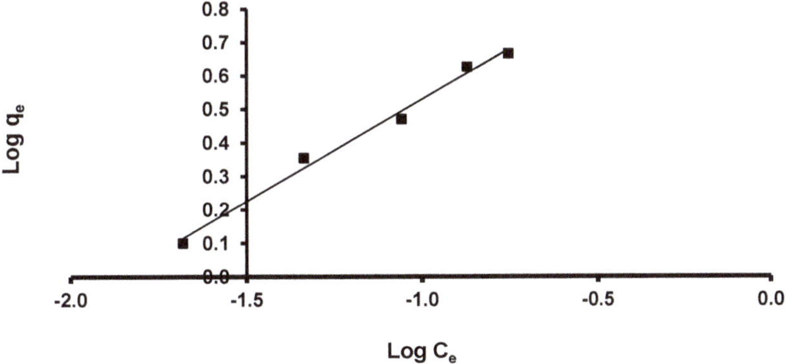

Figure 11. Freundlich isotherm of the IV2R onto the MWCNTs.

3.3.2. Langmuir Isotherm

When the Langmuir isotherm was applied, the obtained results agreed with the data throughout the experiment, with a high correlation coefficient of $R^2 = 0.993$ for the MWCNTs. Furthermore, the IV2R dye showed a high maximal uptake capacity (q_{max}) of 76.92 mg g^{-1} by the MWCNTs (Figure 12). This is consistent with the formation of a complete monolayer on the surface of the adsorbent. The Langmuir constant (b), which is related to the heat of adsorption, was found to be 1.42.

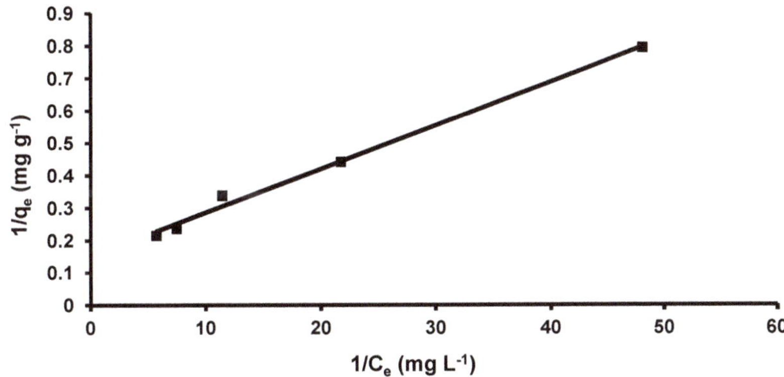

Figure 12. Linear Langmuir isotherm for the sorption of IV2R.

3.3.3. Halsey and Henderson Isotherm

Halsey and Henderson isotherm equations are suitable for multilayer adsorption; in particular, the fitting of these equations can be heteroporous solids [60]. A plot of ln q_e versus ln C_e Halsey and ln$[-\ln(1-C_e)]$ versus ln q_e. Halsey and Henderson adsorption isotherms are given in Figures 13 and 14, respectively. The isotherm constants and correlation coefficients are summarized in Table 2. Halsey showed a correlation coefficient of $R^2 = 0.989$, while Henderson showed an R^2 of 0.988. The results obtained for Halsey and Henderson show that both models are applicable for the adsorption of IV2R onto MWCNTs.

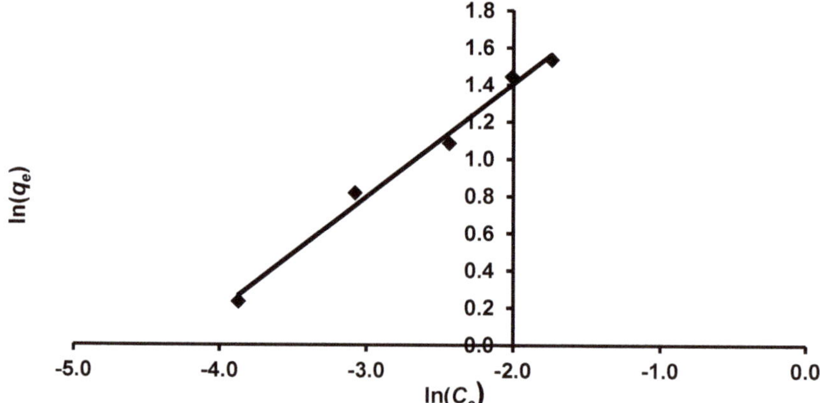

Figure 13. Halsey isotherm of the removal of IV2R.

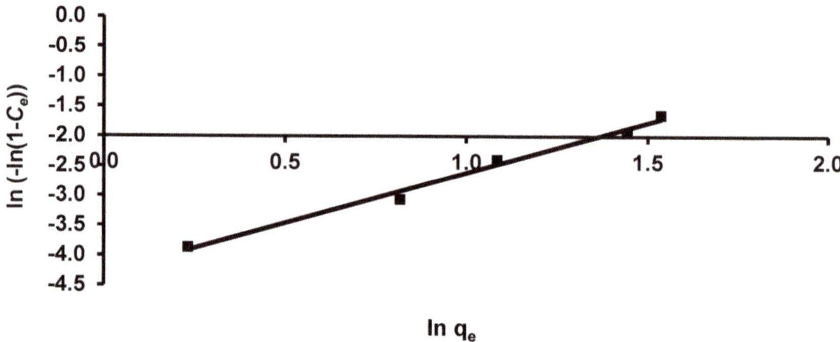

Figure 14. Linear Henderson isotherm of the removal of IV2R.

3.3.4. Harkins–Jura Isotherm

The Harkins–Jura adsorption isotherm can be expressed as Equation (8), which can be solved by a plot of $1/q_e$ versus log C_e, as shown in Figure 15. The Harkins–Jura model can be described as the presence of a distribution of heterogeneous pores and multilayer adsorption. The isotherm constants and correlation coefficients ranged from $R^2 = 0.858$ (Table 2). This may indicate that the Harkins–Jura model is less applicable.

3.3.5. Smith Isotherm

The Smith model is useful for explaining the adsorption isotherm of a biological substance such as cellulose and starch, in addition to being appropriate for heteroporous solids and multilayer adsorption. The Smith model can be solved by a plot of q_e versus ln$(1-C_e)$, as presented in Figure 16. The Smith model adequately represents the sorption isotherms throughout the entire range of water activity. However, the Smith equation was successful in describing the isotherms of IV2R onto the MWCNTs, because the model gave a higher R^2 value (0.964).

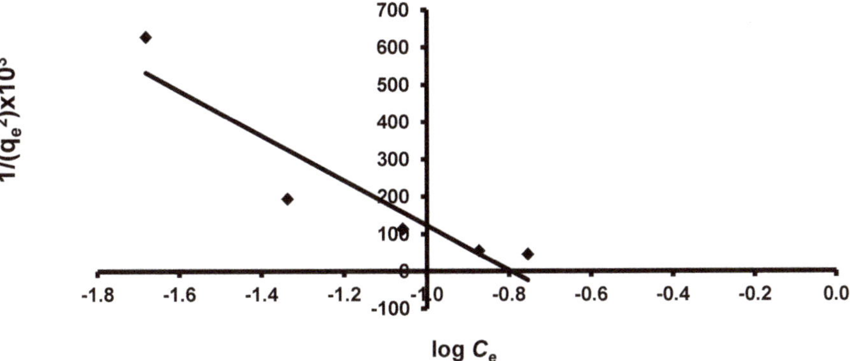

Figure 15. Linear Harkins–Jura isotherm of the removal of IV2R.

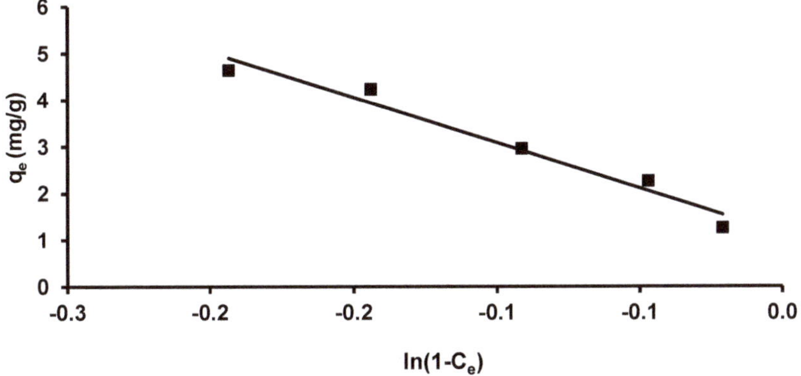

Figure 16. Linear Smith isotherm of the removal of IV2R.

3.3.6. Tempkin Isotherm

The Tempkin isotherm model was applied. Consequently, linear plots of (q_e) vs. (ln C_e) allowed for the calculation of the Tempkin isotherm parameters K_T and b_T from the slope and intercept, respectively, as presented in Figure 17 and Table 2. The data indicate that the Tempkin isotherm model applies to the adsorption of IV2R dye onto MWCNTs, as shown by the high linear regression correlation coefficient (R^2 = 0.965).

3.3.7. Error Function Examination for the Best and Most Appropriate Isotherm Model

The best-fit model for the investigational data was calculated by numerous error functions, such as the average percentage error (APE) equation, chi-square error (χ^2) equation, and the sum of absolute errors (EABS). The data obtained from the different error functions are summarized in Table 3. From the observed data, the best appropriate isotherm models are Tempkin, Smith, Henderson, Freundlich, Langmuir, Halsey, and Harkins–Jura. Nevertheless, the error functions test provided variable data for all models, and the evaluation among the models focused on all error functions individually. For the EABS and χ^2 equation, all the isotherm models can be applied to the investigational data, except the Harkins–Jura isotherm.

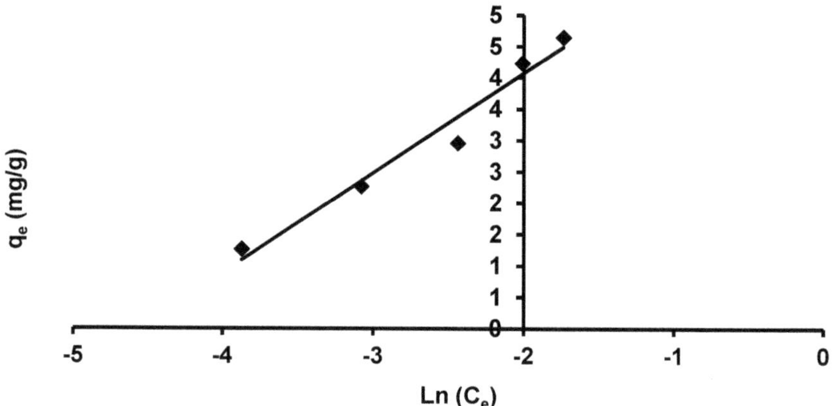

Figure 17. Linear Tempkin isotherm of the removal of IV2R.

Table 2. Factors of the isotherm models from nonlinear and linear solvation.

Isotherm Model	Isotherm Parameter	Value
Freundlich	$1/n$	13.77
	K_F (mg$^{1-1/n}$L$^{1/n}$g^{-1})	0.608
	R^2	0.989
Langmuir	Q_{max} (mg g^{-1})	76.92
	b	1.42
	R^2	0.99
Harkins–Jura	A_{HJ}	1.31
	B_{HJ}	1.73
	R^2	0.858
Halsey	$1/n_H$	0.6089
	K_H	2.714
	R^2	0.989
Henderson	$1/n_h$	1.688
	K_h	0.014
	R^2	0.988
Smith	W_{bs}	1.138
	W_s	19.46
	R^2	0.964
Tempkin	A_T	95.58
	B_T	1.59
	b_T	1558.2
	R^2	0.965

3.4. Adsorption Kinetics

3.4.1. Pseudo-First-Order Kinetics Model

The pseudo-first-order model is the latest well-known equation applied to describe the sorption rate depending on the capacity of adsorption. This model states that the ratio of occupation of adsorption sites is proportional to the number of unoccupied sites [61].

Figure 18A shows the linear figure of log (q_e–q_t) versus t at the original dye concentration of 10 mg/L. The q_e and K_1 values were calculated using the intercept, as well as the slope of the linear plots for removal of the IV2R dye from the MWCNTs.

Table 3. The most appropriate isotherm model to investigate the equilibrium data through various errors functions.

Isotherm Model	APE%	χ^2	EABS
Freundlich	0.031	0	0.034
Langmuir	2.079	0.326	2.236
Harkins–Jura	85.602	551.664	92.065
Halsey	12.38	11.539	13.315
Henderson	0.022	0	0.024
Smith	0.004	0	0.004
Tempkin	0.001	0	0.001

3.4.2. Pseudo-Second-Order Kinetics Model

The pseudo-second-order equation depends on the assumption of the rate-limiting step due to chemical adsorption containing valence forces by exchange and/or sharing of electrons between dye ions and the carbon nanotube (adsorbent). The pseudo-second-order equation is expressed following [62]. Plots of (t/q_t) against (t) should offer a linear correlation from which the data of parameters q_e and k_2 can be calculated from the slope and intercept, respectively. It can be observed in Figure 18B and Table 4 that this model fits well, with a higher correlation coefficient (R^2 = 0.999). Apart from this, it was obvious that the value of the k_2 parameter was greater than the corresponding k_1 parameter value. This is because the pseudo-second-order model presumes that the adsorption rate is proportional to the square of a number of unoccupied sites, as reported by Joseph and David et al. [53].

Table 4 provides the values of K_1, the experimental and calculated values of q_e, and the correlation coefficients for the pseudo-first-order kinetics plots. The theoretical values of q_e do not agree with experimental data obtained. This suggests a poor fit between the kinetics data and the pseudo-first-order model. Thus, the pseudo-first-order equation was ruled out as a result of its correlation coefficients (R^2) for the current experimental results being small for the IV2R dye in MWCNTs.

Table 4. Parameters of the different adsorption kinetics models.

Kinetic Models	Parameters	Value
First-order	q_e (calc.) (mg g^{-1})	29.3
	$k_1 \times 10^3$ (min^{-1})	6.6787
	R^2	0.0797
Second-order	q_e (calc.) (mg g^{-1})	1.066
	$k_2 \times 10^3$ (mg g^{-1} min^{-1})	9300.677
	R^2	0.9993
Intraparticle diffusion	K_{dif} (mg g^{-1} min$^{-0.5}$)	13.434
	C cal (mg g^{-1})	0.862
	R^2	0.867

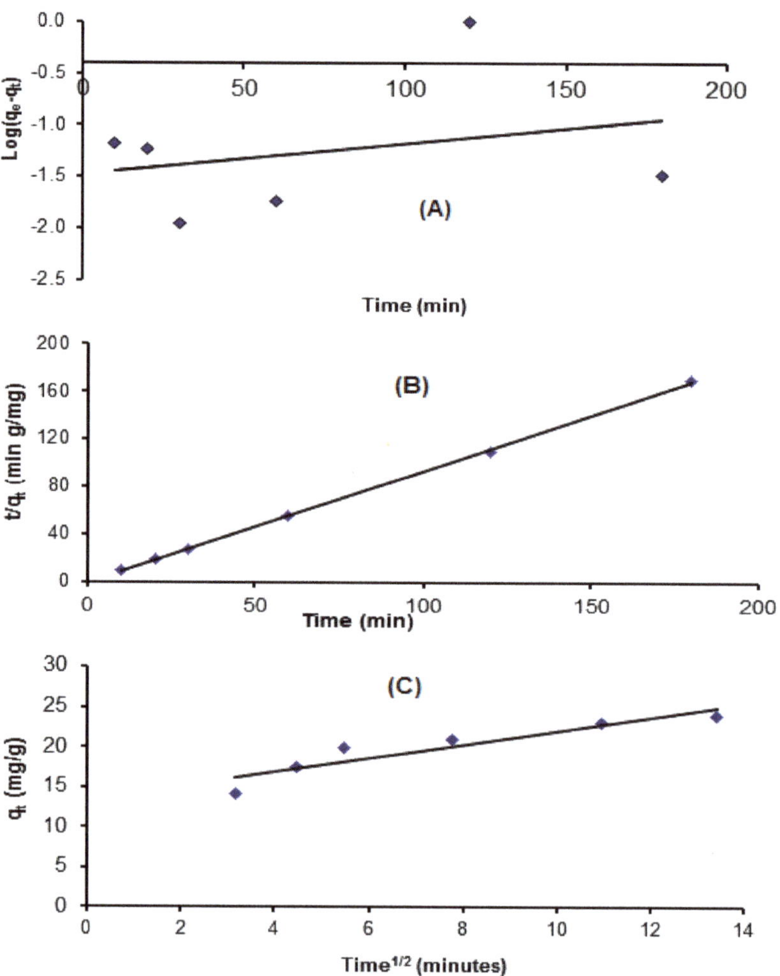

Figure 18. Adsorption kinetics of the pseudo-first-order (**A**), pseudo-second-order (**B**), and intraparticle diffusion equation (**C**) of IV2R adsorption.

3.4.3. The Intraparticle Diffusion Equation

The adsorption method requires numerous steps, including the transport of solute particles from the aqueous part to the external of the solid molecules, followed by diffusion of the solute molecules inside the interior of the cavities, where is likely to be a slow process and a rate-determining step [15]. The figures of q_t inverse $t^{0.5}$ may present a numerous-linearity correlation, which shows that two or more stages happen through the adsorption process (Figure 18C). The rate constant K_{dif} is directly estimated by the slope and the intercept is C, as reported in Table 4. The value of the C factor offers information about the thickness of the border layer, since the resistance to the exterior mass transfer rises as the intercept increases. The low linearity of the plots demonstrates that intraparticle diffusion in the uptake of the IV2R dye onto the MWCNTs with a low correlation coefficient of $R^2 = 0.867$.

3.5. Adsorption Thermodynamics

The thermodynamic factors of the IV2R adsorption onto the MWCNTs are given in Table 5. The great negative value of ΔG° confirms that the sorption of dye was spontaneous and feasible. Nevertheless, the ΔG° values in Table 5 show that the free energy values increased with decreasing temperature, indicating that the adsorption method is endothermic [37,41] and suggesting that the spontaneity of the adsorption method reduces at a lower temperature. In general, the range of ΔG° values is between –20 and 0 kJ mol^{-1} for physisorption; however, the chemisorption process is between –80 and –400 kJ mol^{-1} [63]. The observed data of activation energy in this study indicate that the sorption of the IV2R onto the MWCNTs was by physisorption process. The value of ΔH° for this technique is positive, demonstrating an endothermic nature. The negative sign of the ΔS° parameter shows an increase in the affinity of the MWCNT for the dye, as well as in randomness at the solid–liquid interface [37].

Table 5. Thermodynamic factors of the sorption of IV2R onto MWCNTs.

Temperature (°C)	ΔG° (kJ mol^{-1})	ΔH° (kJ mol^{-1})	ΔS° (J mol^{-1})
25	−7.87669		
35	−7.94698	21.877	−98.76
45	−9.763		
55	−10.5637		

3.6. Regeneration and Reusability Study

The regeneration and reuse of the sorbent material are essential for its economic viability. To evaluate the reusability of MWCNTs, adsorption–desorption experiments were carried out using 40 mg L^{-1} of dye solution, shaken at 110 rpm for 2 h. The results showed that the removal efficiency of the MWCNTs for IV2R uptake dye remained nearly unchanged for the three consecutive series. The initial removal efficiency was 95.2% in 120 min, which reduced slightly to 92.1% after the third round, as shown in Figure 19. Hence, it could be concluded that the sorption capacity of MWCNTs remains unaffected with extended regeneration rounds. This decrease shows that the adsorbent represents a good material that can be used in industrial operations. The diminishing adsorption of IV2R dyes over the three cycles was brought about by the destructive influence of stripping agents and the weight loss in the adsorbent [64]. Furthermore, the resident IV2R dyes in the adsorbent (irreversible binding) resulted in a decrease in the number of available binding sites [65,66].

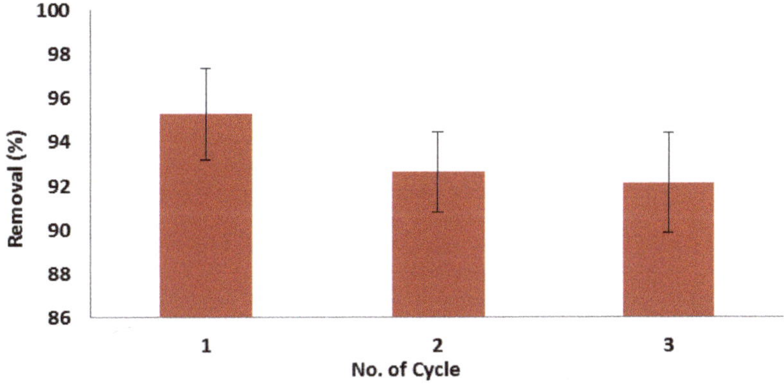

Figure 19. Reusability study for removal of IV2R in various cycles.

Ahmad et al. [67] achieved more than 85% removal of dyes, indicating the reusability of the adsorbent. Moreover, Tunali et al. [68] tested the regenerated adsorbent for up to five adsorption–desorption cycles and showed that the decline in efficiency was less than 5%, highlighting the favorable prospects of these fungi for Pb(II) elimination in industrial operations. Gupta et al. [69] studied the regeneration of MWCNTs for reactive red 3BS, and 92.8% reuse efficiency was achieved after four cycles.

3.7. Applicability on Actual Wastewater

To evaluate the validity of the use of MWCNTs as adsorbents, real wastewater mixed with simulated dye samples was collected to examine the removal of IV2R by the adsorbent under optimization conditions. The results confirmed that the highest removal was significantly altered by changing the kind of water, for which deionized water exhibited the lowest influence on the adsorption method, with color removal of 98.4% at an acidic pH after 120 min. By contrast, real wastewater comprises very high concentrations of interfering ions from numerous pollutants, which had a significant influence on the elimination efficiency of IV2R dye, with color removal of 88% after 120 min. Regardless, the results revealed that MWCNTs adsorbents can be successfully used as a low-cost adsorbent to remove IV2R dye from aqueous mixtures and wastewater. The process showed a typical feature of effluents after the treatment of dye, observed to be 110 mg L^{-1} of TSS and 652 mgL^{-1} of TDS, as presented in Table 6. Moreover, these results show that wastewater containing dye is well treatable through this method, with percentages of color removal of 98.4% and 88.2%; additionally, TSS and TDS removal decreased after the treatment process.

Table 6. Characteristics of the actual wastewater before and after treatment, and standards for the safe disposal of prescribed dye effluents.

Parameters	Before Treatment	After Treatment	Standards for Cotton Textile Industries [66]
pH	9.5	2–6	5.5–9.0
TSS (mg L^{-1})	2050	110	100
TDS (mg L^{-1})	3240	652	500

3.8. Comparative Studies of the Sorption Capacity of MWCNTs

The results of the present study were compared to those of other works on the adsorption capacities of various dyes to show the efficiency of MWCNTs (Table 7).

3.9. Future Research Perspectives and Hallenges

Although a lot of studies have successfully examined the treatment efficiency of MWCNTs and CNT-based adsorbents, more studies should be focused on the following:

(a) Improving CNT filters, films, and sheets for real industrial wastewater purification on an experimental scale.
(b) Creating purified and functionalized compounds of CNT in commercial amounts with little environmental impact at a reasonable price.
(c) Producing CNTs with comparable adsorption capabilities through various techniques such as laser ablation, chemical vapor deposition, and arc discharge processes.
(d) Predicting the adsorption mechanism and dye elimination capability from real industrial wastewater under a range of operating conditions in batch and column processes, as well as on a larger scale.
(e) Examining the toxicity of CNTs to the environment.

Table 7. Summary of the elimination of several dyes from aquatic mixtures by various carbon nanotubes.

CNTs	Dye Adsorbed	q_e (mg g^{-1})	Ref.
MWCNTs	Sufranine O	43.48	[65]
MWCNTs	Methylene blue	35.4	[67]
Oxidize MWCNTs	Bromothymol blue	55	[25]
SWCNT–COOH	Basic red 46	45.33	[1]
MWCNTs	methylene blue	64.7	[67]
MWNTs	Orange II	66.12	[68]
SWCNT	Basic red 46	38.35	[69]
MWNTs	Reactive blue	335.7	[19]
MWNTs	Alizarin red S	161.290	[70]
MWCNTs	Reactive blue 4	502.5	[19]
MWNTs	Methyl orange	52.86	[17]
CNTs/activated carbon fiber	Basic violet 10	220	[71]
MWCNTs/Fe$_3$C	Direct Red 23	172	[72]
Chitosan/Fe$_2$O$_3$/MWCNTs	Methyl orange	66.90	[21]
MWCNTs/Fe$_2$O$_3$	Methylene blue	42.3	[73]
MWCNT		76.9	Present study

4. Conclusions

MWCNTs were prepared, purified, and functionalized in this study. MWCNTs are good alternative adsorbents with which to remove IV2R dye from aqueous solutions and wastewater. MWCNT nanocompounds were examined using SEM, FTIR, XRD, and BET surface analysis. Purified MWCNTs were characterized by FTIR, which showed the existence of functional groups such as N-H, O-H, COOH, C=N, and S-H stretching, thereby increasing their ion exchange properties for the selective adsorption of opposing charged molecules. Furthermore, various factors affecting IV2R dye elimination efficiency were examined by the batch process, including adsorbent dose, agitation time, pH, initial dye concentration, and temperature. The highest operation influences were found to be 0.04 g of MWCNTs, 10 mg L^{-1} of dye, pH 4 at 30 °C, and 120 min of contact time. The percentage removal of the IV2R dyes increased with a rise in the contact time and MWCNT dosage, while it decreased with augmented initial dye concentrations, pH, and temperature. On the contrary, the equilibrium isotherm adsorption results were examined using numerous applied isotherms such as the Freundlich, Langmuir, Harkins–Jura, Halsey, Henderson, Smith, and Tempkin models. A higher adsorption capacity (q_{max}) of 76.92 mg g^{-1} was acquired from the Langmuir model. The best-fitting isotherm model in this study was obtained by different error function models. However, a kinetics study was conducted using the intraparticle diffusion, pseudo-first-order, and pseudo-second-order models. The regression data indicated that the sorption kinetics equation was the best via the pseudo-second-order model (R^2 = 0.999). Additionally, the percentage of removal of the MWCNTs from real wastewater reached 88%. Moreover, the thermodynamic parameters of the sorption process (ΔG°, ΔH°, and ΔS°) were calculated. The adsorption of IV2R dye was endothermic, and was a spontaneous process, and the reaction of adsorption is a physisorption process. The efficiency of the regenerated MWCNTs showed a slight change after three cycles of reuse and regeneration.

Author Contributions: Conceptualization, A.E.A.; methodology, A.E.A. and K.M.A.; software, A.E.A. and K.M.A.; formal analysis, A.E.A., M.A., and A.T.M.; investigation, A.E.A.; resources, A.E.A. and K.M.A.; data curation, A.E.A., M.A. and A.T.M.; writing—original draft preparation, A.E.A., M.A., and A.T.M.; writing—review and editing, A.E.A., M.A., K.M.A., and A.T.M.; visualization, K.M.A.

and M.A.; supervision, K.M.A. project administration, A.E.A.; funding acquisition, K.M.A. All authors have read and agreed to the published version of the manuscript.

Funding: The authors appreciated Taif University Researchers Supporting Project number TURSP-2020/267, Taif University, Taif, Saudi Arabia.

Institutional Review Board Statement: Not applicable.

Informed Consent Statement: Not applicable.

Data Availability Statement: All relevant data are within the paper, and those are available from the corresponding author.

Acknowledgments: The authors appreciated Taif University Researchers Supporting Project number TURSP-2020/267, Taif University, Taif, Saudi Arabia.

Conflicts of Interest: The authors declare no conflict of interest.

References

1. Rajabi, M.; Mahanpoor, K.; Moradi, O. Removal of dye molecules from aqueous solution by carbon nanotubes and carbon nanotube functional groups: Critical review. *RSC Adv.* **2017**, *7*, 47083–47090. [CrossRef]
2. El Moselhy, K.M.; Azzem, M.A.; Amer, A.; AlProl, A.E. Al Adsorption of Cu (II) and Cd (II) from Aqueous Solution by Using Rice Husk Adsorbent. *Phys. Chem. Indian J.* **2017**, *12*, 1–13.
3. Al Prol, A.E.; El-Azzem, M.A.; Amer, A.; El-Metwally, M.E.A.; El-Hamid, H.T.A.; El-Moselhy, K.M. Adsorption of Cadmium (II) Ions (II) from Aqueous Solution onto Mango Leaves. *Asian J. Phys. Chem. Sci.* **2017**, *2*, 1–11. [CrossRef]
4. Saratale, G.D.; Saratale, R.G.; Chang, J.S.; Govindwar, S.P. Fixed-bed decolorization of Reactive Blue 172 by Proteus vulgaris NCIM-2027 immobilized on Luffa cylindrica sponge. *Int. Biodeterior. Biodegrad.* **2011**, *65*, 494–503. [CrossRef]
5. Kishor, R.; Purchase, D.; Saratale, G.D.; Saratale, R.G.; Ferreira, L.F.R.; Bilal, M.; Chandra, R.; Bharagava, R.N. Ecotoxicological and health concerns of persistent coloring pollutants of textile industry wastewater and treatment approaches for environmental safety. *J. Environ. Chem. Eng.* **2021**, *9*, 105012. [CrossRef]
6. Al Prol, A.E. Study of Environmental Concerns of Dyes and Recent Textile Effluents Treatment Technology: A Review. *Asian J. Fish. Aquat. Res.* **2019**, *3*, 1–18. [CrossRef]
7. El-hamid, H.T.A.; Al-Prol, A.E.; Hafiz, M.A. Plackett-Burman and Response Surface Methodology for Optimization of Oily Wastewater Bioremediation by *Aspergillus* sp. *South Asian J. Res. Microbiol.* **2018**, *2*, 1–9. [CrossRef]
8. Miricioiu, M.G.; Niculescu, V.-C. Fly Ash, from Recycling to Potential Raw Material for Mesoporous Silica Synthesis. *Nanomaterials* **2020**, *10*, 474. [CrossRef]
9. Niculescu, V. Mesoporous silica—Efficient media for phenols removal from wastewater. *Sect. Hydrol. Water Resour.* **2017**. [CrossRef]
10. Niculescu, V. Experimental set-up for simultaneously wastewater treatment and carbon dioxide separation through porous materials. In Proceedings of the 17th International Multidisciplinary Scientific GeoConference SGEM2017, Energy and Clean Technologies, Vienna, Austria, 27–29 November 2017. [CrossRef]
11. Nourmoradi, H.; Ghiasvand, A.; Noorimotlagh, Z. Removal of methylene blue and acid orange 7 from aqueous solutions by activated carbon coated with zinc oxide (ZnO) nanoparticles: Equilibrium, kinetic, and thermodynamic study. *Desalanation Water Treat.* **2015**, *55*, 252–262. [CrossRef]
12. Samad, S.A.; Huq, D.; Masum, M.M.A.S.M. Textile Dye Removal from Wastewater Effluents Using Chitosan-ZnO Nanocomposite. *J. Text. Sci. Eng.* **2015**, *5*, 5–8. [CrossRef]
13. Nakkeeran, E.; Varjani, S.J.; Dixit, V.; Kalaiselvi, A. Synthesis, characterization and application of zinc oxide nanocomposite for dye removal from textile industrial wastewater. *Indian J. Exp. Biol.* **2018**, *56*, 498–503.
14. Sampranpiboon, P.; Charnkeitkong, P.; Feng, X. Equilibrium isotherm models for adsorption of zinc (II) ion from aqueous solution on pulp waste. *WSEAS Trans. Environ. Dev.* **2014**, *10*, 35–47.
15. Al Prol, A.E.; El-Metwally, M.E.A.; Amer, A. Sargassum latifolium as eco-friendly materials for treatment of toxic nickel (II) and lead (II) ions from aqueous solution. *Egypt. J. Aquat. Biol. Fish.* **2019**, *23*, 285–299. [CrossRef]
16. Eid, A.; Prol, A. Adsorption and Bioremediation as Technologies of Wastewater Treatment. Ph.D. Thesis, University of Almonifia, Tanta, Egypt, 2017.
17. Zhao, D.; Zhang, W.; Chen, C.; Wang, X. Adsorption of Methyl Orange Dye onto Multiwalled Carbon Nanotubes. *Procedia Environ. Sci.* **2013**, *18*, 890–895. [CrossRef]
18. Saber-Samandari, S.; Saber-Samandari, S.; Joneidi-Yekta, H.; Mohseni, M. Adsorption of anionic and cationic dyes from aqueous solution using gelatin-based magnetic nanocomposite beads comprising carboxylic acid functionalized carbon nanotube. *Chem. Eng. J.* **2017**, *308*, 1133–1144. [CrossRef]
19. Machado, F.M.; Bergmann, C.P.; Lima, E.C.; Royer, B.; de Souza, F.E.; Jauris, I.M.; Calvete, T.; Fagan, S.B. Adsorption of Reactive Blue 4 dye from water solutions by carbon nanotubes: Experiment and theory. *Phys. Chem. Chem. Phys.* **2012**, *14*, 11139–11153. [CrossRef]

20. Mashkoor, F.; Nasar, A. Inamuddin Carbon nanotube-based adsorbents for the removal of dyes from waters: A review. *Environ. Chem. Lett.* **2020**, *18*, 605–629. [CrossRef]

21. Yao, Y.; Bing, H.; Feifei, X.; Xiaofeng, C. Equilibrium and kinetic studies of methyl orange adsorption on multiwalled carbon nanotubes. *Chem. Eng. J.* **2011**, *170*, 82–89. [CrossRef]

22. Shirmardi, M.; Mahvi, A.H.; Hashemzadeh, B.; Naimabadi, A.; Hassani, G.; Niri, M.V. The adsorption of malachite green (MG) as a cationic dye onto functionalized multi walled carbon nanotubes. *Korean J. Chem. Eng.* **2013**, *30*, 1603–1608. [CrossRef]

23. Kareem, A.; Alrazak, N.A.; Aljebori, K.H.; Aljeboree, A.M.; Algboory, H.L.; Alkaim, A.F. Removal of methylene blue dye from aqueous solutions by using activated carbon/ureaformaldehyde composite resin as an adsorbent. *Int. J. Chem. Sci.* **2016**, *14*, 635–648.

24. Machado, F.; Bergmann, C.P.; Lima, E.C.; Adebayo, M.; Fagan, S.B. Adsorption of a textile dye from aqueous solutions by carbon nanotubes. *Mater. Res.* **2014**, *17*, 153–160. [CrossRef]

25. Yu, J.; Zhao, X.; Yang, H.; Chen, X.; Yang, Q.; Yu, L. Aqueous adsorption and removal of organic contaminants by carbon nanotubes. *Sci. Total Environ.* **2014**, *483*, 241–251. [CrossRef]

26. Mohammed, M.I.; Razak, A.A.A.; Al-Timimi, D.A.H. Modified Multiwalled Carbon Nanotubes for Treatment of Some Organic Dyes in Wastewater. *Adv. Mater. Sci. Eng.* **2014**, *2014*, 1–10. [CrossRef]

27. Bahgat, M.; Farghali, A.; El-Rouby, W.; Khedr, M. Synthesis and modification of multi-walled carbon nano-tubes (MWCNTs) for water treatment applications. *J. Anal. Appl. Pyrolysis* **2011**, *92*, 307–313. [CrossRef]

28. El-Mohdy, H.L.A.; Mostafa, T.B. Synthesis of Polyvinyl Alcohol/Maleic Acid Hydrogels by Electron Beam Irradiation for Dye Uptake. *J. Macromol. Sci. Part A Pure Appl. Chem.* **2013**, *50*, 6–17. [CrossRef]

29. Ghoneim, M.M.; El-Desoky, H.S.; El-Moselhy, K.M.; Amer, A.; El-Naga, E.H.A.; Mohamedein, L.I.; Al-Prol, A.E. Removal of cadmium from aqueous solution using marine green algae, Ulva lactuca. *Egypt. J. Aquat. Res.* **2014**, *40*, 235–242. [CrossRef]

30. Langmuir, A.O.D.; Temkin, F.; Radushkevich, D. Isotherms Studies of Equilibrium Sorption of Zn2+ Unto Phosphoric Acid Modified Rice Husk. *IOSR J. Appl. Chem.* **2012**, *3*, 38–45. [CrossRef]

31. Freundlich, H.M.F. Über die adsorption in solution. *J. Phys. Chem* **1906**, *57*, 1100–1107.

32. Langmuir, I. The constitution and fundamental properties of solids and liquids. Part II—Liquids. *J. Frankl. Inst.* **1917**, *184*, 721. [CrossRef]

33. Halsey, G. Physical Adsorption on Non-Uniform Surfaces. *J. Chem. Phys.* **1948**, *16*, 931–937. [CrossRef]

34. Harkins, D.; Jura, G. An adsorption method for the determination of the area of a solid without the assumption of a molecular area, and the area occupied by nitrogen molecules on the surfaces of solids. *J. Chem. Phys.* **1944**, *66*, 1366–1373. [CrossRef]

35. Mall, I.D.; Srivastava, V.C.; Agarwal, N.K. Removal of Orange-G and Methyl Violet dyes by adsorption onto bagasse fly ash—Kinetic study and equilibrium isotherm analyses. *Dye Pigment.* **2006**, *69*, 210–223. [CrossRef]

36. Foo, K.Y.; Hameed, B.H. Insights into the modeling of adsorption isotherm systems. *Chem. Eng. J.* **2010**, *156*, 2–10. [CrossRef]

37. Potgieter, J.H.; Pearson, S.; Pardesi, C. Kinetic and Thermodynamic Parameters for the Adsorption of Methylene Blue Using Fly Ash under Batch, Column, and Heap Leaching Configurations. *Coal Combust. Gasif. Prod.* **2018**, *11*, 22–33. [CrossRef]

38. Kumar, Y.P.; King, P.; Prasad, V.S.R.K. Removal of copper from aqueous solution using *Ulva fasciata* sp.—A marine green algae. *J. Hazard. Mater.* **2006**, *137*, 367–373. [CrossRef] [PubMed]

39. Al-Aoh, H.A.; Yahya, R.; Maah, M.J.; Bin-Abas, M.R. Adsorption of methylene blue on activated carbon fiber prepared from coconut husk: Isotherm, kinetics and thermodynamics studies. *Desalination Water Treat.* **2013**, *52*, 6720–6732. [CrossRef]

40. Özacar, M.; Şengil, I. Application of kinetic models to the sorption of disperse dyes onto alunite. *Colloids Surf. A Physicochem. Eng. Asp.* **2004**, *242*, 105–113. [CrossRef]

41. Babarinde, N.A.A.; Oyesiku, O.O.; Dairo, O.F. Isotherm and thermodynamic studies of the biosorption of copper (II) ions by *Erythrodontium barteri*. *Int. J. Phys. Sci.* **2007**, *2*, 300–304.

42. El-Feky, A.M.; Alprol, A.E.; Heneash, A.M.M.; Abo-Taleb, H.A.; Yousif, M. Evaluation of Water Quality and Plankton for Mahmoudia Canal in Northern West of Egypt. *Egypt. J. Aquat. Biol. Fish.* **2019**, *22*, 461–474. [CrossRef]

43. Thamri, A.; Baccar, H.; Struzzi, C.; Bittencourt, C.; Abdel-Ghani, A.; Llobet, E. MHDA-Functionalized Multiwall Carbon Nanotubes for detecting non-aromatic VOCs. *Nat. Publ. Gr.* **2016**, *6*, 35130. [CrossRef] [PubMed]

44. Zhang, J.; Zou, H.; Qing, Q.; Yang, Y.; Li, Q.; Liu, Z. Effect of Chemical Oxidation on the Structure of Single-Walled Carbon Nanotubes. *J. Chem. Phys. B* **2003**, *107*, 3712–3718. [CrossRef]

45. Dutta, D.P.; Venugopalan, R.; Chopade, S. Manipulating Carbon Nanotubes for Efficient Removal of Both Cationic and Anionic Dyes from Wastewater. *Mater. Sci. Inc. Nanomater. Polym.* **2017**, *2*, 3878–3888. [CrossRef]

46. Salam, M.A.; Burk, R. Synthesis and characterization of multi-walled carbon nanotubes modified with octadecylamine and polyethylene glycol. *Arab. J. Chem.* **2017**, *10*, S921–S927. [CrossRef]

47. Das, R.; Abd Hamid, S.B.; Ali, M.E.; Ramakrishna, S.; Yongzhi, W. Carbon Nanotubes Characterization by X-ray Powder Diffraction—A Review. *Curr. Nanosci.* **2014**, *11*, 23–35. [CrossRef]

48. Maryam, M.; Suriani, A.; Shamsudin, M.S.; Rusop, M.; Mahmood, M.R. BET Analysis on Carbon Nanotubes: Comparison between Single and Double Stage Thermal CVD Method. *Adv. Mater. Res.* **2012**, *626*, 289–293. [CrossRef]

49. Romantsova, I.; Burakov, A.; Kucherova, A.; Neskoromnaya, E.; Babkin, A.; Tkachev, A. Liquid-Phase Adsorption of an Organic Dye on Non-Modified and Nanomodified Activated Carbons: Equilibrium and Kinetic Analysis. *Adv. Mater. Technol.* **2016**, 42–48. [CrossRef]

50. Ngoh, Y.Y.; Leong, Y.-H.; Gan, C.Y. Optimization study for synthetic dye removal using an agricultural waste of Parkia speciosa pod: A sustainable approach for waste water treatment. *Int. Food Res. J.* **2015**, *22*, 2351–2357.
51. Sadik, R.; Lahkale, R.; Hssaine, N. Removal of textile dye by mixed oxide-LDH: Kinetics, isotherms of the adsorption and retention mechanisms studies. *IOSR J. Environ. Sci. Toxicol. Food Technol.* **2014**, *8*, 28–36. [CrossRef]
52. Ananta, S.; Saumen, B.; Vijay, V. Adsorption Isotherm, Thermodynamic and Kinetic Study of Arsenic (III) on Iron Oxide Coated Granular Activated Charcoal. *Int. Res. J. Environ. Sci.* **2015**, *4*, 64–77.
53. Joseph, D.A. The effect of ph and biomass concentration on lead (pb) adsorption by *aspergillus niger* from simulated waste water. *Sci. China Life Sci.* **2008**, *49*, 69–73.
54. Senthil-Kumar, P.; Gayathri, R. Adsorption of Pb^{2+} ions from aqueous solutions onto bael tree leaf powder: Isotherms, kinetics and thermodynamics study. *J. Eng. Sci. Technol.* **2009**, *4*, 381–399.
55. Hashem, F.S. Removal of Methylene Blue by Magnetite-Covered Bentonite Nano-Composite. *Chem. Bull* **2013**, *2013*, 524–529. [CrossRef]
56. Elzain, A.A.; El-Aassar, M.R.; Hashem, F.S.; Mohamed, F.M.; Ali, A.S.M. Removal of methylene dye using composites of poly (styrene-co-acrylonitrile) nanofibers impregnated with adsorbent materials. *J. Mol. Liq.* **2019**, *291*, 111335. [CrossRef]
57. Putra, W.P.; Kamari, A.; Najiah, S.; Yusoff, M.; Ishak, C.F. Biosorption of Cu (II), Pb (II) and Zn (II) Ions from Aqueous Solutions Using Selected Waste Materials: Adsorption and Characterisation Studies. *J. Encapsulation Adsorpt. Sci.* **2014**, *4*, 25–35. [CrossRef]
58. Kumar, P.S.; Ramakrishnan, K.; Kirupha, S.D.; Sivanesan, S. Thermodynamic and kinetic studies of cadmium adsorption from aqueous solution onto rice husk. *Braz. J. Chem. Eng.* **2010**, *27*, 347–355. [CrossRef]
59. Malakootian, M.; Toolabi, A.; Moussavi, S.G.; Ahmadian, M. Equilibrium and Kinetic Modeling of Heavy Metals Biosorption from Three Different Real Industrial Wastewaters onto Ulothrix Zonata Algae. *Aust. J. Basic Appl. Sci.* **2011**, *5*, 1030–1037.
60. Volesky, B. Biosorption: Application Aspects—Process Simulation Tools. *Hydrometallurgy* **2001**, *11*, 69–80.
61. Kooh, M.R.R.; Dahri, M.K.; Lim, L.B. The removal of rhodamine B dye from aqueous solution using *Casuarina equisetifolia* needles as adsorbent. *Cogent Environ. Sci.* **2016**, *2*. [CrossRef]
62. Abdelkarim, S.; Mohammed, H.; Nouredine, B. Sorption of Methylene Blue Dye from Aqueous Solution Using an Agricultural Waste. *Trends Green Chem.* **2017**, *3*, 1–7. [CrossRef]
63. Abdelwahab, O.; Amin, N. Adsorption of phenol from aqueous solutions by Luffa cylindrica fibers: Kinetics, isotherm and thermodynamic studies. *Egypt. J. Aquat. Res.* **2013**, *39*, 215–223. [CrossRef]
64. Nessim, R.B.; Bassiouny, A.R.; Zaki, H.R.; Moawad, M.N.; Kandeel, K.M. Biosorption of lead and cadmium using marine algae. *Chem. Ecol.* **2011**, *27*, 579–594. [CrossRef]
65. Ghaedi, M.; Shojaeipour, E.; Ghaedi, A.M.; Sahraei, R. Isotherm and kinetics study of malachite green adsorption onto copper nanowires loaded on activated carbon: Artificial neural network modeling and genetic algorithm optimization. *Spectrochim. Acta Part. A Mol. Biomol. Spectrosc.* **2015**, *142*, 1–57. [CrossRef]
66. Mondal, P.; Baksi, S.; Bose, D. Study of Environmental Issues in Textile Industries and Recent Wastewater Treatment Technology. *World Sci. News* **2017**, *61*, 98–109.
67. Yao, Y.; Xu, F.; Chen, M.; Xu, Z.; Zhu, Z. Bioresource Technology Adsorption behavior of methylene blue on carbon nanotubes. *Bioresour. Technol.* **2010**, *101*, 3040–3046. [CrossRef]
68. Rodriguez, A.; Ovejero, G.; Sotelo, J.L.; Mestanza, M.; Garcia, J. Adsorption of dyes on carbon nanomaterials from aqueous solutions. *J. Environ. Sci. Health Part A* **2010**, *45*, 1642–1653. [CrossRef] [PubMed]
69. Moradi, O. Adsorption Behavior of Basic Red 46 by Single-Walled Carbon Nanotubes Surfaces. *Full Nanotub. Carbon Nanostruct.* **2013**, *21*, 37–41. [CrossRef]
70. Jamshidi, H.; Ghaedi, M.; Sabzehmeidani, M.M.; Bagheri, A.R. Comparative study of acid yellow 119 adsorption onto activated carbon prepared from lemon wood and ZnO nanoparticles loaded on activated carbon. *Appl. Organomet. Chem.* **2018**, *32*, e4080. [CrossRef]
71. Das, D.; Charumathi, D.; Das, N. Bioaccumulation of the synthetic dye Basic Violet 3 and heavy metals in single and binary systems by Candida tropicalis grown in a sugarcane bagasse extract medium: Modelling optimal conditions using response surface methodology (RSM) and inhibition kinetics. *J. Hazard. Mater.* **2011**, *186*, 1541–1552. [CrossRef]
72. Konicki, W.; Pełech, I.; Mijowska, E.; Jasińska, I. Adsorption of anionic dye Direct Red 23 onto magnetic multi-walled carbon nanotubes-Fe_3C nanocomposite: Kinetics, equilibrium and thermodynamics. *Chem. Eng. J.* **2012**, *210*, 87–95. [CrossRef]
73. Inyinbor, A.A.; Adekola, F.A.; Olatunji, G.A. Kinetics, isotherms and thermodynamic modeling of liquid phase adsorption of Rhodamine B dye onto *Raphia hookerie* fruit epicarp. *Water Resour. Ind.* **2016**, *15*, 14–27. [CrossRef]

applied
sciences

Article

A Simple Device for the On-Site Photodegradation of Pesticide Mixes Remnants to Avoid Environmental Point Pollution

Biagio Esposito [1], Francesco Riminucci [1,2], Stefano Di Marco [3], Elisa Giorgia Metruccio [3], Fabio Osti [3], Stefano Sangiorgi [4] and Elida Nora Ferri [4,*]

1   Proambiente s.c.r.l.—Tecnopolo Bologna CNR, Via P. Gobetti 101, 40129 Bologna, Italy; b.esposito@consorzioproambiente.it (B.E.); f.riminucci@consorzioproambiente.it (F.R.)

2   National Research Council, Institute of Marine Sciences (CNR-ISMAR), Via P. Gobetti 101, 40129 Bologna, Italy

3   National Research Council—Institute of BioEconomy (CNR-IBE), Via P. Gobetti 101, 40129 Bologna, Italy; stefano.dimarco@ibe.cnr.it (S.D.M.); elisa.metruccio@ibe.cnr.it (E.G.M.); fabio.osti@ibe.cnr.it (F.O.)

4   Department of Pharmacy and Biotechnology, University of Bologna, Via S. Donato 15, 40127 Bologna, Italy; stefano.sangiorgi9@unibo.it

*   Correspondence: elidanora.ferri@unibo.it; Tel.: +39-0512095658

Featured Application: The prototype developed in this work can rapidly detoxify agrochemical solutions directly at farms, allowing for their safe disposal in the environment or water reuse.

Abstract: The worldwide increase in the number and use of agrochemicals impacts nearby soil and freshwater ecosystems. Beyond the excess in applications and dosages, the inadequate management of remnants and the rinsing water of containers and application equipment worsen this problem, creating point sources of pollution. Advanced oxidation processes (AOPs) such as photocatalytic and photo-oxidation processes have been successfully applied in degrading organic pollutants. We developed a simple prototype to be used at farms for quickly degrading pesticides in water solutions by exploiting a UV–H_2O_2-mediated AOP. As representative compounds, we selected the insecticide imidacloprid, the herbicide terbuthylazine, and the fungicide azoxystrobin, all in their commercial formulation. The device efficiency was investigated through the disappearance of the parent molecule and the degree of mineralization. The toxicity of the pesticide solutions, before and during the treatment, was assessed by *Vibrio fischeri* and *Pseudokirchneriella subcapitata* inhibition assays. The results obtained have demonstrated a cost-effective, viable alternative for detoxifying the pesticide solutions before their disposal into the environment, even though the compounds, or their photoproducts, showed different sensitivities to physicochemical degradation. The bioassays revealed changes in the inhibitory effects on the organisms in agreement with the analytical data.

Keywords: agrochemicals; photo-reactor; advanced oxidation processes; water pollution; point sources

Citation: Esposito, B.; Riminucci, F.; Di Marco, S.; Metruccio, E.G.; Osti, F.; Sangiorgi, S.; Ferri, E.N. A Simple Device for the On-Site Photodegradation of Pesticide Mixes Remnants to Avoid Environmental Point Pollution. *Appl. Sci.* **2021**, *11*, 3593. https://doi.org/10.3390/app11083593

Academic Editor: Amerigo Capria

Received: 25 March 2021
Accepted: 12 April 2021
Published: 16 April 2021

1. Introduction

Pesticides, herbicides, fertilizers, plant growth regulators, and other agrochemicals play a decisive role in agriculture, protecting crops and improving yields, but the current main challenge is the development of profit-gaining agriculture increasingly geared to environmentally friendly management systems, aimed at reducing the damages produced by agrochemicals to humans and the environment [1].

High intensity agricultural models are characterized by a massive use of pesticides and herbicides contaminating water and soil [2–4]. In 2018, the worldwide quantities of applied pesticides corresponded to 4112.3 metric tons, and in Italy alone to 54.1 metric tons [5]. Animal waste from farms, sediments, organic debris, pesticides, and fertilizers have long proven to be the major causes of ground and surface water pollution [6,7].

Soil and water bodies risk contamination every time pesticides are manipulated [8]. Pesticide drift, percolation, drainage, runoff, and leaching greatly depend on their correct management, both before and after the treatment, including the correct use of the application equipment, i.e., the pneumatic sprayers [9–11]. A known source of point source pollution is represented by incorrect practices in the disposal of spray remnants and tank rinsing water. Directive 2009/128/EC [12] regulates the disposal of agro-wastewater containing pesticides; however, it is quite common for, after the application, the spray remnants in the barrel of the spray-tank to be deposited on the ground or even discarded directly into water bodies, often without any dilution. These point sources of pollution from leftover mixture and cleaning water disposal are a matter of concern since pesticide residues may persist for a very long time; some compounds can be still detected many years after their ban [13,14].

Several solutions have been studied and proposed to remove or inactivate pesticides in water. Most are plants based on water evaporation produced by solar radiation and wind (Heliosec®, Syngenta Agro sas, Guyancourt, France) or by infrared radiation and heat (Osmofilm®, Panteck- France sarl, Montesson France) [15]. Quicker evaporation can be obtained by using electrical resistance, filtering the vapors through activated carbon filters (Evapofit®, Staphytr, Inchy-en-Artois, France) [15]. Otherwise, a chemical pretreatment of coagulation–flocculation can be used to produce decantation of the active ingredient as solid material (Sentinel®, Alba environnement sas, Cluny, France) [15]. Various systems are based on biomass beds, based on microbial degradation (using fungi, bacteria, actinomycetes, and viruses) into less toxic forms [1,9,16]. However, all these solutions require a place and/or a plant where they can be carried out.

More effective treatment systems are required, and the most recent and effective water treatment technologies are based on advanced oxidation processes (AOPs), a set of innovative techniques including UV-based techniques, ultrasound, electron beams, γ rays, and plasma methods, which remove organic contaminants by mineralization [17]. The UV treatments are the easier and less expensive ones, optimized by the addition of a photocatalyst or a radical mediator, such as H_2O_2, which promotes the radical-mediated degradation of the molecules [18–20], according to the main steps illustrated in Scheme 1.

$$H_2O_2 + h\nu \ \rightarrow \ 2\ HO^{\cdot}$$
$$HO^{\cdot} + R \ \rightarrow \ ROH\bullet \qquad \textbf{(a)}$$
$$HO^{\cdot} + RH \ \rightarrow \ H_2O + R^{\cdot} \qquad \textbf{(b)}$$
$$HO^{\cdot} + R \ \rightarrow \ OH^- + R^{\cdot +} \qquad \textbf{(c)}$$

Scheme 1. UV irradiation in the presence of H_2O_2 produces the hydroxyl radicals responsible for the fragmentation of the organic molecules via **(a)** radical addition to double bonds, **(b)** hydrogen abstraction, and **(c)** electron transfer to the molecule.

Exactly like other organic pollutants, agrochemicals can be degraded by light irradiation, especially by UV light. A huge number of studies on the photochemistry of these pollutants have been carried out, determining the photodegradation kinetics, the degradation pathways, and the optimal conditions such as the addition of oxidants and/or catalysts able to enhance the process for the most recalcitrant compounds [21–31].

Our research group is involved in the practical application of AOPs for the remediation of polluted waters to reduce environmental dispersion and restore water quality for further reuse [32–34]. Deeply convinced that the best compliance to ecofriendly management of agro-wastewater can be obtained offering simple, flexible, cost-effective, and time-saving solutions, we designed and built up a small, portable device, a photoreactor effective enough in degrading the active ingredients, which is cheap, extremely easy to use in situ, and not time-consuming concerning the abovementioned solutions. We employed UV irradiation in the presence of hydrogen peroxide to reduce the concentration of the agrochemicals, selecting three pesticides that are widely used in our region, each one representing a different family of agrochemicals: insecticides, herbicides, and fungicides.

Terbuthylazine (TBA), used worldwide as a pre-emergence herbicide in corn farming, shows persistence, toxicity, and endocrine disruption effects on wildlife and humans [35]. Imidacloprid insecticide (IMI), today among the top ten agrochemicals worldwide [36], has been ascribed chronic toxicity to multiple aquatic taxa [37], and negative effects on pollinators [38,39]. Azoxystrobin fungicide (AZO), used since 1996, is very toxic for aquatic organisms and can lead to long-term adverse effects [40]. We tested our degradation prototype on the same solutions used on the farm for treatment. We employed the commercial formulation (CFs) of the compounds, which could behave differently with respect to the pure active ingredients because of the presence of the excipients [41], calculating an average residual volume of solution diluted with a suitable volume of water usually employed to rinse the tank. Since a degradation procedure aims to eliminate the pollutants' negative impact on the environment, the success of the procedure is usually witnessed by the reduction or disappearance of their biotoxicity. We evaluated the biotoxicity of irradiated and non-irradiated solutions by applying two ISO standard biological tests for water quality assessment, based on *Vibrio fischeri* light emission inhibition and *Pseudokirchneriella subcapitata* (now *Raphidocelis subcapitata*) algal growth inhibition, respectively [42,43].

2. Materials and Methods

The commercial formulations of the tested agrochemicals were: TrekP® (ADAMA AGAN, Ltd., Ashdod, Israel) for the herbicide Terbuthylazine, Confidor® (Bayer AG, Leverkusen, Germany) for the insecticide Imidacloprid, and Ortiva® (Syngenta Italia, Milano, Italy) for the fungicide Azoxystrobin. The chemical structure of the active ingredients is reported in Figure 1.

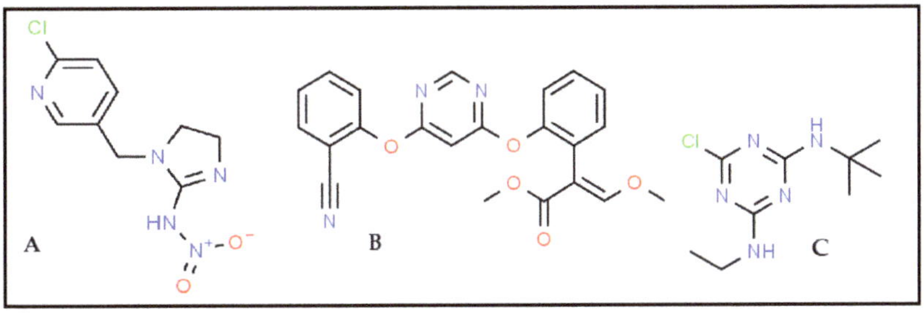

Figure 1. The chemical structure of the compounds under study: (**A**) imidacloprid, (**B**) azoxystrobin, and (**C**) Terbuthylazine.

The H_2O_2 solution (30%) Perhydrol®, used as a catalyst in the AOP experiments, was supplied by Merck Life Science (Milano, Italy). The LCK-385 test-in-cuvette kit for the total organic compound determination was supplied by Hach Lange (Milano, Italy). Lyophilized aliquots of the luminescent bacteria *V. fischeri* were prepared from fresh cultures at our laboratory starting from an original batch supplied by the Pasteur Institute (Paris, France). The 96-wells "Black Cliniplate" microplates were supplied by Thermo Scientific (Vantaa, Finland) and the luminometer was the Victor Light 1420 model from Perkin-Elmer, USA. The Istituto Zooprofilattico Sperimentale of Abruzzo and Molise "G. Caporale" (Teramo, Italy) supplied the freshwater microalga *P. subcapitata* culture. Inorganic salts and nutrient broth components for the bioluminescent bacteria and algal growth were obtained from Sigma-Aldrich. Tap water was chosen as the solvent of the agrochemicals.

A DR5000 spectrophotometer (Hach Lange, Milano, Italy) was employed to evaluate the active ingredients concentration and the total organic compounds (TOC) content.

The photoreactor prototype was assembled at the Proambiente laboratories by using materials currently employed in AOP experiments (UV lamps and stainless steel components).

2.1. The Photodegradation Treatment

The experiments were planned according to the hypothesis that the residual volume of agrochemical solution in a spray tank after the treatment was about 5 L. Agrochemical solutions were prepared at concentrations very close to the recommended ones employed on fields. Accordingly, we decided to dissolve 6.25 mL of each pesticide suspension in 5 L of tap water, and then this volume was brought to 50 L, simulating the water volume used during the rinsing operation of an 800 L tank. Before starting the degradation process, 50 mL of H_2O_2 30% solution was added, obtaining a 10 mM final concentration.

Small specimens of the solution under treatment (5–10 mL) were collected before the start and after 2, 3, 4, 5, 6, 7, and 8 h after the beginning of the irradiation. The lamps were always switched on during the treatment time.

The degradation process was evaluated by following the decrease of the active ingredients' concentration by measuring their absorbance (ABS) at the maximum wavelength for each compound: 242 nm for TBA and AZO and 270 nm for IMI. The mineralization process was followed by spectrophotometrically measuring the TOC value determined by the Hach Lange test kit. The test was performed according to the manufacturer's instructions. Shortly thereafter, the sample was added to the cuvette containing the digestion reagents. After 5 min in the TOC-X5 shaker (Hach Lange, Milano, Italy), a second cuvette containing the revealing reagent was screwed on the digestion cuvette, which was placed in a thermostat for two hours. After cooling at room temperature, the revealing cuvette was read by the spectrophotometer. All experimental results were obtained in at least triplicate and the data were expressed as means $\pm$ SD.

2.2. The Toxicity Assays

2.2.1. Bioluminescence Inhibition Assay

Lyophilized aliquots of *V. fischeri* containing NaCl 3% *w/v* were reconstituted with 1 mL of distilled water and re-suspended in 10–30 mL of the nutrient broth (NaCl 15 g, peptone 2.5 g, yeast extract 1.5 g, glycerol 1.5 mL, HEPES 0.01 M in 500 mL, pH 7). NaCl was added to the treated and untreated solutions to the 3% *w/v*. 200 µL of the bacteria suspension and 100 µL of each sample was dispensed into the microplate wells. The controls consisted of 200 µL of bacteria plus 100 µL of a 3% NaCl solution in tap water. The emitted light was recorded at fixed intervals in the range 0–48 h. A minimum of 5 replicates and a maximum of 12 were tested for each sample and the light emission values, reported as relative luminescence units (RLU), were expressed as means $\pm$ SD.

The bioluminescence inhibition percentage (*I*%) was used to express the toxicity of the samples and calculated according to:

$$I\% = \frac{L_{blank} - L_{sample}}{L_{blank}} \times 100 \tag{1}$$

where *L* is the intensity of the light emitted by the sample or by the control (blank).

2.2.2. Algal Growth Inhibition Assay

The starter culture of *P. subcapitata* was prepared by inoculating in Erlenmeyer flasks 1 mL of microalgae suspension per 100 mL of the Jaworski's culture medium ($Ca(NO_3)_2 \cdot 4H_2O$ 20 g L^{-1}; KH_2PO_4 12.4 g L^{-1}; $MgSO_4 \cdot 7H_2O$ 50 g L^{-1}; $NaHCO_3$ 15.9 g L^{-1}; EDTAFeNa and EDTANa$_2$ both at 2.25 g L^{-1}; H_3BO_3 2.48 g L^{-1}; $[(NH_4)_6Mo_7O_{24} \cdot 4H_2O]$ 1 g L^{-1}; $MnCl_2 \cdot 4H_2O$ 1.4 g L^{-1}, cyanocobalamin, biotin, and thiamine, each one 0.04 g L^{-1}, $NaNO_3$ 80 g L^{-1}; $NaH_2PO_4 \cdot 2H_2O$ 36 g L^{-1}).

The flasks were stopped by a porous cotton plug and illuminated by a white lamp/red lamp Osram daylight 2 $\times$ 36 W plus an Osram Gro-Lux lamp 36 W (8 h light/16 h dark) at room temperature of 20 °C.

To perform the tests, smaller flasks were filled with equal volumes of the treated or untreated samples, added with the suitable amount of Jaworski's medium salts and the algal diluted suspension (approximately 10^5 cells mL^{-1}). The small flasks were kept in the same conditions as the starter culture. Controls were prepared by adding one volume of the Jaworski's salts mixture to one volume of algae suspension. We evaluated the algal density by measuring the ABS at 545.6 nm, an indirect method for cell counting also suggested in the ISO 8692/2004 [44]. Aliquots of carefully hand-shaken samples or controls were measured without dilution, in triplicate, and then the aliquots were poured back into the flask.

3. Results

3.1. The Portable Photoreactor Prototype

The prototype was assembled at the Proambiente laboratories based on the idea of developing a device effective enough to produce significant detoxification of quite large water solution volumes during a reasonable time period (hours) that is robust and easy to be used anywhere a power supply is available. These characteristics of the system, together with the reduced production costs and the simple procedure, should encourage its regular use any time agro-wastewater has to be managed.

To design the irradiating unit (or portable reactor) illustrated in Figure 2, we considered the average capacity of the spray tanks to be about 800 L, the average volume of residual solution to be 5 L, and the washing volume for this tank to be about 50 L (data obtained from personal communication with Italian spray machine sellers and users). The simplest solution, taking into account that the unit had to be easily transportable, was to prepare a stainless steel, cylindrical container (Ø 300 cm, high 1.007 cm) capable of holding up to 70 L of liquid. The UV lamps were selected, among those commercially available, according to the dimensions of the container. We employed four immersion, 20 W, low-pressure mercury UV lamps emitting a monochromatic light at $\lambda = 254$ nm. The lamps were mounted on a circular support to obtain a uniform distribution of the irradiated light inside the sample volume. The support was completed by adding the electric contacts and was placed inside the system case, whose upper part was hermetically closed by a circular cover blocked by a flange to prevent the escape of UV radiation. In addition, the container was equipped with a circular glass window to check the proper functioning of the lamps and with a drain cock to allow for sampling during the irradiation and emptying at the end of the treatment.

To proceed with the photodegradation, the washing water must be transferred inside the prototype. The filling of the prototype with the agrochemical solutions can occur at the spray-tank side using an immersion pump, connected on one side to a tube, capped by a filter, and placed inside the solution into the barrel of the spray-tank. On the other side, it is connected to a tube with an automatic stop dispenser to avoid overfilling the container (Figure 3).

3.2. The Photodegradation and Mineralization Profiles

Of the samples withdrawn from the pesticide solution under treatment, we determined a decrease in both the active ingredient concentration and the degree of mineralization, expressed by the reduction in the ABS and in the TOC value, respectively.

The amount of hydrogen peroxide added to the solutions (50 mL of H_2O_2 at 30% solution) was selected according to our previous experience at the lab scale [33] and the literature data [21]. The pH values of the three solutions after hydrogen peroxide addition were in the range 7.4–7.6, and they were not modified, because this photodegradation procedure was intended for use at farms where a pH adjustment would be an unusual and unlikely operation. Furthermore, those values were suitable for performing the biotoxicity assays both before and after treatment.

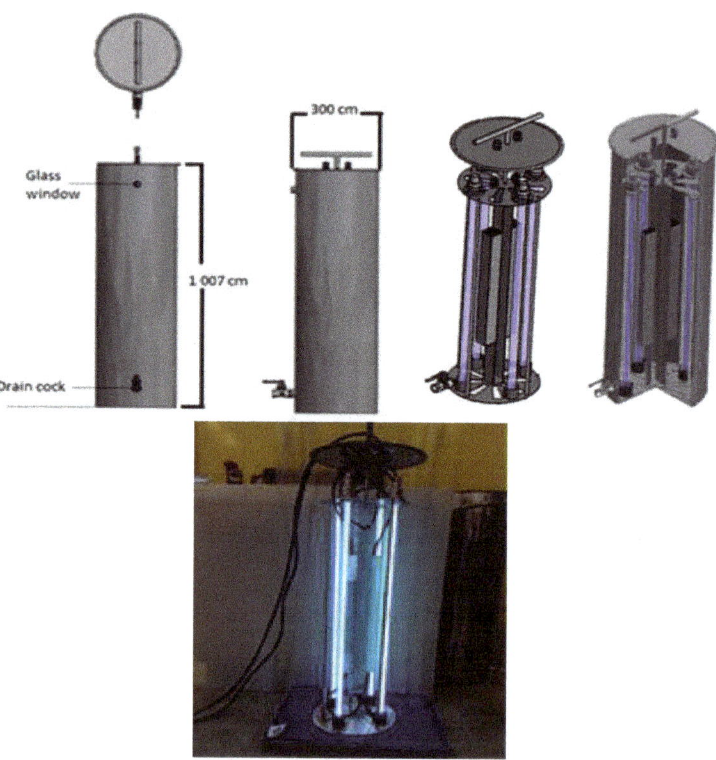

Figure 2. The portable photolytic cell. From left to right: external aspect and dimensions of the system case, scheme of the UV lamps support, the support inside the case system. Below, a photo of the completed lamps' support.

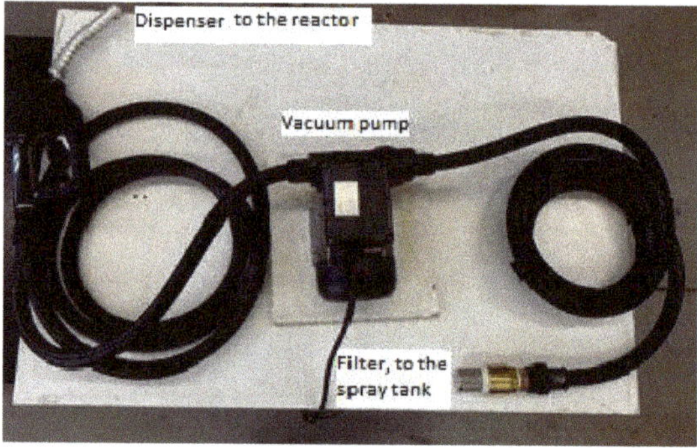

Figure 3. The vacuum pump system employed to quickly and safely transfer the pesticide solution from the spray-tank, or any other container, to the photodegradation unit.

It was out of the scope of this work to follow the photodegradation mechanism by determining the kind and amount of the photoproducts, first of all because our main aim was to develop a more effective device by adapting the UV–H_2O_2 AOP to the out-of-lab conditions encountered at farm locations. On the other hand, several studies on the photochemistry of the selected compounds were already available in the literature, as mentioned in the Introduction.

In Figure 4, for each pesticide, the data obtained for the absorbance and the TOC parameters were reported in a double-axis graph in order to compare the trends of the two processes. Two hours was the first sampling interval chosen to evaluate the degradation. At that time, only the reduction in concentration of TBA and IMI and a modest IMI mineralization were measurable. Indeed, the AOPs proceed through complex pathways, starting with the fragmentation of the parental organic molecules, followed by various transformations leading, in optimal conditions, to the production of CO_2 and carbonates. The disappearance of the parent compound and the reduction of the organic molecule content can follow completely different kinetics.

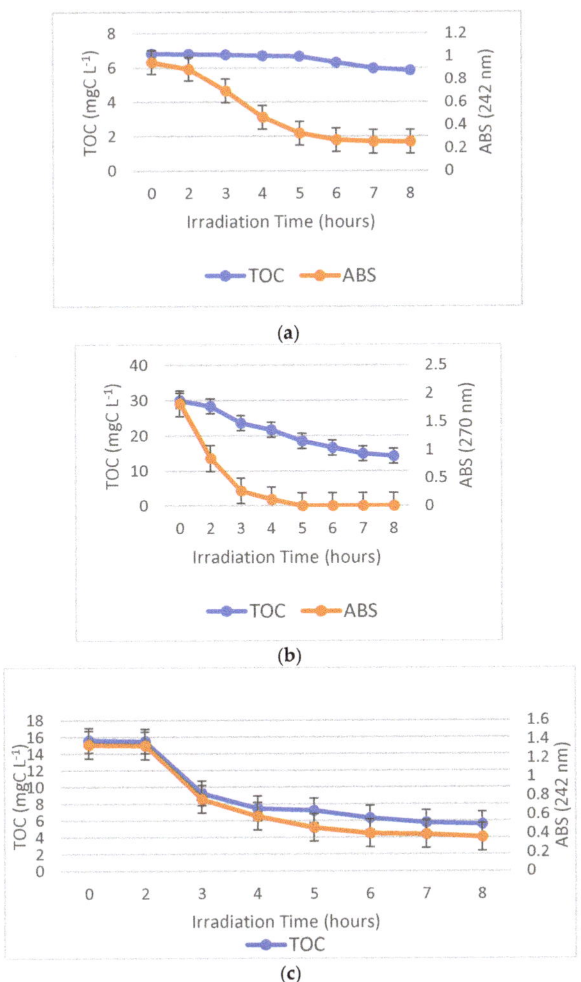

Figure 4. The changes in both the active ingredient concentration and in the TOC content over time with respect to the untreated solutions (irradiation time 0). (**a**) TBA, (**b**) IMI, (**c**) AZO.

Concerning the selected pesticides, these differences were particularly evident for the compound TBA. This molecule, or better its degradation derivatives, resulted in being the most recalcitrant. TBA was degraded at 80% after about 6 h, but the TOC value of the solution was quite unaffected after 8 h of treatment (Figure 4a).

A similar, and definitely less pronounced, discrepancy between the two processes was observed for the insecticide IMI; the parental molecule completely disappeared after 5 h of treatment, while to halve the TOC value, 7 h of irradiation was required (Figure 4b).

On the opposite, the two phenomena showed the same trend in the case of AZO. The disappearance of the active ingredient was strictly followed by the reduction of the TOC value, both halved after 3 h of treatment. After this time, the two curves slightly diverged, probably because of the formation of resistant photoproducts (Figure 4c).

The prototype was able to remove about 80% (mean value) of the active ingredients in an acceptable time (8 h), which can correspond, for example, to an overnight treatment. The mineralization rate was significantly lower, the mean value being about 44%, but it was known that to obtain the disappearance of any organic compound is a rather more difficult task.

3.3. The Toxicity Assays

In order to evaluate in depth the efficiency of remediation technologies such as AOP treatments, the determination of the pollutant degradation and of the mineralization rate must be coupled to various biological methodologies in order to determine the actual changes in toxicity of the treated materials. Toxicity assessment is usually performed by using a set of tests based on simple organisms living in the environment of interest. We applied two of the most used assays to assess the water quality, the bioluminescent bacteria *Vibrio fisheri*, and the freshwater microalga *Pseudokirchneriella subcapitata*, representing the basic level of heterotrophic and autotrophic organisms.

3.3.1. The Bacterial Bioluminescence Inhibition Test

In Figure 5a–c, we reported, separately for the three compounds, the intensities of the bacterial light emission from the controls (blank), the solutions treated for 2 h, and those at the end of the treatment (8 h).

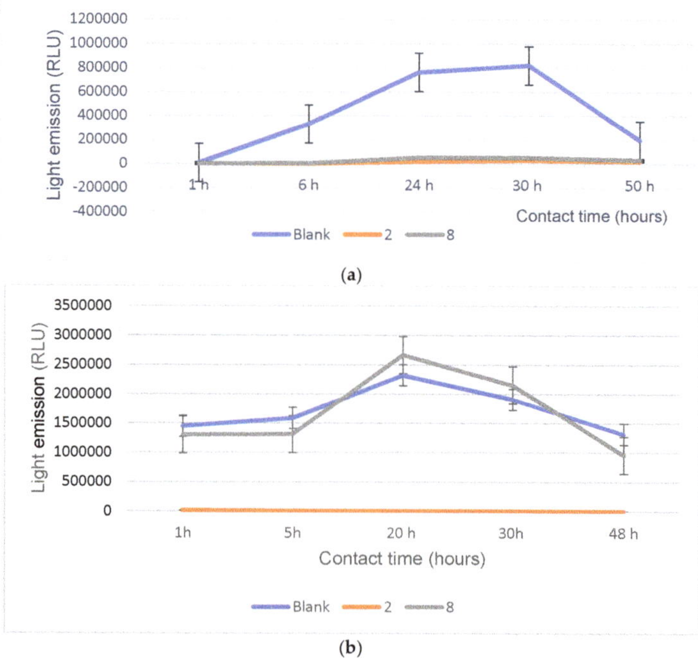

Figure 5. *Cont.*

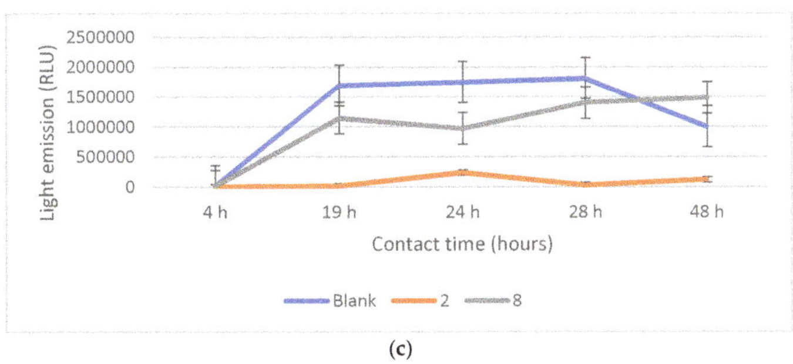

Figure 5. Light emission from bacteria in contact with the untreated solutions (blank) and the solutions after 2 h and 8 h of irradiation. (**a**) TBA, (**b**) IMI, (**c**) AZO.

Both solutions of TBA, at 2 and 8 h, reduced the light emission to zero, confirming that the important decrease in the active ingredient concentration was ineffective in reducing the toxicity of this compound, a characteristic probably completely retained by its photoproducts.

On the contrary, the sample of IMI collected after 2 h of treatment still completely inhibited the light emission, confirming the solution toxicity, but the sample at the end of the treatment showed no inhibition. The light intensity was practically overlapped with that of the control. The incomplete mineralization of the solution did not represent a problem of residual toxicity from the degradation products.

A similar behavior was observed by comparing the control's emission with those of the first and last treated samples of the AZO solution. Again, the sample after 2 h of treatment deeply inhibited the light emission, whereas at the end of the treatment it was possible to observe a reduced, albeit still detectable, toxic effect.

In Figure 6 we reported, for a direct comparison among the three compounds, the mean values of the chronic toxicity, i.e., the % of light emission inhibition produced by all the irradiated solutions and the untreated samples with respect to the control after 24 h of contact with the bacteria. In this way, it was easier to highlight the effects of the H_2O_2–UV treatment on the light emission inhibition over time.

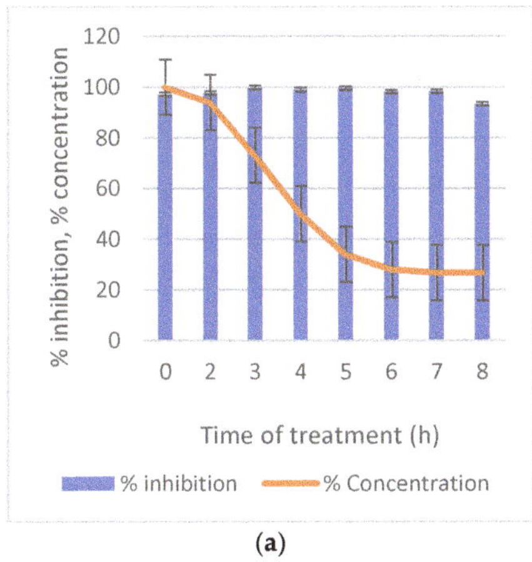

Figure 6. *Cont.*

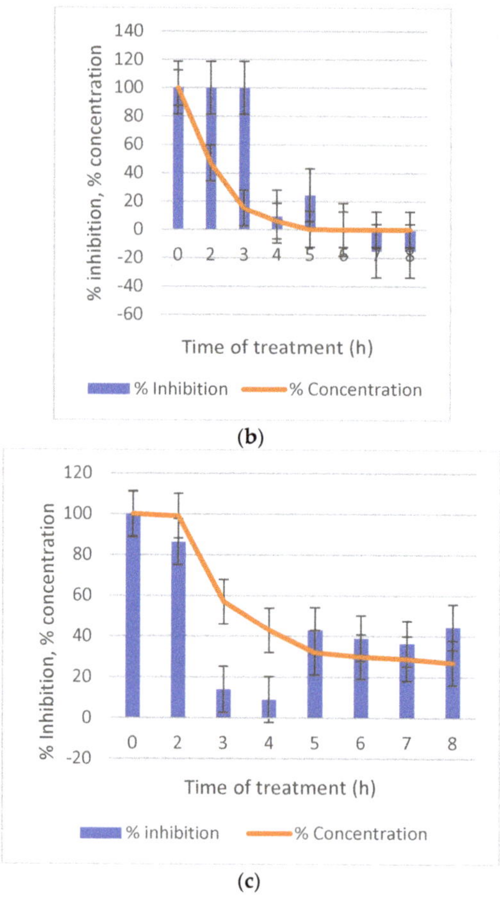

(b)

(c)

Figure 6. The inhibition of light emission produced by the eight solutions of each compound, including the untreated solutions (time of treatment = 0). (**a**) TBA, (**b**) IMI, (**c**) AZO.

Apart from TBA, for which inhibition never decreased significantly, interesting behaviors can be observed. After 4 h of treatment, a sudden reduction in IMI toxicity from 100% to 6% occurred. The negative inhibition values produced by the last three samples indicate that the photoproducts slightly stimulate the wellness of the bacteria, most probably acting as a supplementary carbon source. A significant drop in AZO toxicity occurred after 3 h of treatment (from 83% at 2 h to 14%). Surprisingly, AZO solutions treated for a longer time (5–8 h) partially recovered their toxicity (39%). This result confirmed the partial toxicity represented in Figure 5c, and we can hypothesize that the photoproducts obtained at those times were more toxic than the previous ones.

3.3.2. The Algal Growth Inhibition Assay

The effects of the pesticide solutions on the microalgae growth rate appeared, as expected, different from those on bacteria and, in some cases, more complicated to interpret. The sensitivity of these organisms is surely lower than that of bioluminescent bacteria, and all solutions, treated or not, took the same, quite long, time to influence their growth (Figure 7). At the start of the experiment and still after 4 days of contact, the optical density in the samples (defined as the absorbance at 545.6 nm) and the controls were very similar. Stopping the experiment at 72 h, as frequently done, these compounds could be declared nontoxic. However, after 6 days of contact, the controls continued to grow, whereas all the samples of IMI and TBA (Figure 7a,c) produced a variable drop in the microalgae population. The effect of the various IMI samples was practically the same, i.e., concentration independent. On the contrary, the effects of the herbicide were more specific and dependent on its concentration, and this effect was more evident after 4 days of contact. It was not possible to define a clear toxic effect of AZO solutions at all times, since the counts of samples and controls were in the same range even after 6 days from the beginning, indicating this fungicide as a not harmful compound to these organisms.

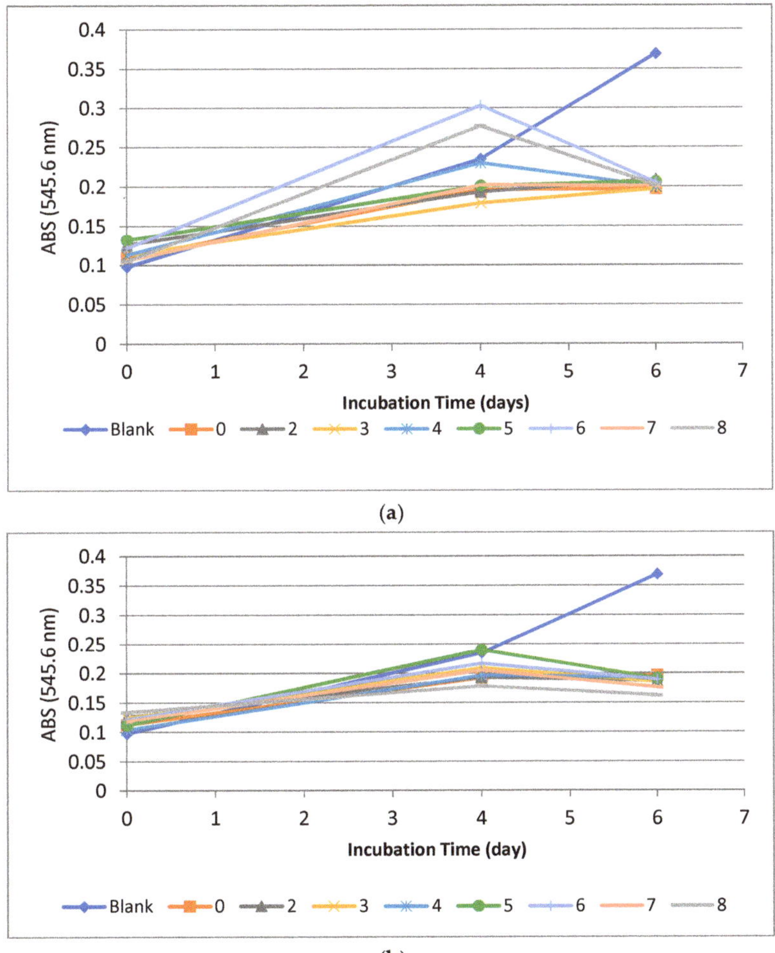

(a)

(b)

Figure 7. *Cont.*

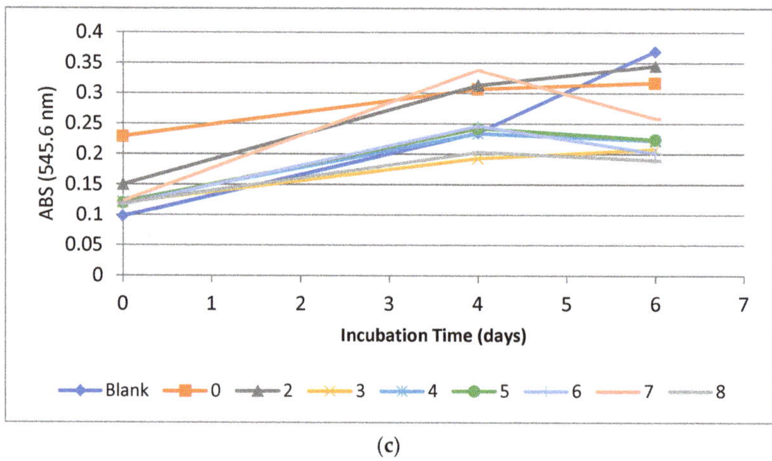

(**c**)

Figure 7. The absorbance values were recorded for the microalgae populations in contact with the various solutions (0–8) and for the control (blank) immediately after the preparation of the samples–algae mixture (incubation time = 0) and after 4 and 6 days. (**a**) TBA, (**b**) IMI, (**c**) AZO.

4. Discussions

Nowadays, the general awareness about the incalculable damages produced by agro-chemicals on all living beings imposes that we take reasonable precautions to ensure that the storage, handling, and disposal of these products do not endanger further human health or the environment. Nevertheless, each crop requires various treatments, and a direct consequence of any phytosanitary treatment is, among others, the generation of a pesticide containing a volume of agro-wastewater from the remnants and rinsing of the treatment machinery. In most cases, the fate of unused tank mixes is to be simply disposed of on the soil of treated fields or at the farm area, since the several developed alternatives to correctly treat these waste results are expensive and time (and space)-consuming, or the threat posed by this action is not completely understood. Efficient, more applicable, and flexible methods are in demand.

To promote the degradation of pollutants by the oxidation of the organic compound is a powerful solution actually applied mainly to the treatment of wastewaters. Among the various possibilities offered by the advanced oxidation processes, we selected one of the most used in its simpler and possibly cheaper formulation, the UV promoted oxidation supported by the addition of hydrogen peroxide as the hydroxyl radicals' source, in a portable, flexible format.

We assembled a photodegradation cell able to contain and treat most parts of the remnant volumes from spraying activities or leftover pesticide waste. The use of this device is elementary; it must simply be filled and connected to the electric power. Its size make it easily transportable and the costs are really reduced. The cost of purchasing it is in the range of €3000–4000 and the maintenance costs concern the substitution of the exhausted UV lamp, in addition to the electric power consumption. Such costs can be supported by any farm, even one of reduced dimensions, and this device can become a small but effective facility, located exactly where the wastewater detoxification is required.

By using the commercial formulation of pesticides with different degrees of suscep-tibility to mineralization, the tap water to prepare the solutions, and the concentrations usually present in the spray-tank washing volumes, we exactly simulated the scenario occurring at farms and planned a suitable treatment time, long enough to be effective but no too long so as to be practically unacceptable.

The chemical determination of the pollutants' degradation and of the mineralization rate was coupled to biological evaluations aimed at determining if the degradation actually resulted in a reduction of the toxic effects on the environment, confirming the effectiveness and usefulness of the device.

The disappearance of the active ingredient can have completely different kinetics with respect to the reduction of the organic matter content, that being the sensitivity of the parental compound and of its derivatives to oxidative degradation, as well as their toxicity, which are often completely independent [32–34].

The degradation of the selected compounds began immediately, except for the fungicide, whose concentration was reduced later than 2 h, but mineralization was not appreciable before 3 h. After 8 h, the degradation of the active ingredients was very satisfying. Overall, for the test, the mineralization rate was significantly lower and at 8 h, less than 50% mineralization had been attained, as an average value. In particular, the photoproducts of TBA are known to be recalcitrant to mineralization [45,46], and this compound was selected just to simulate the worst working conditions. Additionally, the results obtained for the imidacloprid were in agreement with the literature data reporting the formation of several photoproducts during the degradation process of IMI, mainly the well-known 6-chloronicotinic acid, which slows down the mineralization process [25,47].

Nevertheless, these results can be considered extremely positive, and they already seem to offer useful indications for the correct use of the device as such. Indeed, the photoproducts are often more prone to bio- or phytodegradation than their parent compounds, for which fragmentation promotes the different inherent water remediation processes.

It was known that photocatalytic or even more powerful techniques must be applied to obtain exhaustive mineralization processes, but these also increase, in parallel, the costs and/or the skills required to carry out these procedures. Moreover, the complete mineralization of organic pollutants is not always necessary. It is not rare that degradation intermediates show higher biotoxicity than the parental compound [47], suggesting that avoiding the presence of these compounds in the environment (blocking them before they reach soil or water) is the only solution. For example, the toxicity of the fungicide collapses after three hours of treatment, corresponding to a drastic increase in both the degradation and mineralization of the parent compound. However, after four hours, toxicity partially resumes and no longer subsides until the test is complete. It is interesting to note that in precisely the vicinity of the fourth hour and, above all, of the fifth hour, the degradation and mineralization curves of the compound diverge, albeit slightly. It could not be excluded that, from that moment on, a toxic product reluctant to mineralization was formed. The use of the device for AOPs for 3–4 h seems to be preferable both to shorter and longer times (within the limits of what has been tested in this work).

On the other hand, Imidacloprid shown still the 100% toxicity at 3 h, changed to less the 10% after 4 h of treatment. Nevertheless, the IMI concentration was not so different among 3 and 4 h and in both cases very low. It is possible to imagine that the photoproduct(s) retaining the toxic effect underwent definitive degradation only at 4 h, whereas the organic content was still high (low mineralization rate). Additionally, in this case seems that the optimal effect of the device takes place in 4 h.

The assessment of the post-treatment toxicity by the highly sensitive *Vibrio fisheri* test revealed that in only one case, the TBA, did treated and untreated solutions show the same inhibition effect. The difficulties encountered in removing this compound, or better, its derivatives, surely represent a serious, but fortunately not so frequent, obstacle in wastewater remediation.

A detectable impact of the agrochemical solutions on the freshwater microalgae population was observed only after a rather long contact period (144 h). It must be underlined that, for the most part, this kind of assay collects data for shorter periods (72–96 h). In this case, our samples should be considered as not harmful to these organisms.

In conclusion, it is possible to affirm that the overall performance of this first version of the photodegradation portable cell was positive, even though the pollutants' concentration/pH/H_2O_2 ratios were not the optimal ones, which can be employed in laboratory studies to maximize the degradation yields.

Further improvements of the prototype are under study, such as the application of an accessory to mix the solution during the treatment and to enhance the homogeneity of the UV rays' absorbance in the whole volume.

The results obtained from the current studies must be confirmed on a wider selection of agrochemicals and on a mixture of different compounds and organic pollutants to ascertain the effectiveness and feasibility of this device in various conditions and indicate the possibility for scaling it for different applications. The use of the studied device could also improve crop management in terms of bioeconomy, thanks to the increased possibility of recovering contaminated washing water for other agronomic uses [48].

Author Contributions: Conceptualization, B.E., F.R., and S.D.M.; methodology, B.E. and F.R.; formal analysis F.R. and F.O.; investigation, B.E., E.N.F., and S.S.; data curation, F.R. and E.G.M.; writing—original draft preparation, E.N.F. and S.D.M.; writing E.N.F., S.D.M., B.E., and F.R.—review and editing F.O., S.D.M., E.N.F., S.S., F.R., and B.E.; project administration, F.R. All authors have read and agreed to the published version of the manuscript.

Funding: This research received no external funding.

Institutional Review Board Statement: Not applicable.

Informed Consent Statement: Not applicable.

Acknowledgments: We express our sincere gratitude to Giorgio Longino for his fundamental contribution to the realization of the prototype, for his patience during the discussions on its design, and for his great sense of humor.

Conflicts of Interest: The authors declare no conflict of interest.

References

1. De Wilde, T.; Spanoghe, P.; Debae, C.; Ryckeboer, J.; Springael, D.; Jaeken, P. Overview of on-farm bioremediation systems to reduce the occurrence of point source contamination. *Pest Manag. Sci.* **2007**, *63*, 111–128. [CrossRef]
2. Boyd, A.; Ridout, C.; O'Sullivan, D.M.; Leach, J.E.; Leung, H. Plant–pathogen interactions: Disease resistance in modern agriculture. *Trends Genet.* **2013**, *29*, 233–240. [CrossRef]
3. Duru, M.; Therond, O.; Martin, G.; Martin-Clouaire, R.; Magne, M.A.; Justes, E.; Journet, E.P.; Aubertot, J.N.; Savary, S.; Bergez, J.E.; et al. How to implement biodiversity-based agriculture to enhance ecosystem services: A review. *Agron. Sustain. Dev.* **2015**, *35*, 1259–1281. [CrossRef]
4. Sakadevan, K.; Nguyen, M.-L. Chapter Four-Livestock Production and Its Impact on Nutrient Pollution and Greenhouse Gas Emission. *Adv. Agron.* **2017**, *141*, 147–184.
5. FAO. World Food and Agriculture—Statistical Pocketbook 2020. Rome. Available online: https://doi.org/10.4060/cb1521en (accessed on 16 March 2021).
6. Viessman, W., Jr.; Hammer, M.J. *Water Supply and Pollution Control*, 6th ed.; Addison Wesley Longman Inc: Boston, CA, USA, 1998; pp. 876–885.
7. Breach, R.A. UK experiences in developing a catchment scale, risk based, approach to mitigating all routes for pesticide loss to water. In *Proceeding of the International Advances in Pesticide Application, Robinson College, Cambridge, UK, 9–11 January 2008*; Alexander, L.S., Carpenter, P.I., Cooper, S.E., Glass, C.R., Gummer Andersen, P., Magri, B., Robinson, T.H., Stock, D., Taylor, W.A., Thornhill, E.W., Eds.; Association of Applied Biologists: Wellesbourne, UK, 2008; pp. 353–356.
8. Shen, C.; Yang, X.; Wang, Y.; Zhou, J.; Chen, C. Complexation of capsaicin with β-cyclodextrins to improve pesticide formulations: Effect on aqueous solubility, dissolution rate, stability and soil adsorption. *J. Incl. Phenom. Macrocycl. Chem.* **2012**, *72*, 263–274. [CrossRef]
9. Briggs, J.A.; Riley, M.B.; Whitwell, T. Quantification and remediation of pesticides in runoff water from containerized plant production. *J. Environ. Qual.* **1998**, *27*, 814–820. [CrossRef]
10. Vischetti, C.; Capri, E.; Trevisan, M.; Casucci, C.; Perucci, P. Biomassbed: A biological system to reduce pesticide point contamination at farm level. *Chemosphere* **2004**, *55*, 823–828. [CrossRef]
11. Grella, M.; Miranda-Fuentes, A.; Marucco, P.; Balsari, P. Field assessment of a newly-designed pneumatic spout to contain spray drift in vineyards: Evaluation of canopy distribution and off-target losses. *Pest Manag. Sci.* **2020**, *76*, 4173–4191. [CrossRef]
12. Directive 2009/128/EC of the European Parliament and of the Council of 21 October 2009 establishing a framework for Community action to achieve the sustainable use of pesticides (Text with EEA relevance). *OJ L 24.11.* **2009**, *309*, 71–86. Available online: https://eur-lex.europa.eu/legal-content/EN/TXT/?uri=celex%3A32009L0128 (accessed on 16 March 2021).
13. Vischetti, C.; Monaci, E.; Coppola, L.; Marinozzi, M.; Casucci, C. Evaluation of BiomassBed system in bio-cleaning water contaminated by fungicides applied in vineyard. *Int. J. Environ. Anal. Chem.* **2012**, *92*, 949–962. [CrossRef]

14. Özkara, A.; Akyıl, D.; Konuk, M. Pesticides, environmental pollution, and health. In *Environmental Health Risk-Hazardous Factors to Living Species*; Larramendy, M.L., Soloneski, S., Eds.; IntechOpen: London, UK, 2016. [CrossRef]

15. Jouy, L.; Moquet, M.; Yème, P.-V. Rinçage et lavage du pulvérisateur: Au champ ou à la ferme, bien gérer ses effluents phytosanitaires. *Perspect. Agric.* **2009**, *354*, 34–42.

16. Fait, G.; Nicelli, M.; Fragoulis, G.; Trevisan, M.; Capri, E. Reduction of point contamination sources of pesticide from a vineyard farm. *Environ. Sci. Technol.* **2007**, *41*, 3302–3308. [CrossRef] [PubMed]

17. Vagi, M.C.; Petsas, A.S. Recent advances on the removal of priority organochlorine and organophosphorous biorecalcitrant pesticides defined by Directive 2013/39/EU from environmental matrices by using advanced oxidation processes: An overview (2007–2018). *J. Environ. Chem. Eng.* **2020**, *8*, 102940. [CrossRef]

18. Acero, J.L.; Real, F.J.; Benitez, F.J.; Matamoros, E. Degradation of neonicotinoids by UV irradiation: Kinetics and effect of real water constituents. *Sep. Purif. Technol.* **2019**, *211*, 218–226. [CrossRef]

19. Zheng, Q.; Wang, L.; Chen, B.; Chen, Y.; Ma, Y. Understanding and modeling the formation and trasformation of hydrogen peroxide in water irradiated by 254 nm ultraviolet (UV) and 185 nm vacuum UV (VUV): Effects of pH and oxygen. *Chemosphere* **2020**, *244*, 125483. [CrossRef] [PubMed]

20. Reddy, P.V.L.; Kim, K.-H. A review of photochemical approaches for the treatment of a wide range of pesticides. *J. Hazard. Mater.* **2015**, *285*, 325–335. [CrossRef]

21. Rozas, O.; Vidal, C.; Baeza, C.; Jardim, W.F.; Rossner, A.; Mansilla, H.D. Organic micropollutants (OMPs) in natural waters: Oxidation by UV/H2O2 treatment and toxicity assessment. *Water Res.* **2016**, *98*, 109–118. [CrossRef]

22. Fenoll, J.; Garrido, I.; Flores, P.; Hellín, P.; Vela, N.; Navarro, G.; García-García, J.; Navarro, S. Implementation of a new modular facility to detoxify agro-wastewater polluted with neonicotinoid insecticides in farms by solar photocatalysis. *Energy* **2019**, *175*, 722–729. [CrossRef]

23. Díez, A.M.; Sanromán, M.A.; Pazos, M. New approaches on the agrochemical degradation by UV oxidation processes. *Chem. Eng. J.* **2019**, *376*, 120026. [CrossRef]

24. Vela, N.; Fenoll, J.; Garrido, I.; Pérez-Lucas, G.; Flores, P.; Hellín, P.; Navarro, S. Reclamation of agro-wastewater polluted with pesticide residues using sunlight activated persulfate for agricultural reuse. *Sci. Tot. Env.* **2019**, *660*, 923–930. [CrossRef]

25. Kitsiou, V.; Filippidis, N.; Matzavinos, D.; Poulios, I. Heterogeneous and homogeneous photocatalytic degradation of the insecticide imidacloprid in acqueous solutions. *Appl. Catal. B Environ.* **2009**, *86*, 27–35. [CrossRef]

26. Sanlaville, Y.; Guittonneau, S.; Mansour, M.; Feicht, E.A.; Meallier, P.; Kettrup, A. Photosensitized degradation of terbuthylazine in water. *Chemosphere* **1996**, *33*, 353–362. [CrossRef]

27. Wamhoff, H.; Schneider, V. Photodegradation of imidacloprid. *J. Agric. Food Chem.* **1999**, *47*, 1730–1734. [CrossRef]

28. Lányi, K.; Dinya, Z. Photodegradation study for assessing the environmental fate of some triazine-, urea-, and thiolcarbamate-type herbicides. *Microchem. J.* **2005**, *80*, 79–87. [CrossRef]

29. Kurwadkar, S.; Evans, A.; DeWinne, D.; White, P.; Mitchell, F. Modeling photodegradation kinetics of three systemic neonicotinoids-dinotefuran, imidacloprid, and thiamethoxam-in aqueous and soil environment. *Environ. Toxicol. Chem.* **2016**, *35*, 1718–1726. [CrossRef]

30. Boudina, A.; Emmelin, C.; Baaliouamer, A.; Païssé, O.; Chovelon, J.M. Photochemical transformation of azoxystrobin in aqueous solutions. *Chemosphere* **2007**, *68*, 1280–1288. [CrossRef]

31. Katagi, T. Direct photolysis-mechanism of pesticides in water. *J. Pestic. Sci.* **2018**, *43*, 57–72. [CrossRef] [PubMed]

32. Emmi, S.S.; Caminati, S.; Castro, C.; Esposito, B.; Saracino, M.; Sconziano, S. Comparison between Ionizing and Non-Ionizing Radiation Technologies for Wastewater Remediation. Proceeding of the 1st Research Coordination Meeting on "Radiation Treatment of Wastewater for Reuse with Particular Focus on Wastewaters Containing Organic Pollutants", Vienna, Austria, 2–6 May 2011; pp. 82–91. Available online: http://www-naweb.iaea.org/napc/iachem/working_materials/RC-1188-1_report_complete.pdf (accessed on 16 March 2021).

33. Emmi, S.S.; Caminati, S.; Esposito, B.; Saracino, M. About the OH yield in the radiolysis of an aqueous/H_2O_2 system. Its optimisation for water treatment. *Radiat. Phys. Chem.* **2012**, *81*, 1430–1433. [CrossRef]

34. Esposito, B.R.; Capobianco, M.L.; Martelli, A.; Navacchia, M.L.; Pretali, L.; Saracino, M.; Zanelli, A.; Emmi, S.S. Advanced water remediation from ofloxacin by ionizing radiation. *Radiat. Phys. Chem.* **2017**, *141*, 118–124. [CrossRef]

35. Tasca, A.; Puccini, M.; Clematis, D.; Panizza, M. Electrochemical removal of Terbuthylazine: Boron-doped diamond anode coupled with solid polymer electrolyte. *Environ. Pollut.* **2019**, *251*, 285–291. [CrossRef]

36. Simon-Delso, N.; Amaral-Rogers, V.; Belzunces, L.P.; Bonmatin, J.M.; Chagnon, M.; Downs, C.; Furlan, L.; Gibbons, D.W.; Giorio, C.; Girolami, V.; et al. Systemic insecticides (neonicotinoids and fipronil): Trends, uses, mode of action and metabolites. *Environ. Sci. Pollut. Res. Int.* **2015**, *22*, 5–34. [CrossRef] [PubMed]

37. Morrissey, C.A.; Mineau, P.; Devries, J.H.; Sánchez-Bayo, F.; Liess, M.; Cavallaro, M.C.; Liber, K. Neonicotinoid contamination of global surface waters and associated risk to aquatic invertebrates: A review. *Environ. Int.* **2015**, *74*, 291–303. [CrossRef]

38. Bryden, J.; Gill, R.J.; Mitton, R.A.A.; Raine, N.E.; Jansen, V.A.A. Chronic sublethal stress causes bee colony failure. *Ecol. Lett.* **2013**, *16*, 1463–1469. [CrossRef] [PubMed]

39. Sanchez-Bayo, F.; Goulson, D.; Pennacchio, F.; Nazzi, F.; Goka, K.; Desneux, N. Are bee diseases linked to pesticides?—A brief review. *Environ. Int.* **2016**, *89–99*, 7–11. [CrossRef]

40. Hocinat, A.; Boudemagh, A. Biodegradation of commercial Ortiva fungicide by isolated actinomycetes from the acti-vated sludge. *Desalination Water Treat.* **2015**, *57*, 6091–6097. [CrossRef]
41. Banić, N.D.; Šojić, D.V.; Krstić, J.B.; Abramović, B.F. Photodegradation of neonicotinoid active ingredients and their commercial formulations in water by different advanced oxidation processes. *Water Air Soil Pollut.* **2014**, *225*, 1954. [CrossRef]
42. ISO 11348-3:2007. Water Quality—Determination of the Inhibitory Effect of Water Samples on the Light Emission of Vibrio Fischeri (Luminescent Bacteria Test). Available online: https://www.iso.org/standard/40518.html (accessed on 16 March 2021).
43. ISO 14442:2006. Water Quality—Guidelines for Algal Growth Inhibition Tests with Poorly Soluble Materials, Volatile Compounds, Metals, and Wastewater. Available online: https://www.iso.org/standard/34814.html (accessed on 16 March 2021).
44. ISO 8692:2004. Water quality–Freshwater Algal Growth Inhibition Test with Unicellular Green Algae. (Revised by ISO 8692:2012). Available online: https://www.iso.org/standard/29924.html (accessed on 16 March 2021).
45. Lányi, K.; Dinya, Z. Photodegradation study of some triazine–type herbicides. *Microchem. J.* **2003**, *75*, 1–14. [CrossRef]
46. Navarro, S.; Vela, N.; Giménez, M.J.; Navarro, G. Persistence of four s-triazine herbicides in river, sea and groundwater samples exposed to sunlight and darkness under laboratory conditions. *Sci. Total Env.* **2004**, *329*, 87–97. [CrossRef]
47. Žabar, R.; Dolenc, D.; Jerman, T.; Franko, M.; Trebe, P. Photolytic and photocatalytic degradation of 6-chloronicotinic acid. *Chemosphere* **2011**, *85*, 861–868. [CrossRef]
48. Han, B.; Kim, J.K.; Kim, Y.; Zommer, N. Treatment of wastewater for reuse with mobile electron beam plant. In Proceedings of the 1st Research Coordination Meeting on "Radiation Treatment of Wastewater for Reuse with Particular Focus on Wastewaters containing Organic Pollutants", Vienna, Austria, 2–6 May 2011; Available online: http://www-naweb.iaea.org/napc/iachem/working_materials/RC-1188-1_report_complete.pdf (accessed on 16 March 2021).

Article

TiO$_2$-Photocatalyzed Water Depollution, a Strong, yet Selective Depollution Method: New Evidence from the Solar Light Induced Degradation of Glucocorticoids in Freshwaters

Luca Pretali, Angelo Albini, Alice Cantalupi, Federica Maraschi, Stefania Nicolis and Michela Sturini *

Department of Chemistry, University of Pavia, via Taramelli 12, 27100 Pavia, Italy; luca.pretali@gmail.com (L.P.); angelo.albini@unipv.it (A.A.); alice.cantalupi01@universitadipavia.it (A.C.); federica.maraschi@unipv.it (F.M.); stefania.nicolis@unipv.it (S.N.)
* Correspondence: michela.sturini@unipv.it; Tel.: +39-0382-987347

Citation: Pretali, L.; Albini, A.; Cantalupi, A.; Maraschi, F.; Nicolis, S.; Sturini, M. TiO$_2$-Photocatalyzed Water Depollution, a Strong, yet Selective Depollution Method: New Evidence from the Solar Light Induced Degradation of Glucocorticoids in Freshwaters. *Appl. Sci.* **2021**, *11*, 2486. https://doi.org/10.3390/app11062486

Academic Editor: Bin Gao

Received: 14 February 2021
Accepted: 6 March 2021
Published: 11 March 2021

Publisher's Note: MDPI stays neutral with regard to jurisdictional claims in published maps and institutional affiliations.

Abstract: The photodegradation of the most prescribed glucocorticoids (GCs) was studied under relevant environmental conditions in the presence of suspended TiO$_2$. The considered drugs included cortisone (CORT), hydrocortisone (HCORT), betamethasone (BETA), dexamethasone (DEXA), prednisone (PRED), prednisolone (PREDLO), and triamcinolone (TRIAM). The experiments were carried out at concentrations (50 µg L^{-1}) close to the real ones in freshwater samples (tap and river) under simulated and natural sunlight, and their decomposition took place very efficiently under natural sunlight. The reactions were monitored by high-pressure liquid chromatography coupled to electrospray ionization tandem mass spectrometry (HPLC-ESI-MS/MS). According to a pseudo-first-order decay, all drugs underwent degradation within 15 min, following different paths with respect to the direct photolysis. The observed kinetic constants, slightly lower in river than in tap water, varied from 0.29 to 0.61 min^{-1} with modest differences among GCs in the same matrix. Among main matrix macro-constituents, humic acids (HAs) were the most interfering species involved in GCs degradation. The photogenerated primary products were identified by HPLC-ESI-MS/MS, allowing to elucidate the general photochemical path of GCs. Finally, a comparison with literature data obtained using different advanced oxidation processes (AOPs) highlights the treatment efficiency with TiO$_2$/solar light for removing such persistent aquatic contaminants.

Keywords: glucocorticoids; TiO$_2$-solar light degradation; freshwater pollution; photoproducts; matrix constituents

1. Introduction

The role of glucocorticoids in biochemistry is difficult to exaggerate. The many possible configurations and the various chemical functions present explain the variety of reactions they catalyze in the cell and the sophisticated machinery that, starting from the regulation of glucocorticoid secretion by the hypothalamic-pituitary-adrenal (HPA), governs virtually all of the physiological processes, such as metabolism, immune and cardiovascular functions, skeletal growth, reproduction, and cognition [1,2]. Specifically, glucocorticoids (GCs) can inhibit immunological, inflammatory, and allergic processes in our body in response to an outer stimulus. Since their coming in the 1950s, many synthetic molecules have been synthesized to increase pharmacological activity and reduce side effects [3].

Due to their therapeutic properties and low cost, GCs are currently the most frequently used class of drugs to treat rheumatic and inflammatory diseases. As recently reported, the yearly number of prescribed GCs largely exceeds that of estrogens and androgens [4], although their use as doping agents and prophylactic in husbandry was banned by the World Anti-Doping Agency and by the Council Directive 96/22/EC, respectively [5,6]. Their occurrence in freshwaters, mainly as the initial drugs or slightly metabolized compounds, suggests that urban wastewater treatment plants (WWTPs) cannot remove them quantitatively [7]. Indeed, just as many other drugs, they have been detected worldwide in the aquatic environment. River concentrations are in the ng L^{-1} or sub-ng L^{-1} range in European [4,8–11] and non-European countries [12–16]. On the other hand, significantly higher levels, ranging from few tens to some hundred nanograms per liter, have been measured in WWTPs and hospital wastewaters [17]. In particular, influent WWTPs contain massive GCs amounts, up to 2 μg L^{-1}. Besides, some tens of ng per gram have been found in sewage sludge samples [9,18].

These concentrations are high enough to cause a negative eco-toxicological impact on various aquatic organisms, as demonstrated by the plethora of papers devoted to this issue. Under these conditions, reproduction, growth, and development of aquatic biota are affected by chronic exposure to low levels of GCs [19–22]. It was also reported that GCs act additively [22] and have cumulative effects in mixture with other steroid classes [20]. Depollution is a critical issue because of their negligible volatility and low biodegradability in urban WWTPs [7].

Due to the above, new strategies to abate GCs and those emerging contaminants of the aquatic environment that affect endogenous steroid actions have to be taken into account. In a recent paper [23], we studied the possibility of addressing this issue through direct photolysis in natural water. However, GCs photoreactivity under environmental conditions was found to be almost negligible due to their rigid cyclopetaperhydropenathrene structure and their limited light absorption in such highly diluted solutions, also when a more absorbing moiety, such as cross-conjugated chromophore in ring A, was present in the structure (see Scheme 1, X = H, F, Y = OH, O).

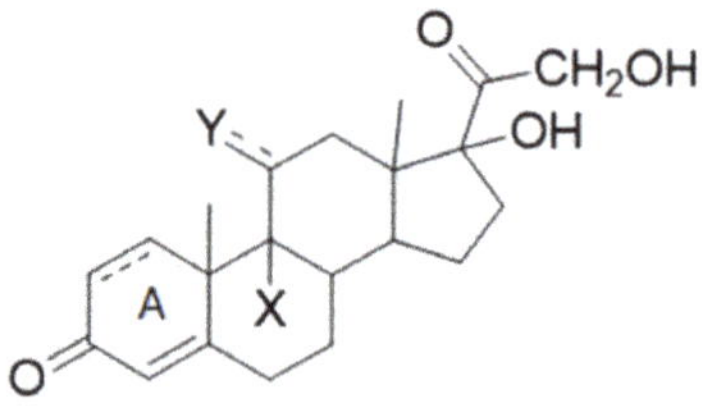

Scheme 1. The general structure of glucocorticoids (GCs).

We then decided to explore the possibility of improving the light-driven depollution process by adding a photocatalyst into the equation. Photocatalytic degradation of organic pollutants, such as drugs, has already demonstrated to be a highly effective process, also compared with other advanced oxidation processes (AOPs), and titania semiconductors are the benchmark in terms of efficiency and efficacy compared with other semiconductor oxides, under both UV-A and solar light ([24–26] and references therein). To date, the effectiveness of TiO_2 photocatalysis for removing GCs under actual conditions, viz. low-level concentrations and natural waters, has not yet been investigated. Only a few studies are available dealing with the TiO_2-photocatalytic degradation of DEXA [27], CORT acetate [28], and PREDLO [29] under experimental conditions (mg per liter concentration, ultrapure water), quite different from the ones mentioned above.

In this work, a range of GCs was chosen among the largely prescribed natural and semisynthetic drugs (see Scheme 2) and were submitted to irradiation using a solar simulator and natural sunlight, in tap and river water at micrograms per liter concentration (50 μg L^{-1}).

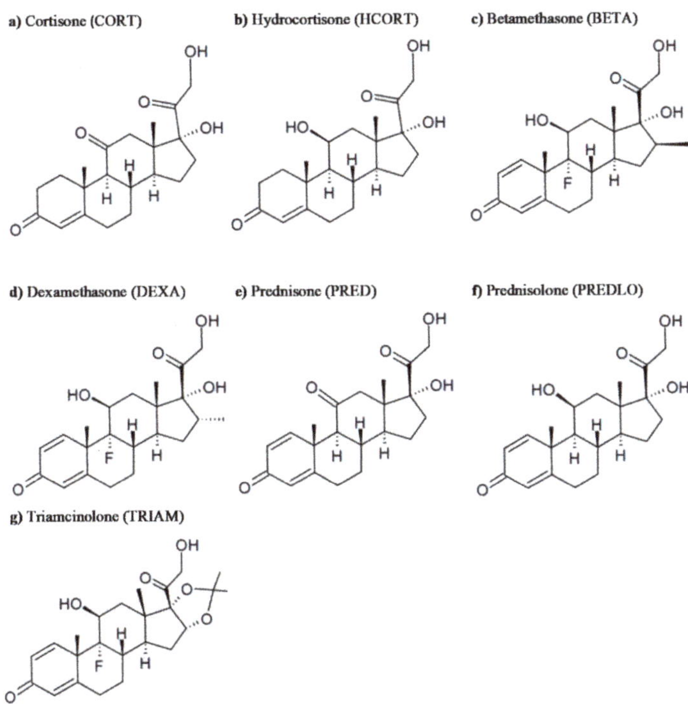

Scheme 2. Structures of the GCs examined in this study.

The potential effect due to some of the main matrix constituents on GCs-TiO$_2$ photore-action was also investigated.

Further experiments were carried out at mg per liter concentration (10 mg L^{-1}) to identify the primary photodegradation products by HPLC-ESI-MS/MS and elucidate the general photochemical path of GCs.

Finally, the sunlight TiO$_2$-photocatalytic efficiency for GCs removal was highlighted and compared with other advanced oxidation processes recently proposed [30–36].

2. Materials and Methods

2.1. Reagents and Materials

All the chemicals were of reagent grade or higher in quality and were used without any further purification. GCs standards (CORT, HCORT, BETA, DEXA, PRED, and PREDLO), acetic acid (99–100%), High-Performance Liquid Chromatography (HPLC) gradient–grade acetonitrile (ACN), CaSO$_4$ (99%), MgNO$_3$ hexahydrate (97%), and Humic acid sodium salt (MW = 100,000–150,000) were purchased by Merck (Milan, Italy). TRIAM was supplied by Farmabios (Gropello Cairoli, Italy). Methanol and water for Liquid Chromatography-Mass Spectrometry (LC/MS) were purchased by Carlo Erba Reagents (Cornaredo, Milano, Italy). GCs stock solutions of 10 mg L^{-1} were prepared in tap water and stored in the dark at 4 °C for a maximum of a week. The working solution of 50 µg L^{-1} was prepared daily. Aereoxide P25 titanium dioxide (Evonik, Resource Efficiency GmbH, Germany) was used for the photocatalytic experiments.

Nylon syringe filters (0.2 µm, 13 mm, Whatman, Milano, Italy) were used immediately after sample collection and before HPLC injection.

Tap water was from the Pavia municipal waterworks, while river water sample was collected from the Staffora River at 30–50 cm-depth in amber glass bottles. All the samples were stored in the dark (4 °C) before use. The physico-chemical parameters are reported in Table 1.

Table 1. Physico-chemical characterization of tap and river water samples.

Parameters/Ions		Tap Water	River Water
pH		7.7	7.9
Conductivity at 20 °C	$\mu S\ cm^{-1}$	271	293
TOC *	$mg\ L^{-1}$	<2	6.9
Cl^-	$mg\ L^{-1}$	5.0	4.0
NO_3^-	$mg\ L^{-1}$	0.6	1.5
SO_4^{2-}	$mg\ L^{-1}$	5.0	13
HCO_3^-	$mg\ L^{-1}$	182	195
Ca^{2+}	$mg\ L^{-1}$	35	54
Mg^{2+}	$mg\ L^{-1}$	10	7.5
Na^+	$mg\ L^{-1}$	12	5.5

* TOC total organic carbon.

2.2. Analytical Determination

HPLC systems with different sensitivity and analysis modes were used. For the kinetic experiments at 10 mg L^{-1}, the HPLC-UV system consisted of a Shimadzu (Shimadzu Corporation, Milano, Italy) LC-20AT solvent delivery module equipped with a DGU-20A3 degasser and interfaced with a SPD-20A UV detector. The wavelength selected for analysis was 238 nm for all GCs. Each sample was diluted (30% *v/v*) with MeOH and injected (20 µL) into a 250 × 4.6 mm, 5 µm GraceSmart RP18 (Sepachrom) coupled with a similar guard-column. The mobile phase was ultrapure water and ACN mixture, 70:30 for CORT, HCORT, PRED, and PREDLO; 65:35 for BETA, DEXA and TRIAM at a flow rate of 1.0 mL. The isocratic elution was maintained for 20 min, and then, after washing with 100% ACN for 5 min, the initial conditions were re-established. The instrumental quantification limits were 0.3 mg L^{-1} for CORT, HCORT, and DEXA, 0.09 mg L^{-1} for BETA, 0.2 mg L^{-1} for PRED, PREDLO, and TRIAM.

For the kinetic experiments at 50 µg L^{-1}, the HPLC-ESI-MS/MS system consisted of an Agilent (Cernusco sul Naviglio, Milano, Italy) HPLC apparatus 1260 Infinity coupled with an Agilent 6460C ESI-MS/MS spectrometer. Each sample (5 µL) was injected into an Agilent 120 EC-C18 Poroshell column (50 × 3 mm, 2.7 µm) with a similar guard-column. The column temperature was maintained at 50 ± 1 °C. The mobile phases were (A) ultrapure water (0.1% *v/v* acetic acid) and (B) MeOH (0.1% *v/v* acetic acid). A linear gradient from 40% to 84% B was applied in 12 min, followed by a column re-equilibration time of 8 min. The flow rate was 0.6 mL min^{-1}.

GCs identification and quantification were performed in negative electrospray ionization (ESI). Source parameters were set as follows: drying gas temperature 300 °C (N_2); drying gas flow 5 L min^{-1} (N_2); nebulizer 45 psi; sheath gas temperature 250 °C; sheath gas flow 11 L min^{-1}; capillary voltage at 3500 V (positive mode) and 3000 V (negative mode); nozzle voltage 500 V positive, 0 V negative; electron multiplier voltage (EMV) 0 V for both polarities. MRM conditions (precursor ion $[M + AcO]^-$ adduct) for each compound are reported in Table S1. MassHunter Software from Agilent was used for data processing.

The HPLC-ESI-MS/MS analyses for photoproducts identification were performed by using a surveyor HPLC system (Thermo Finnigan, San Jose, CA, USA), equipped with a 150 × 2.0 mm, 4 µm Jupiter 4U Proteo column (Phenomenex). The mobile phase consisted in: (A) ultrapure water (0.1% *v/v* formic acid) and (B) ACN (0.1% *v/v* formic acid). The starting concentration of eluent B was 2%, increased to 100% by 40 min with a linear gradient; this concentration was maintained for 5 min to wash the column. The flow rate was 0.2 mL min^{-1}. The MS/MS system consisted of an LCQ ADV MAX iontrap mass spectrometer, with an ESI ion source operating in ion-positive mode with the following instrument conditions: source voltage 5.0 kV; capillary voltage 46 V; capillary temperature 210 °C; tube lens voltage 55 V. Xcalibur 2.0.7 SP1 Software (Thermo Finnigan, San Jose, CA, USA) was used for spectra processing.

2.3. Irradiation Experiments

Irradiation experiments under simulated sunlight were carried out using a solar simulator (Solar Box 1500e, CO.FO.ME.GRA, Milano, Italy) set at a power factor 250 W m^{-2} (equipped with a UV outdoor filter of soda-lime glass IR-treated, and with a BST temperature sensor), and under natural sunlight. The incident power of natural sunlight (Pavia, 45°11′ N, 9°09′ E, June–July 2017, 12.00 a.m.–3.00 p.m., 27–30 °C) was measured by means of an HD 9221 (Delta OHM, 450–950 nm) and a Multimeter (CO.FO.ME.GRA, 295–400 nm) pyranometers and resulted in the range 400–501 W m^{-2} (Vis) and 22–34 W m^{-2} (UV). The experiments were performed in a closed glass container containing 100 mL of untreated tap or river water samples (depth 40 mm, exposed surface 9500 mm^2) at native pH, enriched with 50 μg L^{-1} of each drug, separately dissolved and magnetically stirred. Aliquots (0.5 mL) of each sample were withdrawn at planned times, filtered, and injected in HPLC-ESI-MS/MS system (multiple reaction monitoring mode, MRM) to follow the degradation profiles.

Each kinetic experiment was performed in triplicate, and the degradation kinetic constant (k_{deg}) was calculated by using dedicated software (Fig P application, Fig P Software Corporation, version 2.2a, BIOSOFT, Cambridge, UK).

Before irradiation, each suspension (50 μg L^{-1} of each GC, 0.5 g L^{-1} TiO$_2$) was magnetically stirred in the dark for 20 min to promote the drug adsorption on the catalyst surface. For each GC, a control solution with no catalyst was also measured. No changes in GCs concentration were detected in the control samples.

To investigate matrix effects, a set of experiments, as described above, were carried out with the addition of salts (SO$_4$$^{2-}$ 50 mg L^{-1}, NO$_3$$^-$ 20 mg L^{-1}) and humic acids (HAs, 10 and 5 mg L^{-1}).

For identifying the photoproducts, 100 mL tap water samples were spiked with 10 mg L^{-1} of each GC (separately dissolved) and irradiated under simulated solar light as described above. Aliquots (0.5 mL), treated as above, were injected into the HPLC-UV system before performing the HPLC-ESI-MS/MS analysis (in full scan, zoom scan, and MS/MS mode).

3. Results and Discussion

We investigated TiO$_2$ photocatalytic degradation of seven most prescribed GCs (see Scheme 2) under relevant environmental conditions. To better mimic the realistic GCs degradation, a set of experiments were carried out on freshwater samples (not filtered tap and river water samples, native pH) enriched with 50 μg L^{-1} of each GC under simulated and natural sunlight. Photoproducts structures were determined by HPLC-ESI-MS/MS.

Results were discussed and compared to the most recently reported data obtained employing different advanced degradation techniques [30–36].

3.1. Kinetic Degradation in Actual Samples

The degradation profiles of the investigated GCs (50 μg L^{-1}) in tap (a) and river (b) water samples under simulated solar light are shown in Figure 1.

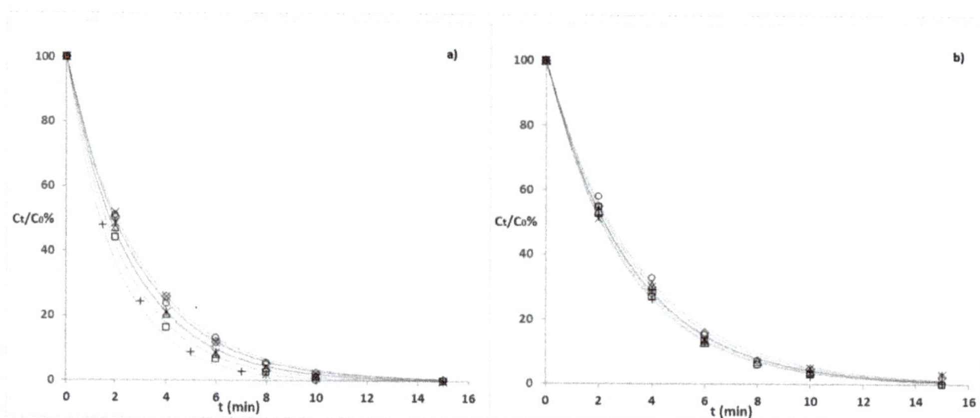

Figure 1. Evolution profiles of each GC (50 µg L^{-1}) in tap (**a**) and river water (**b**) under simulated sunlight: CORT ($\diamond$), HCORT ($*$), BETA ($\times$), DEXA ($\triangle$), PRED ($\square$), PREDLO ($+$), and TRIAM ($\bigcirc$) (see Section 2.3 for the irradiation conditions).

A pseudo-first-order equation (see Equation (1)) satisfactorily fitted all of the experimental data:

$$\frac{C_t}{C_0} = e^{-k_{deg}t} \tag{1}$$

C_0 is the initial GC concentration, C_t is GC concentration at time t, and k_{deg} is the kinetic degradation constant.

Table 2 shows the photolytic and photocatalytic degradation constants observed under the different experimental conditions described above.

Table 2. Observed GCs degradation constants (k_{deg}) (50 µg L^{-1} GCs, 0.5 g L^{-1} TiO$_2$, tap and river water, simulated, and natural sunlight).

	Photolysis Simulated Solar Light		Photocatalysis Simulated Solar Light		Natural Sunlight
Compound	**Tap Water** k_{deg} **(min^{-1})**	**River Water** k_{deg} **(min^{-1})**	**Tap Water** k_{deg} **(min^{-1})**	**River Water** k_{deg} **(min^{-1})**	**River Water** k_{deg} **(min^{-1})**
CORT	0.00106(5)	0.00128(3)	0.35(3)	0.300(8)	0.39(1)
HCORT	0.00246(9)	0.0033(2)	0.38(1)	0.315(8)	0.38(2)
BETA	0.0115(6)	0.0114(1)	0.336(9)	0.32(1)	0.35(2)
DEXA	0.0117(1)	0.0097(6)	0.386(9)	0.314(7)	0.61(1)
PRED	0.0185(9)	0.0186(6)	0.431(9)	0.33(1)	0.474(9)
PREDLO	0.024(1)	0.0199(4)	0.51(1)	0.336(4)	0.52(1)
TRIAM	0.0139(3)	0.0099(4)	0.355(9)	0.29(1)	0.38(1)

In brackets, the standard deviation values (sd), R^2 values in the range 0.992–0.999.

The most relevant result is that GCs direct photolysis occurred slowly, particularly for those GCs that scarcely absorb solar light (CORT and HCORT), with no significant difference between tap and river waters. On the contrary, TiO$_2$-assisted photocatalysis appeared to be independent of the chemical structure of the GCs. It was much faster than the direct photolysis, one order of magnitude for the less persistent studied GCs (PRED and PREDLO) and up to two orders of magnitude for the more persistent ones (CORT and HCORT). Both observations are well in accord with the drug initial adsorption on the TiO$_2$ granule surface, which is known to happen mostly through the carbonyl function (see Scheme 3), with no considerable effect on the rest of the molecule.

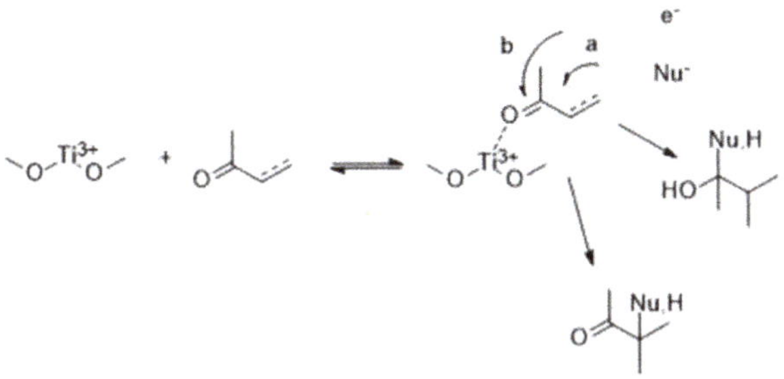

Scheme 3. Initial adsorption of the drugs on the TiO_2 granule surface through the carbonyl function. As it is generally assumed, within TiO_2 crystals the titanium nuclei are bound to six oxygen anions in a bipyramid. However, the outermost titanium ions are not fully compensated by the oxygen charge and bear some positive charge. Complexation then involves a Ti^{3+} ion on the surface of the TiO_2 granule and the n orbital in the carbonyl function, as it happens in the acid catalysis of the addition to α, β-carbonyl onto the β (a) or carbonyl (b) functions [37,38].

As previously reported in the literature [29], GCs degradation is then expected to proceed mainly via the photocatalytic oxidation of the adsorbed molecules on the catalyst surface. This reaction is initiated on the surface of the catalyst, but the generated reactive intermediates are susceptible to a cascade of oxidative processes, which may happen both at the catalyst surface and in the nearby solution where photoproduced OH radicals also play a role. Likewise, the decrease in the decomposition rate, observed when a fluorine atom is introduced on the GC scaffold, well fits with the proposed mechanistic scheme (Scheme 3).

No significant difference in the order of reactivity was observed in the photocatalytic GCs degradation at mg per liter concentration; a quantitative GCs removal ($\geq$95%) under actual conditions (natural pH, untreated water sample, and simulated sunlight) occurred in 15 min for all GCs. The observed kinetic degradation constants (k_{deg}) calculated for each GC are shown in Table 3.

Table 3. Degradation constants (k_{deg}) for each GC (10 mg L^{-1}) in tap water under simulated sunlight (see Paragraph 2.3 for the irradiation conditions) and comparison with other AOPs.

GC	AOP	Matrix	Kinetic Constant, % Removal, Degradation Time	Ref
DEXA 2-50 mg L^{-1}	Gamma ray, gamma ray with H_2O_2 or Fenton	Pure water, pH 7.2	$5 \times 10^{-4} - 4.7 \times 10^{-3}$ Gy^{-1}, $8 \times 10^{-4} - 1.6 \times 10^{-3}$ Gy^{-1}	[35]
PRED 50 mg L^{-1}	Electrochemical oxidation process 20 mA cm^{-2}	Pure water, pH 11, Na_2SO_4 1 g L^{-1}	0.1052 mg h^{-1}, 78% removal 4 h	[33]
PRED acetate 5 mg L^{-1}	O_3 50 mg min^{-1}	Pure water, pH 3	0.0595 min^{-1}, 90% removal 30 min	[32]
PREDLO 3.6 mg L^{-1}	UV/chlorine, 254 nm	Phosphate buffer pH 7, artificial fresh water pH 6	5.53×10^{-3} s^{-1}	[31]
PREDLO 100 mg L^{-1}	UV-Fenton, 360 nm	Pure water, pH 3	Quantitative removal 15 min	[30]
DEXA 5 mg L^{-1}	ZrO_2 1.5 g L^{-1}, 365 nm WO_3 0.5 g L^{-1}, >380 nm	Pure water, pH 3	ZrO_2 0.0078 min^{-1}, 76% removal 180 min WO_3 0.0277 min^{-1}, 100% removal 80 min	[34]

Table 3. *Cont.*

GC	AOP	Matrix	Kinetic Constant, % Removal, Degradation Time	Ref
BETA phosphate 30 mg L^{-1}	ZnO 0.44 g L^{-1}, 254 nm ZnO 0.44 g L^{-1}, persulfate, 254 nm	Pure water, pH 9	Removal 63%, 180 min Removal 98%, 180 min	[36]
PREDLO 25 mg L^{-1}	TiO$_2$ P25 1 g L^{-1}, 365 nm or solar light	Pure water, pH 6.7	Removal 94%, 1 h solar light Removal 73%, 1 h 365 nm	[29]
CORT acetate 10 mg L^{-1}	TiO$_2$ P25, 375 nm, air saturated, TiO$_2$ P25, 375 nm, persulfate air saturated	Different buffer solution, air saturated	0.040 min^{-1}, quantitative removal 100 min, 0.071 min^{-1}, quantitative removal 30 min	[28]
DEXA 10 mg L^{-1}	TiO$_2$ P25 0.2 g L^{-1}, simulated solar light	Pure water, air saturated	Quantitative removal 15 min	[27]
CORT, HCORT, BETA, DEXA, PRED, PREDLO, TRIAM 10 mg L^{-1}	TiO$_2$ P25 0.5 g L^{-1}, simulated solar light	Tap water, natural pH	0.184(5), 0.230(7), 0.19(1), 0.206(5), 0.177(7), 0.224(3), 0.24(1) min^{-1} Removal >95% 15 min	[This work]

These results are particularly informative for evaluating the validity of TiO$_2$ remediation in real matrices under UV-A or solar light compared with other advanced oxidation processes, such as ZnO [36] WO$_3$ and ZrO$_2$ [34] semiconductors, ozone [32], gamma irradiation [35], UV/chlorine process [31], photo-Fenton [30] and electrochemical oxidation process [33]. For example, ZnO has a low stability in aqueous solutions at natural pH [36]; WO$_3$ and ZrO$_2$ require strong acidic conditions to work [34]. Other AOPs, such radiolysis [35] or photo-Fenton [30], need the addition of a sacrificial oxidizing agent (H$_2$O$_2$, S$_2$O$_8^{2-}$, etc.) to improve their photocatalytic performance; ozonization [32] and electrochemical oxidation process [33] require acidic and alkaline solutions, respectively, and UV/chlorine process [31] occurs under UV-C irradiation.

None of the above-mentioned AOPs is competitive in terms of GCs degradation rate if compared to TiO$_2$, and, importantly, they have never been applied to GCs degradation at μg L^{-1} levels in environmental conditions.

3.2. Matrix Effects: Salts and Humic Acids

As opposed to photolysis, GCs TiO$_2$-mediated photodegradation took place slightly slower in river samples than in tap water samples. It was faster under natural sunlight in river water than simulated light (see Table 2), proving the process efficiency despite changeable environmental factors, such as incident power [39]. Despite the similar qualitative composition of the two considered matrices (see Table 1), a decrease in k_{deg} was observed for all GCs in river water samples than in tap water samples under the same experimental conditions (see Table 2). We attempted to evaluate the potential matrix contribution to this effect, although, in general, this is a challenging task, especially in real matrices where the many species present may interact with each other and also with several other xenobiotics [40].

Among the various macro-constituents, nitrate and sulfate were investigated because they are more abundant than others in the considered river water sample (see Table 1) and, at the same time, are known to affect drug photodegradation [40], including GCs. Indeed, some previous works reported the influence of SO$_4^{2-}$ and NO$_3^-$ on the AOPs degradation rate of GCs. In particular, SO$_4^{2-}$ was found to reduce the photodegradation rate of BETA [36], PREDLO [31], and PRED [32], acting directly as hydroxyl radical scavenger to form the respective anion radicals; NO$_3^-$ was reported to slow down the decomposition rate of DEXA decreasing OH radical concentration indirectly via NO$_2^-$ [35]. Other important interferents naturally present in river water are humic acids (HAs); these ubiquitous species can act both as photo-sensitizers or compete in absorbing solar light [40,41]. Recent studies examined the inhibition effect of natural organic matter (NOM) on the degradation rate of PREDLO [31] and showed that increasing the NOM concentration, k_{deg} decreased.

Based on the k_{deg} values reported in Table 2 and on the GCs structures, three model GCs, viz. CORT, PREDLO, and TRIAM were selected, and their photodegradation was investigated in tap water fortified with the above-mentioned inorganic salts and HAs, following the procedure described in Section 2.3. Salt concentrations (some tens of mg per liter) higher than those measured in the river water samples were specifically added to evidence their potential impact on the GCs degradation rate, while HAs levels, corresponding to 7.5 mg L^{-1} of total organic carbon (TOC), were added to mimic the natural organic matter present in the river sample (see Table 1). Irradiation was carried out in tap water by a solar light simulator. The degradation rates were calculated and reported in Table 4.

Table 4. Photodegradation constants (k_{deg}) determined in tap water under simulated sunlight in the presence of inorganic salts and humic acids (HAs).

Compound	CORT	PREDLO	TRIAM
Matrix	k_{deg} (min^{-1})		
tw	0.35 (3)	0.51 (1)	0.355 (9)
tw + SO$_4$$^{2-}$ (50 mg L^{-1})	0.41 (3)	0.539 (5)	0.31 (1)
tw + NO$_3$$^-$ (20 mg L^{-1})	0.28 (1)	0.45 (1)	0.37 (1)
tw + HAs (10 mg L^{-1})	0.10 (1)	0.10 (1)	0.08 (1)

As showed in Table 4, the added SO_4^{2-} (50 mg L^{-1}) had a small or no effect on the degradation rates of the chosen GCs, while NO_3^- (20 mg L^{-1}) slightly affected the degradation rates, in the order CORT > PREDLO > TRIAM. On the contrary, HAs (10 mg L^{-1}) had the most significant impact on all drugs, from 4- to 5-fold decrease of the rate. Interestingly, no significant differences in k_{deg} were observed in a more diluted HAs solution (5 mg L^{-1}) for the considered GCs. As expected, competitive light absorption by dissolved organic matter deviated significantly the process from the photocatalytic path.

3.3. Identification of Photoproducts

As mentioned above, all the studied GCs underwent a fast decomposition both in tap and river water.

HPLC-ESI-MS/MS qualitative analysis revealed many photoproducts (see Supplementary Materials), and some trends can be identified based on the structure of the starting GCs. Six main reaction pathways can be described, involving either the hydroxyacetyl side chain or the steroid ring (see Scheme 4) and two of them, viz. (iv), cyclohexadienone isomerization, only available for those GCs containing a second conjugated double bond, and (v), the photooxidation of C17 side chain, have been already described as the main reaction paths in the direct photolysis of glucocorticoids ([23] and references therein). Nonetheless, the products coming from these two routes are now a minority among the identified degradation mechanisms, and we were able to characterize byproducts from path (iv) only for TRIAM (see Supplementary Materials), while the importance of path (v) is highly reduced with respect to direct photolysis, especially for those GCs having a second conjugated double bond on ring A.

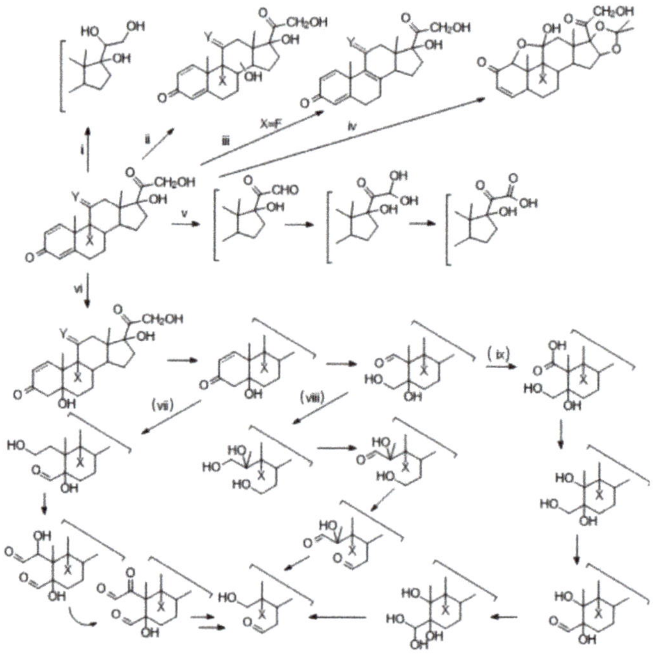

Scheme 4. Paths followed in the photocatalytic degradation of GCs.

Among the remaining four main pathways, route (i) is a minority process only observed for CORT, HCORT, and PRED, and it comes from the formal reduction of the carbonyl at C17. The same is true for path (iii), which led to the reductive dehalogenation of DEXA. The latter reaction has been previously described for DEXA-phosphate photochemistry in the presence of a reducing agent [42]. Both path (i) and (ii) can be related to the free electrons generated on the catalyst surface upon irradiation reacting with adsorbed GCs at the catalyst surface. Nevertheless, most of the identified photoproducts are peculiar of the photocatalytic process, not produced under direct photolysis [23], and can be directly linked to OH radicals photo-generated from the irradiated TiO_2 catalyst. This powerful oxidant can then attack the steroid backbone either on C4=C5 double bond on ring A, path (vi), leading to double bond hydration followed by ring-opening, as already reported in the literature for PREDLO [29], then starting a cascade reaction (path vii, viii, and ix), ultimately yielding highly degraded products, or via OH insertion on the internal rings, (ii) producing hydroxylated byproducts [27]. The preference toward one or more of these degradation paths for the seven studied GCs can be rationalized by looking at their molecular structure, where the not-conjugated CORT and HCORT prefer to undergo path (ii) or (v), so reacting either at the C17 hydroxyacetyl side chain or via OH insertion on the inner steroid rings, while the GCs bearing a second double bond on ring A shift the preferred OH attack to this site (path vi), resulting in a C5–OH pivotal intermediate which then further reacts (path vii, viii, and ix) forming several different oxidized byproducts. Nonetheless, path (ii) and (v) do not completely disappear but contribute to the overall photodegradation process to a lesser extent. In accordance with the proposed mechanism, we saw that the addition of NO_3^- (20 mg L^{-1}) to tap water impacts the degradation rates in the order CORT > PREDLO > TRIAM; this can be tentatively rationalized based on NO_3^- scavenging effects on both OH radicals and photogenerated e_{aq}^-, involving mostly the single conjugated CORT chemistry where the lack of the second double bond impairs competitive degradation paths, viz. double bond hydration (vi) and cyclohexadienone isomerization (iv). On the contrary, the effect of HAs is independent of the chemical structure of the studied GCs and can be explained with HAs competing both with TiO_2 for light absorption and with GCs for the adsorption sites on the catalyst surface. As previously mentioned, other secondary processes are still available for those GCs having particular structures, viz. reductive defluorination for DEXA, cyclohexadienone isomerization for TRIAM.

4. Conclusions

The present study proves that TiO_2-assisted photocatalysis effectively removes various GCs present in freshwaters at µg per liter levels. Degradations occurred within 15 min for all GCs, and they were found to be from one to two orders of magnitude higher than direct photolysis, also in the case of CORT and HCORT that scarcely absorb solar light. The photoproducts formed during the first steps of the photocatalytic process, with a lifetime comparable to that of their parent compounds, were identified and allowed to elucidate the general mechanistic path. As indicated above, two main paths, different from those observed in the direct photolysis experiments, dominate the GCs photooxidation, involving either the unsaturated ketone moiety in ring A, especially for those GCs having a second conjugated double bond, or the C-H bonds on the steroid ring, with a limited number of isomers, which ultimately demonstrate that the process initiates from the TiO_2-drug complex.

In contrast to what was seen from photolytic experiments, we found that GCs photodegradation in the presence of TiO_2 proceeds slightly faster in tap water than in river water due to natural interferents such as HAs and inorganic anions more abundant in the latter. Nonetheless, TiO_2 photodegradation remained effective and faster than direct photolysis, also in the presence of matrix constituents, such as natural organic matter and salts, ubiquitous species in freshwaters. Furthermore, the reaction was shown to be efficient under natural solar light thanks to the photocatalyst ability to absorb in a wide range of wavelengths. The degradation mechanism is dominated by OH radicals generated on the catalyst surface, which then react with adsorbed GCs following different paths based on the particular chemical structure of each drug, while products from direct photolysis are still present but as a minority. Furthermore, even if the kinetic reaction constants are comparable across all the studied GCs, from an environmental point of view, it is important to underline that those compounds bearing a second conjugated double bond on ring A have access to a wider range of degradation paths, leading in the end to the formation of more degraded byproducts which mostly lost the pharmacologically active steroid ring.

Once again, photocatalysis was shown to be the method of choice for water depollution. The mechanism may change somewhat, but the great stability of the titania crystals under irradiation, coupled with the good adsorption properties and catalytic activity, overcome any other method.

Supplementary Materials: The following are available online at https://www.mdpi.com/2076-3417/11/6/2486/s1, Table S1: Optimized MRM conditions for the HPLC-ESI-MS/MS analysis, Table S2: Fragmentation of photolytic products of CORT ($[M + 1]^+ = 361$), Table S3: Fragmentation of photolytic products of HCORT ($[M + 1]^+ = 363$), Table S4: Fragmentation of photolytic products of BETA ($[M + 1]^+ = 393$). Table S5: Fragmentation of photolytic products of DEXA ($[M + 1]^+ = 393$), Table S6: Fragmentation of photolytic products of PRED ($[M + 1]^+ = 359$), Table S7: Fragmentation of photolytic products of PREDLO ($[M + 1]^+ = 361$), Table S8: Fragmentation of photolytic products of TRIAM ($[M + 1]^+ = 435$).

Author Contributions: Conceptualization, M.S., A.A., L.P. and F.M.; formal analysis, M.S.; investigation, A.C. and S.N.; writing—original draft preparation, M.S., L.P. and A.A.; writing—review and editing, M.S., L.P., F.M. and S.N.; visualization, M.S., L.P., A.C. and F.M. All authors have read and agreed to the published version of the manuscript.

Funding: This research received no external funding.

Institutional Review Board Statement: Not applicable.

Informed Consent Statement: Not applicable.

Data Availability Statement: The data presented in this study are available on request from the corresponding author.

Conflicts of Interest: The authors declare no conflict of interest.

References

1. Ramamoorthy, S.; Cidlowski, J.A. Corticosteroids. *Rheum. Dis. Clin. N. Am.* **2016**, *42*, 15–31. [CrossRef] [PubMed]
2. Oakley, R.H.; Cidlowski, J.A. The biology of the glucocorticoid receptor: New signaling mechanisms in health and disease. *J. Allergy Clin. Immunol.* **2013**, *132*, 1033–1044. [CrossRef] [PubMed]
3. Bolster, M.B. Corticosteroids: Friends and Foes. *Rheum. Dis. Clin. N. Am.* **2016**, *42*, xv–xvi. [CrossRef] [PubMed]
4. Jia, A.; Wu, S.; Daniels, K.D.; Snyder, S.A. Balancing the Budget: Accounting for Glucocorticoid Bioactivity and Fate during Water Treatment. *Environ. Sci. Technol.* **2016**, *50*, 2870–2880. [CrossRef]
5. Duclos, M. Glucocorticoids: A Doping Agent? *Endocrinol. Metab. Clin. N. Am.* **2010**, *39*, 107–126. [CrossRef]
6. EEC council directive No. 96/22/EC. *Off. J. Eur. Commun. L* **1996**, *125*. Available online: https://eur-lex.europa.eu/legal-content/EN/ALL/?uri=CELEX%3A31996L0023 (accessed on 14 February 2021).
7. Weizel, A.; Schlüsener, M.P.; Dierkes, G.; Wick, A.; Ternes, T.A. Analysis of the aerobic biodegradation of glucocorticoids: Elucidation of the kinetics and transformation reactions. *Water Res.* **2020**, *174*, 115561. [CrossRef] [PubMed]
8. Kugathas, S.; Williams, R.J.; Sumpter, J.P. Prediction of environmental concentrations of glucocorticoids: The River Thames, UK, as an example. *Environ. Int.* **2012**, *40*, 15–23. [CrossRef]
9. Herrero, P.; Borrull, F.; Pocurull, E.; Marcé, R.M. Determination of glucocorticoids in sewage and river waters by ultra-high performance liquid chromatography–tandem mass spectrometry. *J. Chromatogr. A* **2012**, *1224*, 19–26. [CrossRef]
10. Weizel, A.; Schlüsener, M.P.; Dierkes, G.; Ternes, T.A. Occurrence of Glucocorticoids, Mineralocorticoids, and Progestogens in Various Treated Wastewater, Rivers, and Streams. *Environ. Sci. Technol.* **2018**, *52*, 5296–5307. [CrossRef]
11. Speltini, A.; Merlo, F.; Maraschi, F.; Sturini, M.; Contini, M.; Calisi, N.; Profumo, A. Thermally condensed humic acids onto silica as SPE for effective enrichment of glucocorticoids from environmental waters followed by HPLC-HESI-MS/MS. *J. Chromatogr. A* **2018**, *1540*, 38–46. [CrossRef]
12. Wu, S.; Jia, A.; Daniels, K.D.; Park, M.; Snyder, S.A. Trace analysis of corticosteroids (CSs) in environmental waters by liquid chromatography–tandem mass spectrometry. *Talanta* **2019**, *195*, 830–840. [CrossRef]
13. Fan, Z.; Wu, S.; Chang, H.; Hu, J. Behaviors of Glucocorticoids, Androgens and Progestogens in a Municipal Sewage Treatment Plant: Comparison to Estrogens. *Environ. Sci. Technol.* **2011**, *45*, 2725–2733. [CrossRef] [PubMed]
14. Chang, H.; Hu, J.; Shao, B. Occurrence of Natural and Synthetic Glucocorticoids in Sewage Treatment Plants and Receiving River Waters. *Environ. Sci. Technol.* **2007**, *41*, 3462–3468. [CrossRef] [PubMed]
15. Shen, X.; Chang, H.; Sun, Y.; Wan, Y. Determination and occurrence of natural and synthetic glucocorticoids in surface waters. *Environ. Int.* **2020**, *134*, 105278. [CrossRef]
16. Gong, J.; Lin, C.; Xiong, X.; Chen, D.; Chen, Y.; Zhou, Y.; Wu, C.; Du, Y. Occurrence, distribution, and potential risks of environmental corticosteroids in surface waters from the Pearl River Delta, South China. *Environ. Pollut.* **2019**, *251*, 102–109. [CrossRef] [PubMed]
17. Schriks, M.; van Leerdam, J.A.; van der Linden, S.C.; van der Burg, B.; van Wezel, A.P.; de Voogt, P. High-Resolution Mass Spectrometric Identification and Quantification of Glucocorticoid Compounds in Various Wastewaters in The Netherlands. *Environ. Sci. Technol.* **2010**, *44*, 4766–4774. [CrossRef] [PubMed]
18. Guedes-Alonso, R.; Santana-Viera, S.; Montesdeoca-Esponda, S.; Afonso-Olivares, C.; Sosa-Ferrera, Z.; Santana-Rodríguez, J.J. Application of microwave-assisted extraction and ultra-high performance liquid chromatography–tandem mass spectrometry for the analysis of sex hormones and corticosteroids in sewage sludge samples. *Anal. Bioanal. Chem.* **2016**, *408*, 6833–6844. [CrossRef]
19. Willi, R.A.; Faltermann, S.; Hettich, T.; Fent, K. Active Glucocorticoids Have a Range of Important Adverse Developmental and Physiological Effects on Developing Zebrafish Embryos. *Environ. Sci. Technol.* **2018**, *52*, 877–885. [CrossRef] [PubMed]
20. Willi, R.A.; Castiglioni, S.; Salgueiro-González, N.; Furia, N.; Mastroianni, S.; Faltermann, S.; Fent, K. Physiological and Transcriptional Effects of Mixtures of Environmental Estrogens, Androgens, Progestins, and Glucocorticoids in Zebrafish. *Environ. Sci. Technol.* **2020**, *54*, 1092–1101. [CrossRef] [PubMed]
21. Xin, N.; Jiang, Y.; Liu, S.; Zhou, Y.; Cheng, Y. Effects of prednisolone on behavior and hypothalamic–pituitary–interrenal axis activity in zebrafish. *Environ. Toxicol. Pharmacol.* **2020**, *75*, 103325. [CrossRef]
22. Willi, R.A.; Salgueiro-González, N.; Carcaiso, G.; Fent, K. Glucocorticoid mixtures of fluticasone propionate, triamcinolone acetonide and clobetasol propionate induce additive effects in zebrafish embryos. *J. Hazard. Mater.* **2019**, *374*, 101–109. [CrossRef] [PubMed]
23. Cantalupi, A.; Maraschi, F.; Pretali, L.; Albini, A.; Nicolis, S.; Ferri, E.N.; Profumo, A.; Speltini, A.; Sturini, M. Glucocorticoids in freshwaters: Degradation by solar light and environmental toxicity of the photoproducts. *Int. J. Environ. Res. Public Health* **2020**, *17*, 8717. [CrossRef] [PubMed]
24. Gmurek, M.; Olak-Kucharczyk, M.; Ledakowicz, S. Photochemical decomposition of endocrine disrupting compounds – A review. *Chem. Eng. J.* **2017**, *310*, 437–456. [CrossRef]

25. Serpone, N.; Artemev, Y.M.; Ryabchuk, V.K.; Emeline, A.V.; Horikoshi, S. Light-driven advanced oxidation processes in the disposal of emerging pharmaceutical contaminants in aqueous media: A brief review. *Curr. Opin. Green Sustain. Chem.* **2017**, *6*, 18–33. [CrossRef]
26. Pretali, L.; Maraschi, F.; Cantalupi, A.; Albini, A.; Sturini, M. Water Depollution and Photo-Detoxification by Means of TiO$_2$: Fluoroquinolone Antibiotics as a Case Study. *Catalysts* **2020**, *10*, 628. [CrossRef]
27. Calza, P.; Pelizzetti, E.; Brussino, M.; Baiocchi, C. Ion trap tandem mass spectrometry study of dexamethasone transformation products on light activated TiO$_2$ surface. *J. Am. Soc. Mass Spectrom.* **2001**, *12*, 1286–1295. [CrossRef]
28. Romão, J.S.; Hamdy, M.S.; Mul, G.; Baltrusaitis, J. Photocatalytic decomposition of cortisone acetate in aqueous solution. *J. Hazard. Mater.* **2015**, *282*, 208–215. [CrossRef] [PubMed]
29. Klauson, D.; Pilnik-Sudareva, J.; Pronina, N.; Budarnaja, O.; Krichevskaya, M.; Käkinen, A.; Juganson, K.; Preis, S. Aqueous photocatalytic oxidation of prednisolone. *Open Chem.* **2013**, *11*, 1620–1633. [CrossRef]
30. Díez, A.M.; Ribeiro, A.S.; Sanromán, M.A.; Pazos, M. Optimization of photo-Fenton process for the treatment of prednisolone. *Environ. Sci. Pollut. Res.* **2018**, *25*, 27768–27782. [CrossRef] [PubMed]
31. Yin, K.; He, Q.; Liu, C.; Deng, Y.; Wei, Y.; Chen, S.; Liu, T.; Luo, S. Prednisolone degradation by UV/chlorine process: Influence factors, transformation products and mechanism. *Chemosphere* **2018**, *212*, 56–66. [CrossRef]
32. He, X.; Huang, H.; Tang, Y.; Guo, L. Kinetics and mechanistic study on degradation of prednisone acetate by ozone. *J. Environ. Sci. Health Part A* **2020**, *55*, 292–304. [CrossRef] [PubMed]
33. Welter, J.B.; da Silva, S.W.; Schneider, D.E.; Rodrigues, M.A.S.; Ferreira, J.Z. Performance of Nb/BDD material for the electrochemical advanced oxidation of prednisone in different water matrix. *Chemosphere* **2020**, *248*, 126062. [CrossRef] [PubMed]
34. Ghenaatgar, A.; Tehrani, R.M.A.; Khadir, A. Photocatalytic degradation and mineralization of dexamethasone using WO$_3$ and ZrO$_2$ nanoparticles: Optimization of operational parameters and kinetic studies. *J. Water Process Eng.* **2019**, *32*, 100969. [CrossRef]
35. Guo, Z.; Guo, A.; Guo, Q.; Rui, M.; Zhao, Y.; Zhang, H.; Zhu, S. Decomposition of dexamethasone by gamma irradiation: Kinetics, degradation mechanisms and impact on algae growth. *Chem. Eng. J.* **2017**, *307*, 722–728. [CrossRef]
36. Giahi, M.; Taghavi, H.; Habibi, S. Photocatalytic degradation of betamethasone sodium phosphate in aqueous solution using ZnO nanopowder. *Russ. J. Phys. Chem. A* **2012**, *86*, 2003–2007. [CrossRef]
37. Thomas, A.G.; Syres, K.L. Adsorption of organic molecules on rutile TiO$_2$ and anatase TiO$_2$ single crystal surfaces. *Chem. Soc. Rev.* **2012**, *41*, 4207. [CrossRef] [PubMed]
38. Sushko, M.L.; Gal, A.Y.; Shluger, A.L. Interaction of organic molecules with the TiO$_2$ (110) surface: Ab inito calculations and classical force fields. *J. Phys. Chem. B* **2006**, *110*, 4853–4862. [CrossRef] [PubMed]
39. Sturini, M.; Speltini, A.; Maraschi, F.; Profumo, A.; Pretali, L.; Fasani, E.; Albini, A. Photochemical Degradation of Marbofloxacin and Enrofloxacin in Natural Waters. *Environ. Sci. Technol.* **2010**, *44*, 4564–4569. [CrossRef] [PubMed]
40. Petala, A.; Mantzavinos, D.; Frontistis, Z. Impact of water matrix on the photocatalytic removal of pharmaceuticals by visible light active materials. *Curr. Opin. Green Sustain. Chem.* **2021**, *28*, 100445. [CrossRef]
41. Vione, D. Photochemical reactions in sunlit surface waters: Influence of water parameters, and implications for the phototransformation of xenobiotic compounds. *Photochemistry* **2016**, *44*, 348–363.
42. Rasolevandi, T.; Naseri, S.; Azarpira, H.; Mahvi, A.H. Photo-degradation of dexamethasone phosphate using UV/Iodide process: Kinetics, intermediates, and transformation pathways. *J. Mol. Liq.* **2019**, *295*, 111703. [CrossRef]

Article

Ecotoxicological Evaluation of Methiocarb Electrochemical Oxidation

Annabel Fernandes [1,*], Christopher Pereira [1], Susana Coelho [2,3], Celso Ferraz [2,3], Ana C. Sousa [4,*], M. Ramiro Pastorinho [3,5,6], Maria José Pacheco [1], Lurdes Ciríaco [1] and Ana Lopes [1]

[1] FibEnTech-UBI, Department of Chemistry, Universidade da Beira Interior, 6201-001 Covilhã, Portugal; christophersp1992@gmail.com (C.P.); mjap@ubi.pt (M.J.P.); lciriaco@ubi.pt (L.C.); analopes@ubi.pt (A.L.)
[2] Health Sciences Research Centre (CICS), Universidade da Beira Interior, 6200-506 Covilhã, Portugal; susana_coelho_1@hotmail.com (S.C.); celsoferraz16@gmail.com (C.F.)
[3] NuESA-Health and Environment Study Unit, Faculty of Health Sciences, Universidade da Beira Interior, 6200-506 Covilhã, Portugal; rpastorinho@uevora.pt
[4] CICECO-Aveiro Institute of Materials, Department of Chemistry, Universidade de Aveiro, 3810-193 Aveiro, Portugal
[5] Department of Biology, Universidade de Évora, 7002-554 Évora, Portugal
[6] Comprehensive Health Research Centre (CHRC), 7002-554 Évora, Portugal
* Correspondence: annabelf@ubi.pt (A.F.); anasousa@ua.pt (A.C.S.); Tel.: +351275241329 (A.F.); +351234370200 (A.C.S.)

Received: 21 September 2020; Accepted: 21 October 2020; Published: 22 October 2020

Abstract: The ecotoxicity of methiocarb aqueous solutions treated by electrochemical oxidation was evaluated utilizing the model organism *Daphnia magna*. The electrodegradation experiments were performed using a boron-doped diamond anode and the influence of the applied current density and the supporting electrolyte (NaCl or Na_2SO_4) on methiocarb degradation and toxicity reduction were assessed. Electrooxidation treatment presented a remarkable efficiency in methiocarb complete degradation and a high potential for reducing the undesirable ecological effects of this priority substance. The reaction rate followed first-order kinetics in both electrolytes, being more favorable in a chloride medium. In fact, the presence of chloride increased the methiocarb removal rate and toxicity reduction and favored nitrogen removal. A 200× reduction in the acute toxicity towards *D. magna*, from 370.9 to 1.6 toxic units, was observed for the solutions prepared with NaCl after 5 h treatment at 100 A m^{-2}. An increase in the applied current density led to an increase in toxicity towards *D. magna* of the treated solutions. At optimized experimental conditions, electrooxidation offers a suitable solution for the treatment and elimination of undesirable ecological effects of methiocarb contaminated industrial or agricultural wastewaters, ensuring that this highly hazardous pesticide is not transferred to the aquatic environment.

Keywords: acute toxicity; *Daphnia magna*; electrochemical oxidation; emerging contaminants; methiocarb

---

1. Introduction

Emerging contaminants including some pharmaceuticals, personal care products and pesticides are a cause of concern due to their ubiquity in the aquatic environment and their potential for causing deleterious ecological effects [1–4]. This has led the European Union to establish a surveillance mechanism for the monitoring of substances that show a possible risk to the environment. In 2018, the Commission Implementing Decision (EU) 2018/840 was published, establishing a watch list of substances for Union-wide monitoring in the field of water policy, which includes hormones, antibiotics and pesticides [5].

One of the priority substances identified by the Commission Implementing Decision (EU) 2018/840 watch list is methiocarb (MC), $C_{11}H_{15}NO_2S$, a carbamate pesticide employed in agriculture worldwide, which has been detected in natural waters and in wastewaters [5–9]. Due to MC and some of its metabolites' high toxicity, the World Health Organization classified MC as a highly hazardous pesticide, being also strictly regulated by the European Union [10–12]. Although the presence of methiocarb in waters and wastewaters at concentrations ranging from ng L^{-1} to µg L^{-1} is usually associated with its use, the industrial production of pesticides and herbicides generates a large volume of wastewaters with a high concentration of these pollutants that, if not properly treated, will also contaminate the water bodies [13,14].

According to the literature, conventional wastewater treatments are uncapable of removing pollutants such as MC [6,11]. Thus, new treatment technologies have been studied to achieve the complete degradation of these contaminants to environmentally benign, non-toxic and biodegradable products and prevent their discharge in the aquatic environment [15–19]. Methiocarb oxidation through ClO_2, NaOCl and NH_2Cl has been studied by Qiang et al. [20] and Tian et al. [21,22]. The authors found that, despite the effective elimination of MC, a significant increase in toxicity due to the formation of more toxic degradation products occurred. By applying a vacuum ultraviolet/ultraviolet (VUV/UV) process in the degradation of several pesticides including MC, Yang et al. [23] observed a reduction in the toxicity of the solutions although the pesticides' mineralization rates were considerably lower than their degradation rates. Cruz-Alcalde et al. [11] studied the removal of MC by ozonation and observed that during the degradation experiments, direct ozone reaction was effective to remove MC but not its formed intermediates whereas hydroxyl radicals could oxidize all species. In fact, treatment technologies based on oxidation through hydroxyl radicals have been widely studied because hydroxyl radicals are one of the most powerful oxidizing species that are able to react unselectively and instantaneously with the surrounding organic pollutants [24]. Among these technologies, electrochemical processes have received great attention due to their versatility, ease in operation and, specially, their high efficiency in the degradation of recalcitrant compounds [17,25]. Several authors have reported the removal of pesticides and herbicides using different electrochemical methods including electrochemical oxidation (EO), electro-Fenton (EF), photoelectro-Fenton (PEF) and electrochemical peroxidation [13,14,26–28]. In the degradation of solutions containing 100 mg L^{-1} of the herbicide clopyralid, Santos et al. [14] found that, among EO with electrogenerated hydrogen peroxide, EF and PEF processes, PEF led to the most attractive results with almost a complete transformation of clopyralid in inorganic ions. In a different study, Silva et al. [26] applied an EO process utilizing dimensionally stable anodes (DSA) to degrade the insecticide imidacloprid at a concentration of approximately 25 mg L^{-1} and achieved its complete oxidation (100% mineralization). According to these authors, the imidacloprid degradation occurred via direct anodic oxidation and Cl_2-mediated oxidation and the breakdown of the insecticide's molecular structure and formation of short chain compounds occurred before its complete mineralization. In a study performed by Martínez-Huitle et al. [28], the performance of different anode materials, Pb/PbO_2, Ti/SnO_2 and a boron-doped diamond anode (BDD), was evaluated during the electrochemical oxidation of the pesticide methamidophos at a concentration of 50 mg L^{-1}. A few authors have found that, using Pb/PbO_2 and Ti/SnO_2, formaldehyde appeared as a product of the reaction, giving evidence of an indirect mineralization mechanism. However, when BDD was used, formaldehyde production was not detected and the formation of phosphate was observed instead, indicating the complete mineralization of methamidophos. In fact, EO using a BDD anode has been regarded as the electrochemical process of choice with outstanding degradation results that are due to the higher oxidation ability of BDD compared with other anode materials [29–33].

Although the application of the EO process in the treatment of waters and wastewaters appears as an attractive alternative for the degradation of biorefractory or priority pollutants such as pesticides and herbicides, the influence of the EO treatment in the toxicity of the treated solutions is not well known because the removal of the contaminants might not always correspond to a quantitative reduction of the toxic effects. The physicochemical and biological parameters usually determined to evaluate the quality of wastewaters are not sufficient to assess the toxicity to organisms where ecotoxicological tests with aquatic organisms are required to better characterize the toxicity removal and prevent undesirable ecological effects in the aquatic environment caused by wastewater discharge [34]. Hence, the present work aims to fill this gap by evaluating the ecotoxicity of methiocarb solutions treated by EO using the freshwater crustacean *Daphnia magna* as the model species, as endorsed by international organizations including the American Society for Testing and Materials (ASTM, USA), the Organization for Economic Cooperation and Development (OECD, France) and the International Organization for Standardization (ISO, Switzerland). In the particular case of the highly toxic pesticide methiocarb, there are no studies addressing its degradation by electrochemical oxidation with a subsequential ecotoxicological evaluation. Thus, electrochemical oxidation experiments utilizing a BDD anode were performed with MC aqueous solutions at a concentration of 20 mg L^{-1}. In order to assess the influence in the toxicity results of two major experimental conditions that greatly affect the EO performance, MC degradation was carried out at two different applied current densities (j), 100 and 300 A m^{-2}, utilizing two different supporting electrolytes, NaCl and Na_2SO_4. Although the MC concentrations reported for waters and wastewaters are in the range of ng L^{-1} to μg L^{-1}, the concentration of 20 mg L^{-1} was selected for this study with the aim of evaluating the ability of the EO process to reduce the high toxicity of this priority substance in an extreme-case scenario and attending to the concentrations found in the literature for electrochemical studies involving pesticides and herbicides. In addition, in the cases where membrane filtration processes are utilized to purify the effluents, the concentrations in the permeate may reach values much higher than those usually found at the entrance of the effluent treatment plants.

2. Materials and Methods

2.1. Methiocarb Aqueous Solutions Composition and Characterization

The composition and characterization of the methiocarb aqueous solutions used in this study are described in Table 1. MC solutions were prepared in ultrapure water obtained with a Milli-Q system using the methiocarb PESTANAL®, analytical standard (CAS Number 2032-65-7), purchased from Sigma-Aldrich. As the MC aqueous solution did not have enough electrical conductivity to pursue with the electrochemical experiments at the applied current densities, the addition of a background electrolyte was required. Two different salts were studied as background electrolytes, sodium sulfate anhydrous (CAS Number 7757-82-6) of 99.7% purity and sodium chloride ACS reagent (CAS Number 7647-14-5), purity $\geq$ 99.0%. Both were purchased from Sigma-Aldrich. The concentration of added salt was set as the minimum required to perform the EO assays at the highest applied current density studied (300 A m^{-2}), which was found to be 250 mg L^{-1}, corresponding to an increase of the solution's electrical conductivity to approximately 500 μS cm^{-1}.

Table 1. Composition and characterization of the methiocarb aqueous solutions used in the experiments.

Solution		MC + Na_2SO_4	MC + NaCl
Composition	Methiocarb	20	20
mg L^{-1}	Na_2SO_4	250	
	NaCl	–	250
Chemical oxygen demand/mg L^{-1}		43 ± 1	43 ± 1
Total organic carbon/mg L^{-1}		11.6 ± 0.4	11.8 ± 0.5
Total nitrogen/mg L^{-1}		1.23 ± 0.03	1.24 ± 0.04
pH		5.45 ± 0.03	5.28 ± 0.03
Conductivity/μS cm^{-1}		437 ± 5	514 ± 5
Acute toxicity/	EC_{50} (48 h)	0.21%	0.27%
	TU [1]	470.4	370.9

[1] TU—toxic units (TU = $(1/EC_{50}) \times 100$, TU > 100: highly toxic, 10 < TU < 100: very toxic, 1 < TU < 10: toxic, TU < 1 non-toxic [35]).

2.2. Electrodegradation Experiments

The EO experiments aimed to study the influence of the composition of the supporting electrolyte, of the applied current density and of the electrolysis duration/applied electric charge on the toxicity reduction towards *Daphnia magna* and on the removal rate of methiocarb, total organic carbon (TOC) and total nitrogen (TN). Thus, the EO assays were conducted at room temperature (22–25 °C) in batch mode with stirring (250 rpm), using an open, undivided and cylindrical glass cell (250 mL of capacity) and 200 mL of solution. A commercial Si/BDD anode purchased from Neocoat and a stainless-steel cathode, each one with an immersed area of 10 cm^2, were utilized as electrodes. They were placed in parallel with an inter-electrode gap of 0.3 cm and were centered in the electrochemical cell. The experiments, utilizing the solutions described in Table 1, MC + Na_2SO_4 and MC + NaCl, were performed at two different applied current densities, 100 and 300 A m^{-2}, which was the electric current supplied by a GW, Lab DC, model GPS-3030D (0–30 V, 0–3 A) power source. For the applied current density of 100 A m^{-2}, 5 h and 6 h duration assays were performed, corresponding to applied charges of 1.80 and 2.16 kC, respectively. For the experiments run at 300 A m^{-2}, the assays had 3 h duration, corresponding to 3.24 kC of applied charge. Samples were collected each hour for physicochemical determinations to monitor the experiments. Additionally, in the experiments run at 300 A m^{-2}, a sample was collected after 20 min of starting the assay corresponding to an applied charge of 0.36 kC, which was the same reached after 1 h by the experiments run at 100 A m^{-2}. Toxicological assays were only performed for the initial (MC + Na_2SO_4 and MC + NaCl) and final samples (MC + Na_2SO_4–100 A m^{-2}–1.8 kC (5 h), MC + Na_2SO_4–100 A m^{-2}–2.16 kC (6 h), MC + Na_2SO_4–300 A m^{-2}–3.24 kC (3 h), MC + NaCl–100 A m^{-2}–1.8 kC (5 h), MC + NaCl–100 A m^{-2}–2.16 kC (6 h) and MC + NaCl–300 A m^{-2}–3.24 kC (3 h)). For the kinetic study, assays with 1 h duration were run with samples being collected every 15 min.

All of the EO assays were performed at least in duplicate and results presented are mean values of all of the determinations performed.

2.3. Toxicological Assays

The freshwater crustacean *Daphnia magna* Strauss, Clone K6, was used in the toxicological assays. The daphnids were obtained from a stock culture maintained in the laboratory under standardized conditions. The animals were reared at a constant temperature (20 ± 1 °C) and photoperiod (16:8 h light/dark) in ASTM (American Society of Testing Materials) hard water. The culture medium was changed every other day during weekdays and was supplemented with seaweed extract. The daphnids were fed daily with the green algae *Raphidocelis subcapitata* (3.0 × 10^5 cells mL^{-1}). From this stock culture, individual daphnids were transferred to 100 mL beakers prior to the beginning of the tests. Neonates with less than 24 h from the 3rd to the 6th broods were used. The evaluation of daphnia fitness and the validation of the test were achieved by performing the 24 h acute toxicity test with $K_2Cr_2O_7$ as recommended by the OECD Guideline 202 and fully described by Fernandes et al. [36,37].

The acute toxicity tests were performed following the OEDC Guideline 202 [36]. For each treatment and respective controls, five replicates with five neonates each were tested. Daphnids were exposed to different dilutions of the treated and untreated solutions in a six well plate. For each replicate, 10 mL of the test solutions were used, corresponding to 2 mL per daphnid as recommended by the OECD Guideline [36]. The number of immobilized daphnia was registered after 24 and 48 h and the median effective concentration (EC_{50}) was calculated using the software GraphPad Prism8. All of the toxicity tests were performed within 1 month of the EO treatment and the samples were preserved under dark conditions at −20 °C. Prior to the beginning of the tests, the samples were left to unfreeze at 4 °C.

2.4. Physicochemical Determinations

The characterization of samples was performed by measuring TOC and TN measured in a Shimadzu TOC-V CPH analyzer combined with a TNM-1 unit and by the determination of chemical oxygen demand (COD) using a closed reflux and titrimetric method according to standard procedures [38]. High performance liquid chromatography (HPLC) was also performed for MC determinations using a Shimadzu 20A Prominence HPLC system equipped with a SPD-M20A diode array detector, a CTO-20AC column oven and a LC-20AD pump and using a Purospher STAR RP18 endcapped column (250 × 4 mm (i.d.), 5 μm). The elution was performed isocratically with a mixture of formic acid aqueous solution (0.05%) and acetonitrile, 50:50 (*v/v*), at a flow rate of 1 mL min^{-1} and 35 °C. The injection volume was 20 μL and the detection wavelength was 262 nm. The reagents were Sigma-Aldrich HPLC grade and ultrapure water (Milli-Q system) was utilized to prepare the solutions.

pH (HANNA, HI 931400) and conductivity (Mettler Toledo, SevenEasy S30K) were measured along the assays.

3. Results and Discussion

Figure 1 presents the normalized variation of MC and TOC concentrations with applied charge during the EO assays accomplished with different supporting electrolyte and applied current densities. When the influence of the supporting electrolyte was analyzed, the MC and TOC removal rates were higher for the solutions containing NaCl. In fact, in the presence of chloride, MC was completely degraded during the initial period of the assays for both *j* whereas in the presence of Na$_2$SO$_4$, the MC removal rate was lower and decreased along the assays. This faster degradation in the presence of chloride has been described in the literature for other contaminants and, according to the authors, is due to the additional indirect oxidation in the bulk of the solution by active chlorine species generated from the oxidation of chloride ions at the BDD [33,39].

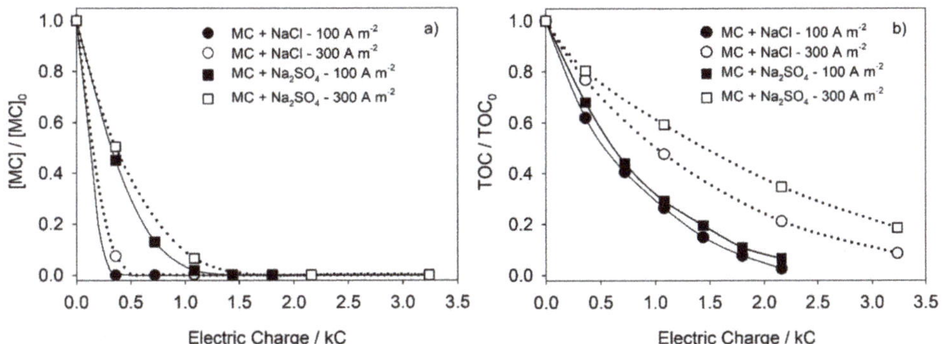

Figure 1. Variation with applied charge of the normalized (**a**) methiocarb (MC) concentration and (**b**) total organic carbon (TOC) for the electrochemical oxidation (EO) experiments performed at 100 and 300 A m^{-2} using Na$_2$SO$_4$ or NaCl as the supporting electrolyte.

To elucidate the reaction kinetics in both electrolytes, the decay in time of the MC concentration determined by HPLC for the first hour of the assays was analyzed. The MC degradation showed a first-order reaction mechanism with the following kinetic constants: k = 0.28 × 10^{-3} s^{-1} (Na$_2$SO$_4$_100 A m^{-2}); k = 0.77 × 10^{-3} s^{-1} (Na$_2$SO$_4$_300 A m^{-2}); k = 1.6 × 10^{-3} s^{-1} (NaCl_100 A m^{-2}); k = 1.7 × 10^{-3} s^{-1} (NaCl_300 A m^{-2}). This showed that, in the chloride medium, the oxidation happened mainly in the bulk of the solution and, because the MC concentration was low, even at the lowest applied current density the chlorine active species formed were enough to oxidize MC at a very good reaction rate. Regarding oxidation in the presence of sulfate, the oxidation mechanism must be either by hydroxyl or sulfate radicals with a lower lifetime, thus giving more importance to MC diffusion towards the anode's surface.

Regarding the TOC, for the highest j, the difference between the removal rates when NaCl or Na_2SO_4 were used was more pronounced than at 100 A m^{-2}. Regardless of that, MC and TOC removals were energetically more efficient at the lowest j. These results indicated that at 300 A m^{-2}, MC and by-product oxidation through hydroxyl radicals were reduced and the oxidation by active chlorine species was enhanced, which was due to the higher amount of active chlorine species available at this j and to the lower amount of hydroxyl radicals available, caused by the augmentation of secondary reactions such as the oxidation of the electrolyte or oxygen and hydrogen evolution [40]. Although the mentioned reactions were always present, their rate increased with applied current densities.

The different degradation mechanisms occurring when NaCl or Na_2SO_4 were used were also evidenced in pH variation along the assays (Figure 2a). When NaCl was used, an increase in pH during the initial period of the assays was found, followed by a decrease to below the initial value and ending in a plateau. The initial increase in pH observed was probably due to the chloride oxidation/water reduction and the subsequent decrease could be a result of the formation of by-products such as short chain carboxylic acids [17]. In the presence of Na_2SO_4, a decrease in pH during the initial period of the assays was observed with a subsequent tendency to a plateau.

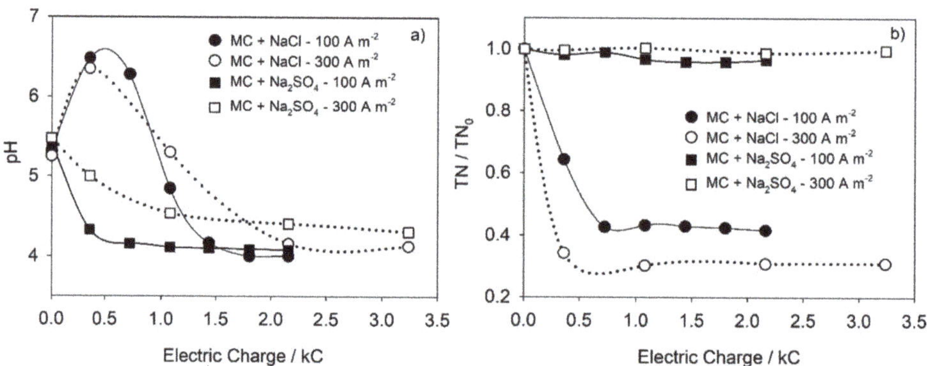

Figure 2. Variation with applied charge of the (**a**) pH and (**b**) normalized total nitrogen (TN) for the EO experiments performed at 100 and 300 A m^{-2} using Na_2SO_4 or NaCl as the supporting electrolyte.

As an MC molecule has nitrogen in its structure, TN variation during the assays was also assessed. A normalized TN variation with an applied charge, presented in Figure 2b, showed that in the presence of Na_2SO_4 no nitrogen removal occurred. However, when NaCl was used, TN concentration decreased by more than 50%. According to the literature, organic nitrogen is converted to NH_4^+ in the EO using a BDD anode, which in turn is oxidized mainly to nitrate and nitrogen gas as NH_4^+ oxidation takes place by indirect oxidation through active chlorine species [41,42]. It was also observed that, for the solutions containing chloride, TN removal was energetically more efficient at the highest current density. These results agreed with the enhancement of the electrolyte oxidation reaction at a higher j because more active chlorine species available lead to a higher TN removal rate.

As MC is a highly toxic compound, the ecotoxicological evaluation of the treatment methods applied to its degradation is mandatory. Thus, the ecotoxicity of the treated samples obtained in this study as well of the initial samples was evaluated towards *Daphnia magna*. As it can be perceived from Figure 3 and Figure S1 (Supplementary Material), the electrochemical oxidation treatment of methiocarb solutions led to a remarkable decrease in the acute toxicity towards *Daphnia magna*. In both MC solutions containing Na_2SO_4 or NaCl, the initial toxicity was very high although solutions prepared with NaCl were slightly less toxic (EC_{50} = 0.21%, 470.4 TUs against EC_{50} = 0.27%, 370.9 TUs for solutions prepared with Na_2SO_4 and NaCl, respectively). Methiocarb is an acetylcholinesterase inhibitor pesticide widely used in agriculture as a molluscicide, acaricide and avicide and its toxicity towards aquatic invertebrates has already been reported [43]. Figure 3a presents the EC_{50} results after 48 h of exposure for the different experimental conditions studied. The higher EC_{50} values were attained for the experiments performed at 100 A m^{-2} using NaCl as the supporting electrolyte. At these experimental conditions, an increase in EC_{50} from 0.27% to 62% was achieved after a 5 h assay (1.8 kC). However, when the assay duration was extended for one more hour to 6 h (2.16 kC) in total, the EC_{50} decreased to 50%. In solutions containing chloride, the increase in toxicity during the electrochemical treatment is usually attributed to the formation of chlorinated compounds [44,45], which exhibit a high octanol/water coefficient (Kow), thus affecting biological membranes and presenting higher toxicity for living organisms [37]. Furthermore, the formation of perchlorate at the final stage of the electrochemical process when using a BDD anode is well described in the literature [46–48]. According to Lacasa et al. [45], the hydroxyl radicals formed in large quantities during the electrolysis of aqueous solutions when a BDD anode is used in the presence of chlorides can be oxidized successively to different oxochlorinated compounds according to Equations (1)–(4).

$$Cl^- + OH^\bullet \rightarrow ClO^- + H^+ + e^-, \tag{1}$$

$$ClO^- + OH^\bullet \rightarrow ClO_2^- + H^+ + e^- \tag{2}$$

$$ClO_2^- + OH^\bullet \rightarrow ClO_3^- + H^+ + e^- \tag{3}$$

$$ClO_3^- + OH^\bullet \rightarrow ClO_4^- + H^+ + e^- \tag{4}$$

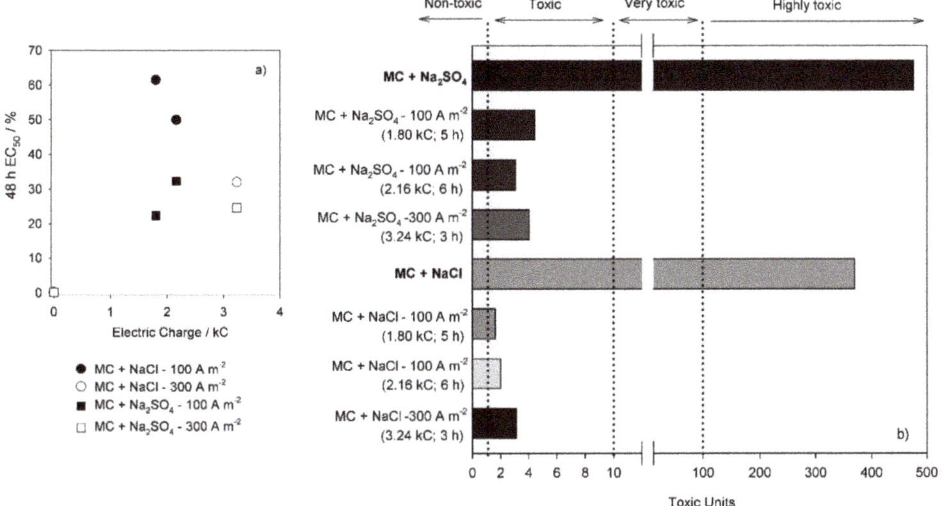

Figure 3. (**a**) EC_{50} results obtained for the treated solutions at 100 and 300 A m^{-2} using Na_2SO_4 or NaCl as the supporting electrolyte. (**b**) Comparison of the toxicity of the different methiocarb solutions in terms of toxic units before and after treatment at the different experimental conditions.

Attending that, in the last hour of the assay, the organic load in the solution is low and its oxidation is under mass transport control, the occurrence of the reactions described by Equations (1)–(4) will be enhanced and the perchlorate concentration in the solution, as the final oxidation product, will increase, explaining the increase in ecotoxicity found. The possible formation of organochlorinated compounds in the last hour of the assay, although in low concentration because the TOC was less than 1 mg L^{-1} at this point, could also be a possible explanation for the increased ecotoxicity.

Both perchlorate and organochlorinated compound formation can also explain the lower ecotoxicity reduction presented by the chloride-containing solutions treated at 300 A m^{-2}. At the applied charge of 3.24 kC (3 h assay at 300 A m^{-2}), EC$_{50}$ increased only to 32%. At higher applied current densities, chloride oxidation and hydroxyl radical formation were enhanced, augmenting the occurrence of the reactions described by Equations (1)–(4) and the possible formation of organochlorinated compounds.

For the solutions prepared with sulfate at 100 A m^{-2}, EC$_{50}$ increased from 0.21% to 22% after a 5 h assay (1.8 kC) and to 32% after 6 h (2.16 kC). These results agreed with the lower MC degradation rate observed for the solutions containing sulfate and indicated that at 1.8 kC there were in the solution toxic MC degradation products that were being further oxidized to less or non-toxic compounds, decreasing the solution toxicity at 2.16 kC. Nevertheless, the increase in applied current density and charge did not result in a further decrease in toxicity. At the applied charge of 3.24 kC (3 h assay at 300 A m^{-2}), EC$_{50}$ increased only to 25% in the sulfate-containing solution. This result can be explained by the energy efficiency loss when j was increased as observed for the TOC removal, mainly due to mass transport limitations.

In terms of toxic units (Figure 3b), it can be seen that, using NaCl as the supporting electrolyte, it was possible to reduce the toxicity more than 200× from 370.9 TUs to 1.6 TUs with 5 h treatment at 100 A m^{-2}, whereas for the same conditions but using Na$_2$SO$_4$, the reduction was only of 107× from 470.4 TUs to 4.4 TUs. At the higher applied current density of 300 A m^{-2}, the difference between the supporting electrolytes was less pronounced, with a similar reduction in toxicity of 120× versus 115× for NaCl and Na$_2$SO$_4$ containing solutions, respectively.

Overall, the results obtained clearly demonstrated the remarkable efficiency of EO treatment for MC complete degradation and its high potential for reducing the toxic ecological effects of this highly hazardous pesticide even when present in concentrations far above those currently reported for waters and wastewaters. According to Pablos et al. [35], the MC solutions went from being classified as highly toxic before treatment to toxic and very close to be considered non-toxic after EO treatment. Although the present study only addressed experiments at a laboratory scale, it offers a useful contribution to further develop the application of the electrochemical oxidation process at an industrial level. In fact, according to the results achieved, EO can be considered suitable for the remediation of contaminated industrial or agricultural wastewaters, enabling the elimination of MC and its adverse effects before they reach the aquatic environment.

4. Conclusions

Electrochemical oxidation with a BDD anode can effectively degrade methiocarb and its transformation by-products, reducing drastically the acute toxicity towards *Daphnia magna*. The MC electrodegradation mechanism depends on the species present in the solution and on the applied current density:

- In the presence of chloride, MC is rapidly transformed as result of its additional indirect oxidation by active chlorine species. These active chlorine species are responsible for the nitrogen elimination, above 50%.
- In the presence of sulfate, the electrooxidation mechanism involves mainly hydroxyl radicals formed at the BDD surface leading to a slow MC elimination because the oxidation rate depends on the MC diffusion to the anode's surface. No nitrogen removal was observed using this supporting electrolyte.
- An increase in current density enhances secondary reaction rates, thus decreasing the energy efficiency. An exception is nitrogen removal in chloride-containing solutions, which is promoted by the increased amount of active chlorine species.

- Chloride-containing solutions lead to higher toxicity reductions than sulfate-containing solutions most likely due to the faster MC degradation and of its transformation products when NaCl is used.
- In chloride-containing solutions, the increase in applied electric charge due to the increase in assay duration or applied current density leads to a lower decrease in ecotoxicity probably due to the formation of perchlorate and organochlorinated compounds.

At optimized experimental conditions, the EO process offers a suitable solution for the treatment and elimination of undesirable ecological effects of MC contaminated industrial or agricultural wastewaters, ensuring that this priority substance is not transferred to the aquatic environment.

Supplementary Materials: The following are available online at http://www.mdpi.com/2076-3417/10/21/7435/s1, Figure S1: Number of immobilized daphnids after 24 and 48h of exposure to the different treatments. Immobilization refers to the inability of animals that are not able to swim within 15 seconds, after gentle agitation of the test vessel [1].

Author Contributions: Conceptualization, A.L.; methodology, A.F. and A.C.S.; validation, A.F. and A.C.S.; investigation, C.P., S.C. and C.F.; resources, M.R.P. and L.C.; writing—original draft preparation, A.F. and A.C.S.; writing—review and editing, M.R.P., M.J.P., L.C. and A.L.; visualization, M.J.P.; supervision, A.L.; project administration, A.F. All authors have read and agreed to the published version of the manuscript.

Funding: This research was funded by Fundação para a Ciência e a Tecnologia, FCT, through UID Fiber Materials and Environmental Technologies (FibEnTech), project UIDB/00195/2020. Further financial support was provided by funds from the Health Sciences Research Center (CICS-UBI) through National Funds by FCT (UID/Multi/00709/2019).

Acknowledgments: A.C.S. acknowledges University of Aveiro for funding in the scope of the framework contract foreseen in the numbers 4, 5 and 6 of the article 23, of the Decree-Law 57/2016, of August 29, changed by Law 57/2017, of July 19 and project CICECO-Aveiro Institute of Materials, UIDB/50011/2020 & UIDP/50011/2020, financed by national funds through the FCT/MEC and when appropriate co-financed by FEDER under the PT2020 Partnership Agreement.

Conflicts of Interest: The authors declare no conflict of interest. The funders had no role in the design of the study; in the collection, analyses, or interpretation of data; in the writing of the manuscript or in the decision to publish the results.

References

1. Ebele, A.J.; Abdallah, M.A.-E.; Harrad, S. Pharmaceuticals and personal care products (PPCPs) in the freshwater aquatic environment. *Emerg. Contam.* **2017**, *3*, 1–16. [CrossRef]
2. Westlund, P.; Yargeau, V. Investigation of the presence and endocrine activities of pesticides found in wastewater effluent using yeast-based bioassays. *Sci. Total Environ.* **2017**, *607*, 744–751. [CrossRef] [PubMed]
3. Krieg, S.A.; Shahine, L.K.; Lathi, R.B. Environmental exposure to endocrine-disrupting chemicals and miscarriage. *Fertil. Steril.* **2016**, *106*, 941–947. [CrossRef] [PubMed]
4. Puckowski, A.; Mioduszewska, K.; Łukaszewicz, P.; Borecka, M.; Caban, M.; Maszkowska, J.; Stepnowski, P. Bioaccumulation and analytics of pharmaceutical residues in the environment: A review. *J. Pharm. Biomed. Anal.* **2016**, *127*, 232–255. [CrossRef] [PubMed]
5. Commission Implementing Decision (EU) 2018/840. *Off. J. Eur. Union* **2018**, L141/9–L141/12. Available online: https://eur-lex.europa.eu/legal-content/EN/TXT/PDF/?uri=CELEX:32018D0840&rid=7 (accessed on 12 June 2020).
6. Campo, J.; Masiá, A.; Blasco, C.; Picó, Y. Occurrence and removal efficiency of pesticides in sewage treatment plants of four Mediterranean River Basins. *J. Hazard. Mater.* **2013**, *263*, 146–157. [CrossRef] [PubMed]
7. Masiá, A.; Ibáñez, M.; Blasco, C.; Sancho, J.; Picó, Y.; Hernández, F.H. Combined use of liquid chromatography triple quadrupole mass spectrometry and liquid chromatography quadrupole time-of-flight mass spectrometry in systematic screening of pesticides and other contaminants in water samples. *Anal. Chim. Acta* **2013**, *761*, 117–127. [CrossRef] [PubMed]
8. Fytianos, K.; Pitarakis, K.; Bobola, E. Monitoring of N-methylcarbamate pesticides in the Pinios River (central Greece) by HPLC. *Int. J. Environ. Anal. Chem.* **2006**, *86*, 131–145. [CrossRef]

9. Squillace, P.J.; Scott, J.C.; Moran, M.J.; Nolan, B.T.; Kolpin, D.W. VOCs, pesticides, nitrate, and their mixtures in groundwater used for drinking water in the United States. *Environ. Sci. Technol.* **2002**, *36*, 1923–1930. [CrossRef]

10. World Health Organization. *The WHO Recommended Classification of Pesticides by Hazard and Guidelines to Classification 2009*; World Health Organization: Geneva, Switzerland, 2010; ISBN 978-92-4-154796-3.

11. Cruz-Alcalde, A.; Sans, C.; Esplugas, S. Exploring ozonation as treatment alternative for methiocarb and formed transformation products abatement. *Chemosphere* **2017**, *186*, 725–732. [CrossRef]

12. Commission Implementing Regulation (EU) 2019/1606. *Off. J. Eur. Union* **2019**, L250/53–L250/55. Available online: https://eur-lex.europa.eu/legal-content/EN/TXT/PDF/?uri=CELEX:32019R1606&from=GA (accessed on 12 June 2020).

13. Camacho, F.G.; De Souza, P.A.L.; Martins, M.L.; Benincá, C.; Zanoelo, E.F. A comprehensive kinetic model for the process of electrochemical peroxidation and its application for the degradation of trifluralin. *J. Electroanal. Chem.* **2020**, *865*, 114163. [CrossRef]

14. Santos, G.D.O.; Eguiluz, K.I.; Salazar-Banda, G.R.; Saez, C.; Rodrigo, M.A. Testing the role of electrode materials on the electro-Fenton and photoelectro-Fenton degradation of clopyralid. *J. Electroanal. Chem.* **2020**, *871*, 114291. [CrossRef]

15. Ferreira, M.B.; Souza, F.; Muñoz-Morales, M.; Sáez, C.; Cañizares, P.; Martínez-Huitle, C.; Rodrigo, M.A. Clopyralid degradation by AOPs enhanced with zero valent iron. *J. Hazard. Mater.* **2020**, *392*, 122282. [CrossRef]

16. Vagi, M.C.; Petsas, A.S. Recent advances on the removal of priority organochlorine and organophosphorus biorecalcitrant pesticides defined by Directive 2013/39/EU from environmental matrices by using advanced oxidation processes: An overview (2007–2018). *J. Environ. Chem. Eng.* **2020**, *8*, 102940. [CrossRef]

17. Moreira, F.C.; Boaventura, R.A.; Brillas, E.; Vilar, V.J. Electrochemical advanced oxidation processes: A review on their application to synthetic and real wastewaters. *Appl. Catal. B Environ.* **2017**, *202*, 217–261. [CrossRef]

18. Barbosa, M.O.; Moreira, N.F.; Ribeiro, A.R.; Pereira, M.F.; Silva, A.M.T. Occurrence and removal of organic micropollutants: An overview of the watch list of EU Decision 2015/495. *Water Res.* **2016**, *94*, 257–279. [CrossRef]

19. Petrovic, M.; Gonzalez, S.; Barceló, D. Analysis and removal of emerging contaminants in wastewater and drinking water. *TrAC Trends Anal. Chem.* **2003**, *22*, 685–696. [CrossRef]

20. Qiang, Z.; Tian, F.; Liu, W.; Liu, C. Degradation of methiocarb by monochloramine in water treatment: Kinetics and pathways. *Water Res.* **2014**, *50*, 237–244. [CrossRef]

21. Tian, F.; Qiang, Z.; Liu, W.; Ling, W. Methiocarb degradation by free chlorine in water treatment: Kinetics and pathways. *Chem. Eng. J.* **2013**, *232*, 10–16. [CrossRef]

22. Tian, F.; Qiang, Z.; Liu, C.; Zhang, T.; Dong, B. Kinetics and mechanism for methiocarb degradation by chlorine dioxide in aqueous solution. *Chemosphere* **2010**, *79*, 646–651. [CrossRef]

23. Yang, L.; Li, M.; Li, W.; Jiang, Y.; Qiang, Z. Bench- and pilot-scale studies on the removal of pesticides from water by VUV/UV process. *Chem. Eng. J.* **2018**, *342*, 155–162. [CrossRef]

24. Gligorovski, S.; Strekowski, R.; Barbati, S.; Vione, D. Environmental Implications of Hydroxyl Radicals (•OH). *Chem. Rev.* **2015**, *115*, 13051–13092. [CrossRef] [PubMed]

25. Martínez-Huitle, C.A.; Panizza, M. Electrochemical oxidation of organic pollutants for wastewater treatment. *Curr. Opin. Electrochem.* **2018**, *11*, 62–71. [CrossRef]

26. Silva, L.M.; Dos Santos, R.P.A.; Morais, C.C.O.; Vasconcelos, C.L.; Martínez-Huitle, C.A.; Castro, S.S.L. Anodic Oxidation of the Insecticide Imidacloprid on Mixed Metal Oxide (RuO2-TiO2 and IrO2-RuO2-TiO2) Anodes. *J. Electrochem. Soc.* **2017**, *164*, E489–E495. [CrossRef]

27. Dhaouadi, A.; Adhoum, N. Degradation of paraquat herbicide by electrochemical advanced oxidation methods. *J. Electroanal. Chem.* **2009**, *637*, 33–42. [CrossRef]

28. Martínez-Huitle, C.A.; De Battisti, A.; Ferro, S.; Reyna, S.; Cerro-López, M.; Quiro, M.A. Removal of the Pesticide Methamidophos from Aqueous Solutions by Electrooxidation using Pb/PbO2, Ti/SnO2, and Si/BDD Electrodes. *Environ. Sci. Technol.* **2008**, *42*, 6929–6935. [CrossRef]

29. Dao, K.C.; Yang, C.-C.; Chen, K.-F.; Tsai, Y.-P. Recent Trends in Removal Pharmaceuticals and Personal Care Products by Electrochemical Oxidation and Combined Systems. *Water* **2020**, *12*, 1043. [CrossRef]

30. Pueyo, N.; Ormad, M.P.; Miguel, N.; Kokkinos, P.; Ioannidi, A.; Mantzavinos, D.; Frontistis, Z. Electrochemical oxidation of butyl paraben on boron doped diamond in environmental matrices and comparison with sulfate radical-AOP. *J. Environ. Manag.* **2020**, *269*, 110783. [CrossRef]

31. Ganiyu, S.O.; Martínez-Huitle, C.A. Nature, Mechanisms and Reactivity of Electrogenerated Reactive Species at Thin-Film Boron-Doped Diamond (BDD) Electrodes During Electrochemical Wastewater Treatment. *ChemElectroChem* **2019**, *6*, 2379–2392. [CrossRef]

32. Martínez-Huitle, C.A.; Brillas, E.; Einaga, Y.; Farrell, J. Trends in Synthetic Diamond for Electrochemical Applications. *ChemElectroChem* **2019**, *6*, 4330–4331. [CrossRef]

33. Pereira, G.F.; Rocha-Filho, R.C.; Bocchi, N.; Biaggio, S.R. Electrochemical degradation of bisphenol A using a flow reactor with a boron-doped diamond anode. *Chem. Eng. J.* **2012**, *198*, 282–288. [CrossRef]

34. Ma, M.; Li, J.; Wang, Z. Assessing the Detoxication Efficiencies of Wastewater Treatment Processes Using a Battery of Bioassays/Biomarkers. *Arch. Environ. Contam. Toxicol.* **2005**, *49*, 480–487. [CrossRef] [PubMed]

35. Pablos, M.V.; Martini, F.; Fernández, C.; Babín, M.; Herraez, I.; Miranda, J.; Martínez, J.; Carbonell, G.; San-Segundo, L.; García-Hortigüela, P.; et al. Correlation between physicochemical and ecotoxicological approaches to estimate landfill leachates toxicity. *Waste Manag.* **2011**, *31*, 1841–1847. [CrossRef] [PubMed]

36. *OECD Guideline for Testing of Chemicals—Daphnia so., Acute Immobilisation Test*; OECD: Paris, France, 2004.

37. Fernandes, A.; Pastorinho, M.R.; Sousa, A.C.; Silva, W.; Silva, R.; Nunes, M.J.; Rodrigues, A.; Pacheco, M.; Ciríaco, L.; Lopes, A. Ecotoxicological evaluation of electrochemical oxidation for the treatment of sanitary landfill leachates. *Environ. Sci. Pollut. Res.* **2018**, *26*, 24–33. [CrossRef] [PubMed]

38. Eaton, A.; Clesceri, L.; Rice, E.; Greenberg, A.; Franson, M.A. *Standard Methods for Examination of Water and Wastewater*, 21st ed.; American Public Health Association: Washington, DC, USA, 2005.

39. Murugananthan, M.; Yoshihara, S.; Rakuma, T.; Shirakashi, T. Mineralization of bisphenol A (BPA) by anodic oxidation with boron-doped diamond (BDD) electrode. *J. Hazard. Mater.* **2008**, *154*, 213–220. [CrossRef] [PubMed]

40. Panizza, M.; Cerisola, G. Direct And Mediated Anodic Oxidation of Organic Pollutants. *Chem. Rev.* **2009**, *109*, 6541–6569. [CrossRef]

41. Fernandes, A.; Coelho, J.; Ciríaco, L.; Pacheco, M.; Lopes, A. Electrochemical wastewater treatment: Influence of the type of carbon and of nitrogen on the organic load removal. *Environ. Sci. Pollut. Res.* **2016**, *23*, 24614–24623. [CrossRef]

42. Perez, G.; Saiz, J.; Ibañez, R.; Urtiaga, A.; Ortiz, I. Assessment of the formation of inorganic oxidation by-products during the electrocatalytic treatment of ammonium from landfill leachates. *Water Res.* **2012**, *46*, 2579–2590. [CrossRef]

43. Arena, M.; Auteri, D.; Barmaz, S.; Brancato, A.; Brocca, D.; Bura, L.; Cabrera, L.C.; Chiusolo, A.; Civitella, C.; Marques, D.C.; et al. Conclusion on the peer review of the pesticide risk assessment of the active substance methiocarb. *EFSA J.* **2018**, *16*, e05429. [CrossRef]

44. Costa, C.R.; Botta, C.M.; Espíndola, E.L.; Olivi, P. Electrochemical treatment of tannery wastewater using DSA®electrodes. *J. Hazard. Mater.* **2008**, *153*, 616–627. [CrossRef]

45. Gotsi, M.; Kalogerakis, N.; Psillakis, E.; Samaras, P.; Mantzavinos, D. Electrochemical oxidation of olive oil mill wastewaters. *Water Res.* **2005**, *39*, 4177–4187. [CrossRef]

46. Lacasa, E.; Llanos, J.; Cañizares, P.; Rodrigo, M.A. Electrochemical denitrificacion with chlorides using DSA and BDD anodes. *Chem. Eng. J.* **2012**, *184*, 66–71. [CrossRef]

47. Sánchez-Carretero, A.; Sáez, C.; Cañizares, P.; Rodrigo, M.A. Electrochemical production of perchlorates using conductive diamond electrolyses. *Chem. Eng. J.* **2011**, *166*, 710–714. [CrossRef]

48. Polcaro, A.; Vacca, A.; Mascia, M.; Ferrara, F. Product and by-product formation in electrolysis of dilute chloride solutions. *J. Appl. Electrochem.* **2008**, *38*, 979–984. [CrossRef]

Publisher's Note: MDPI stays neutral with regard to jurisdictional claims in published maps and institutional affiliations.

MDPI

St. Alban-Anlage 66

4052 Basel

Switzerland

Tel. +41 61 683 77 34

Fax +41 61 302 89 18

www.mdpi.com

Applied Sciences Editorial Office

E-mail: applsci@mdpi.com

www.mdpi.com/journal/applsci